*Principles of*
# PHYSICAL
# SEDIMENTOLOGY

*Principles of*
# PHYSICAL SEDIMENTOLOGY

## J.R.L. ALLEN
*Department of Geology, University of Reading*

London
GEORGE ALLEN & UNWIN
Boston          Sydney

**George Allen & Unwin (Publishers) Ltd,**
**40 Museum Street, London, WC1A 1LU, UK**

George Allen & Unwin (Publishers) Ltd,
Park Lane, Hemel Hempstead, Herts HP2 4TE, UK

Allen & Unwin Inc.,
8 Winchester Place, Winchester, Mass 01890, USA

George Allen & Unwin Australia Pty Ltd,
8 Napier Street, North Sydney, NSW 2060, Australia

First published in 1985

---

**British Library Cataloguing in Publication Data**

Allen, John R. L.
    Principles of physical sedimentology: a primer.
    1. Sedimentation and deposition
    I. Title
    551.3'04    QE571
    ISBN 0-04-551095-4
    ISBN 0-04-551096-2 Pbk

---

**Library of Congress Cataloguing in Publication Data**

Allen, John R. L.
    Principles of physical sedimentology.
Includes bibliographies and indexes.
1. Sedimentation and deposition. 2. Hydraulic engineering. I. Title.
QE571.A44 1985    551.3    85–6006
ISBN 0-04-551095-4 (alk. paper)
ISBN 0-04-551096-2 (pbk. : alk. paper)

---

Set in 9 on 11 point Times by
Mathematical Composition Setters Ltd, Salisbury, Wiltshire
and printed in Great Britain by William Clowes Limited,
Beccles and London

*In memoriam* Peter Allen

## Cover illustration

*Front* (Upper)
Pattern of streamlines round an equatorial section through a sphere in a
Hele — Shaw cell.

*Front* (lower)
Current sorted and rounded debris formed from the coralline alga
*Lithothamnium*, Connemara, Republic of Ireland.

*Back* (upper)
Large wave-related ripples in pebbly, shelly very coarse sand, Rocquaine
Bay, Guernsey.

*Back* (lower)
Steeply climbing current ripple cross-lamination in vertical streamwise
profile, Uppsala Esker, Sweden.

# Preface

My aim in this book is simple. It is to set out in a logical way what I believe is the minimum that the senior undergraduate and beginning postgraduate student in the Earth sciences should nowadays know of general physics, in order to be able to understand (rather than form merely a descriptive knowledge of) the smaller-scale mechanically formed features of detrital sediments. In a sense, this new book is a second edition of my earlier *Physical processes of sedimentation* (1970), which continues to attract readers and purchasers, inasmuch as time has not caused me to change significantly the essence of my philosophy about the subject. Time has, however, brought many welcome new practitioners to the discipline of sedimentology, thrown up a multitude of novel and exciting results and problems and, on the personal side, materially altered and (hopefully) sharpened my appreciation of this field. I could not therefore have prepared a second edition in the traditional sense but have instead written *Principles of physical sedimentology* as an entirely new work. It is similar in scope to *Physical processes of sedimentation* but, as my overriding aim is to give a well founded account of general physical principles, the book in no way attempts to be exhaustive as regards subject matter. Thus there is no separate chapter on wind-related features and I have omitted altogether any discussion of glacial phenomena. There is instead a new chapter on mass movements and their analysis, and I have placed considerable emphasis on, amongst other things, turbulence and vortex phenomena, and on the mechanisms and processes relating to muddy sediments.

Physical sedimentology is essentially an observational science. It is important not only to look hard at and think about sedimentary rocks and modern sediments and processes in the field, but also to strengthen one's intuitions and resolve one's uncertainties by frequently making experiments in the laboratory. In the firm belief that the user of this book will find them helpful, instructive and worth the time to be spent on them, I have therefore described and in many cases illustrated a large number of simple laboratory experiments. These are further elaborated in a companion laboratory handbook, *Experiments in physical sedimentology*, in which many additional experiments will also be found. It cannot be too strongly emphasized, and especially to those readers beginning in sedimentology, that elaborate apparatus is generally not required for the making of useful sedimentological experiments. Most of the equipment needed for those I describe can be found in the kitchen, bathroom or general laboratory, and the materials most often required − sand, clay and flow-marking substances − are cheaply and widely available. As described, the experiments are for the most part purely qualitative, but many can with only little modification be made the subject of a rewarding quantitative exercise. The reader is urged to try out these experiments and to think up additional ones. Experimentation should be as natural an activity and mode of enquiry for a physical sedimentologist as the wielding of spade and hammer.

Although a quantitative treatment played an important role in *Physical processes of sedimentation*, I very largely ducked in that book the issue of the derivation of equations, preferring to shelter behind such evasions as 'it can be shown that . . .'. This I now believe was wrong and, while it might have painlessly provided him or her with useful formulae, gave the reader no key to understanding or basis for attacking new situations from first principles. This nettle has been firmly grasped in *Principles of physical sedimentology*, in which I have included an introductory chapter setting out the essence of mechanics and fluid mechanics, and in which I derived as many relationships as possible from first principles, writing out all but the most glaringly obvious steps; only in the case of certain essential aspects of water waves are the mathematical requirements beyond the level of this book. The user with a knowledge of elementary algebra and calculus should therefore have no difficulty in following my developments and, by example, should emerge confident that he or she can tackle new situations. The mathematics introduced is intended to do no more than to serve the requirements of the particular physical problem; the reader should not allow himself or herself to be overawed by the symbols and equations, but should in turn make them his or her servants. A word of caution is nonetheless due. The price of a mathematically simple analysis is with some problems a degree of simplification that might horrify a specialist researcher. I make no apology for this, so long as the point is noted, as I believe it is better to see the truth indistinctly than not at all! It is best, of course, to see the truth clearly and vividly.

No book of this kind can be prepared without help of many kinds from many people. I am deeply indebted to Mr Roger Jones of George Allen & Unwin Ltd, who encouraged me to bring out a revision of my earlier work, and who contributed materially, along with Mr G. D. Palmer, to whom I am also grateful, to the character of the present book. Dr Iaakov Karcz, Dr Michael Leeder and Dr John Southard made many helpful and important suggestions for the structure of the book in its early stages, and to them I express my gratitude. I am also greatly indebted to many friends and colleagues who made suggestions for the improvement of particular chapters or sections and who generously made available illustrative material, in some cases previously unpublished. My final scientific indebtedness is to the many sedimentologists and researchers in other disciplines whose work I cannot directly acknowledge, on account of the need to keep the 'Readings' to a reasonable length. Mrs Audrey Conner, Mrs Gillian Coward, Mrs Mary Downing and Mrs Dorothy West are warmly thanked for their enduring patience and skilful typing. It is a pleasure to thank Mr J. L. Watkins for his help with the photographic work and for his general advice on photographic matters.

John R. L. Allen
Reading

# Contents

# Tables

# Notation

Physical sedimentology draws on so many fields based on general physics, each with its own hallowed conventions (in some instances more than one) regarding the representation of physical quantities, that it has been impossible for me in this book to give the available Roman and Greek symbols a unique meaning in every case. Fortunately, the different usages occur for the most part quite separately, and I have therefore been able to follow at least the more important conventions. Changes of meaning have been clearly indicated in the text. Note that the symbols used least frequently are excluded from the following list.

$a$    a coefficient; long axis of an irregular sediment particle (m); long semi-axis of an ellipsoid (m); acceleration (m s$^{-2}$)

$A$    flow cross-sectional area (m$^2$)

$b$    a coefficient; intermediate axis of an irregular sediment particle (m); intermediate semi-axis of an ellipsoid (m)

$c$    short axis of an irregular sediment particle (m); phase velocity of water waves (m s$^{-1}$); cohesion (N m$^{-2}$)

$C$    constant of integration; fractional volume concentration of granular solids (non-dimensional)

$C_D$    drag coefficient for a particle in relative motion with a fluid (non-dimensional)

$C_{D,0}$    drag coefficient for a solitary particle in relative motion with an infinite fluid (non-dimensional)

$C_{ref}$    reference fractional volume concentration of granular solids (non-dimensional)

$d$    a distance (m); lag distance (m, in some cases non-dimensional); horizontal diameter of water particle orbit beneath water waves (m)

$D$    sediment particle diameter (m, $\mu$m)

$D_A$    diameter of circle with same area as a sediment particle in projection (m, $\mu$m)

$D_m$    mean particle diameter of a mixture (m, $\mu$m)

$D_V$    diameter of sphere of same volume as a sediment particle (m, $\mu$m)

$D_{50}$    median particle diameter of a mixture (m, $\mu$m)

$E$    energy per unit volume (J m$^{-3}$)

$f$    Darcy–Weisbach friction coefficient for flow in a pipe or open channel (non-dimensional); frequency (s$^{-1}$)

$F$    force (N); sediment flux (kg m$^{-2}$ s$^{-1}$)

$Fr$    Froude number = $U/(gh)^{1/2}$ (non-dimensional)

$g$    acceleration due to gravity (m s$^{-2}$)

$h$    flow thickness (m)

$h_{cr}$    critical flow thickness, e.g. in Froude number or wave equations (m)

$H$    hydraulic head (m); height of bedform (m); height of water wave (m)

$I$    unit sediment transport rate on basis of immersed weight (J m$^{-2}$ s$^{-1}$); sediment particle stability number (various definitions) (non-dimensional)

$J$    unit sediment transport rate on dry-mass basis (kg m$^{-1}$ s$^{-1}$)

$k$    a coefficient; specific permeability ($\mu$m$^2$)

$L$    a length (m); wavelength of bedform, water wave, or channel sinuosity (m)

$m$    an exponent; mass (kg); sediment load (normally immersed-weight basis) (N m$^{-2}$)

$M$    mass (Kg); momentum flux (N m$^{-2}$)

$n$    a number or exponent

$p$    fluid pressure (N m$^{-2}$)

$P$    sediment fractional porosity (non-dimensional)

$q$    volumetric fluid discharge per unit width of flow (m$^2$ s$^{-1}$)

$Q_v$    total volumetric discharge (m$^3$ s$^{-1}$) (the subscript v is commonly dropped)

$Q_m$    total mass discharge (kg s$^{-1}$)

$r$    radical distance or radius (m)

$R$    radius (m); sediment dry-mass deposition or erosion rate (rate of transfer) (kg m$^{-2}$ s$^{-1}$)

$Re$    Reynolds number = (length × velocity × density)/viscosity (non-dimensional)

$S$    slope of water surface and/or bed in open channel (non-dimensional)

$Sr$    Strouhal non-dimensional frequency = $fD/V_0$

$t$    time (s)

$T$    time for settlement of liquidized bed (s); surface tension of a liquid (N m$^{-1}$)

$T_b$    non-dimensional burst period

$u$    instantaneous flow velocity measured in $x$-direction (m s$^{-1}$)

$u'$    fluctuating component of velocity measured in $x$-direction (m s$^{-1}$)

$U$    local flow velocity (laminar case) or local time-

$U_{bm}$    averaged velocity (turbulent case) measured in $x$-direction (m s$^{-1}$)

$U_{bm}$    celerity of bedform (m s$^{-1}$)

$U_{cr}$    critical value of flow velocity or sediment erosion threshold (m s$^{-1}$)

$U_m$    local depth-averaged flow velocity in $x$-direction (m s$^{-1}$)

$U_{max}$    maximum $x$-directed velocity shown in a velocity profile, or maximum horizontal velocity observed on a near-bed water particle orbit beneath water waves (m s$^{-1}$)

$U_0$    velocity of undisturbed $x$-directed stream outside boundary layer (m s$^{-1}$)

$U_{om}$    overall mean $x$-directed velocity (m s$^{-1}$)

$U_s$    mean sediment particle transport velocity (m s$^{-1}$)

$U_\tau$    shear velocity $= (\tau/\varrho)^{1/2}$ (m s$^{-1}$)

$v$    instantaneous flow velocity measured in $y$-direction (m s$^{-1}$)

$v'$    fluctuating component of velocity measured in $y$-direction (m s$^{-1}$)

$V$    volume (m$^3$); local time-averaged flow velocity measured in $y$-direction (m s$^{-1}$); terminal fall velocity of particle in a fluid, measured relative to ground (m s$^{-1}$)

$V_0$    terminal fall velocity of a solitary particle in an infinite fluid, measured relative to ground (m s$^{-1}$)

$V_{rel}$    relative velocity between a particle and a fluid (m s$^{-1}$)

$V_s$    superficial velocity (m s$^{-1}$)

$V_{s,cr}$    superficial velocity necessary for fluidization of a granular aggregate (m s$^{-1}$)

$w$    flow width (m); instantaneous flow velocity measured in $z$-direction (m s$^{-1}$)

$w'$    fluctuating component of velocity measured in $z$-direction (m s$^{-1}$)

$W$    local time-averaged flow velocity measured in $z$-direction (m s$^{-1}$); vertical velocity of interface associated with settling of a particle dispersion (m s$^{-1}$)

$x$    distance measured in general direction of flow (m)

$y$    distance measured normal to flow boundary or free surface (m)

$Y$    non-dimensional distance normal to flow boundary $= \varrho y U_\tau/\eta$

$z$    distance measured parallel to flow boundary or free surface but perpendicular to $x$-direction (m)

$Z$    non-dimensional transverse distance $= \varrho z U_\tau/\eta$

$\alpha$    (alpha) an angle

$\beta$    (beta) an angle (generally bed or water-surface slope)

$\gamma$    (gamma) sediment bulk density (kg m$^{-3}$)

$\delta$    (delta) a small increment; boundary-layer thickness (m)

$\Delta$    (delta) the difference between two values of a quantity

$\epsilon$    (epsilon) eddy diffusion coefficient (m$^2$ s$^{-1}$)

$\zeta$    (zeta) angle of bedform climb

$\eta$    (eta) dynamic (molecular) viscosity of a fluid (N s m$^{-2}$, Pa s)

$\eta_a$    apparent viscosity (N s m$^{-2}$)

$\theta$    (theta) an angle; non-dimensional shear stress (various definitions)

$\theta_{cr}$    critical value of non-dimensional stress

$\varrho$    (rho) fluid density (kg m$^{-3}$)

$\sigma$    (sigma) standard deviation of particle size distribution (m, $\mu$m); solids density (kg m$^{-3}$); normal stress (N m$^{-2}$)

$\tau$    (tau) shear stress (N m$^{-2}$)

$\tau_{cr}$    critical value of shear stress (N m$^{-2}$)

$\tau_s$    shear strength (N m$^{-2}$)

$\phi_i$    (phi) angle of initial yield of cohesionless grains

$\phi_r$    residual angle after shearing of cohesionless grains

$\omega$    (omega) power of a fluid stream (W m$^{-2}$)

# 1 Concepts and rules of the game

> Matter — forces — flow rate of a fluid — laws of conservation of mass, momentum and energy — energy losses during fluid flow — Newton's laws of motion — fluid viscosity — boundary layers — flow separation — application of concepts to a water body affected by a wind.

## 1.1 Matter and influences

Taking a commonsense view, the world we live in is composed of quantities of matter which influence each other physically in a variety of ways, depending on the nature of the matter, on whether the lumps are in relative motion and on the nature of their motion. Matter is presented to us in three partly interchangeable forms. We describe as solids those materials which are hard and heavy and capable of retaining their shape. Many solids turn out to have a highly ordered or crystalline atomic structure. Experience teaches us to classify as fluids those materials which readily flow and which, speaking generally, are light or very light in comparison with solids. Amongst fluids we can readily distinguish between those capable of being poured into a vessel, where they fit themselves to its shape and form a horizontal free surface, and those which invariably expand to fill the container provided. The first kind of fluid we call liquids. These possess a disordered atomic or molecular structure overall, but extremely locally, and for very brief periods, a crystalline structure can be present. Gases, the other and lighter kind of fluid, lack any orderly atomic or molecular structure.

Essentially, physical sedimentology is about (1) the interplay that occurs within the Earth's natural environments between solid transportable particles and transporting (or potentially transporting) fluids, and (2) the sedimentary consequences of that interplay.

Of the environments in which the interplay occurs, those of greatest direct significance to man, and those most readily studied, are to be found at the Earth's surface in the shape of, for example, river channels, beaches and tidal flats, and the floors of the oceans. We largely aim this book towards these. But it is as well to remember that contained in the Earth's crust are magma chambers where igneous rocks are forming, and these furnish another kind of sedimentological environment. One of the current tasks of petrology is to work out a 'sedimentology' of igneous rocks.

The solid matter involved in the interplay has various origins. In surface environments, the vast majority of sedimentary particles originate either directly by the weathering of rocks or indirectly by the mediation of organisms that make hard tissues from atmospheric carbon dioxide and/or substances dissolved in natural waters. Amongst the products of rock weathering are pieces of mineral matter. An organism such as a bivalve mollusc also creates during its life a piece of mineral matter, in the form of a structured shell. Plants produce tissues in which mineral matter may be combined with organic compounds with limited potential for survival after the death of the organism. The solids generated in magma chambers are mainly crystals of individual silicate minerals, for example, members of the feldspar and pyroxene families. Mineral crystals capable of being transported by currents can arise in surface environments where natural waters experience sufficient evaporation.

The two most important fluids involved in the interplay are water (a liquid), which forms the rivers, lakes and oceans, and the atmospheric air (a mixture of gases), which shrouds the whole Earth. The fluids operating in magma chambers are complex silicate melts containing dissolved gases (including water vapour) at a high temperature and pressure.

Thus all three states of matter are represented in the sedimentological interplay between transportable solids and transporting fluids. How should we set about exploring the interplay in any particular situation? We shall first of all want to state how much of each kind of matter is present. This can be done in one way by measuring the total mass of each kind, where we define a mass as the amount of matter present in a substance. Under the International System of Units (SI), the unit of mass is taken to be the kilogram, abbreviated to kg (Table 1.1). It is equal to the mass of the international prototype of the kilogram. Mass should never be confused with weight, which is the attraction to the Earth of a mass. In what follows, we shall use either $m$ or $M$ as symbols to denote mass. But we can also specify the amount of each kind of matter by measuring the amount of space occupied by each substance, that is, its

volume. Since fluids and solids exist in three dimensions, volume must be measured as the product of one length by a second length by a third length. Under the International System of Units, the unit of length is the metre, abbreviated to m, which is equal to a specified number of wavelengths of a specified kind of atomic radiation (Table 1.1). Hence the unit of volume is the cubic metre, abbreviated to $m^3$ (Table 1.1).

In distinguishing between the different forms or states of matter, we spoke of solids as generally heavy and of fluids as generally light or very light. What we mean by this commonsense appreciation is that in a heavy solid a large mass is present in a small volume, whereas in a comparable volume of a light or very light fluid only a small or very small mass exists. This is the concept of the density of a substance, defined as its mass divided by its volume. As mass is given in kilograms, and volume

**Table 1.1** Chief quantities encountered in phyical sedimentology (see also Notation, pp. xiv – xv).

| Quantity | Unit | Symbol |
|---|---|---|
| Fundamental | | |
| mass | kilogram (kg) | $m, M$ |
| length | metre (m) | various |
| time | second (s) | $t$ |
| Kinetics and continuity | | |
| area | $m^2$ | $A$ (commonly) |
| volume | $m^3$ | $V$ (commonly) |
| density | $kg\,m^{-3}$ | $\varrho, \sigma$ |
| velocity | $m\,s^{-1}$ | $U$ (and many others) |
| temporal acceleration | $m\,s^{-2}$ | $a$ |
| spatial acceleration | $s^{-1}$ | no special symbol |
| local mass discharge (flux) | $kg\,m^{-2}\,s^{-1}$ | no special symbol |
| unit mass discharge | $kg\,m^{-1}\,s^{-1}$ | $J$ (in case of sediment transport rate) |
| total mass discharge | $kg\,s^{-1}$ | no special symbol |
| local volumetric discharge (flux) | $m\,s^{-1}$ | no special symbol |
| unit volumetric discharge | $m^2\,s^{-1}$ | $q$ |
| total volumetric discharge | $m^3\,s^{-1}$ | $Q, Q_v$ |
| Dynamics | | |
| force | newton (N, $kg\,m\,s^{-2}$) | $F$ |
| normal stress (pressure) | $N\,m^{-2}$, Pa | $p, \sigma$ |
| tangential (shear) stress | $N\,m^{-2}$ | $\tau$ |
| momentum | $N\,s$ | no special symbol |
| momentum flux | $N\,m^{-2}$ | $M$ |
| work | joule (J, N m) | no special symbol |
| power | watt (W, $J\,s^{-1}$) | no special symbol |
| stream power | $W\,m^{-2}$ | $\omega$ |
| energy | $J$ | no special symbol |
| unit fluid energy | $J\,m^{-3}$ | $E$ |
| viscosity | $N\,m^{-2}\,s$, Pa s | $\eta$ |

is in cubic metres, density is measured in kilograms per cubic metre, abbreviated to $kg\ m^{-3}$ (Table 1.1). The density of a substance is one of its most important material properties. That of water is approximately $1 \times 10^3\ kg\ m^{-3}$, whereas that of air at normal temperature and pressure is only $1.3\ kg\ m^{-3}$. The magmas which form minerals and rocks on cooling have densities of the order of $3 \times 10^3\ kg\ m^{-3}$. As far as possible, we shall represent the density of a solid by the Greek letter $\sigma$ (sigma) and that of a fluid by the Greek letter $\varrho$ (rho).

Are substances of constant density? Solids and liquids can be compressed, but to any significant degree only when subjected to almost unimaginably huge pressures, and most do expand on being heated, but only to a slight extent in response to a large temperature rise. As the temperature and pressure at the Earth's surface change but little, solids and liquids in surface environments can be treated as incompressible and of constant density. This is a valuable simplification, because it allows the strict interchange of mass and volume through the definition of density. But what about gases? Our everyday experience teaches us that gases are easily compressed – try squeezing the air in a bicycle pump – and that they vary noticeably in volume when either slightly heated or cooled. Fortunately, in sedimentology, gases can also be treated as incompressible (with some exceptions), but this simplification is never acceptable to either meteorologists or aircraft designers.

What are the influences through which transportable solids and transporting fluids affect each other in natural environments? These influences are called forces, but the the difficulty with them is that we cannot capture a force in a bottle for examination. Force is therefore defined in terms of its effects or consequences, and is said to be that which changes, or tends to change, the state of rest or uniform motion of a mass. Under the International System of Units, the unit of force is the newton, abbreviated to N (Table 1.1). Wherever appropriate in this book, the symbol $F$ will be used to denote a force measured in newtons.

The motion of a brick sliding over a table can be explored mechanically by imagining that each of the forces involved is summed up at, and acts through, the centre of the brick. However, in problems involving fluids, and particularly fluids in motion, we are generally much more interested in the distributed forces, that is, the forces acting over each unit area of, say, the surface of a solid particle immersed in a fluid or the wetted surface of a channel. As the metre is the unit of length, the appropriate unit in which to measure such a distributed force or stress is the newton per square metre,

abbreviated to $N\ m^{-2}$. Clearly, such a force may act either normal to or parallel to the area in question, although the unit of measurement is in each case the same. The first kind of distributed force is a normal stress or pressure, and is sometimes measured using the special unit called the pascal (Pa), 1 Pa being equal to $1\ N\ m^{-2}$ (Table 1.1). We shall denote normal stresses by either the symbol $p$ or, in special cases, by the Greek letter $\sigma$ (sigma). A distributed force that acts parallel with, i.e. tangential to, a surface is called a shear stress (Table 1.1). It has no special unit of measurement, and will be denoted by the Greek letter $\tau$ (tau).

## 1.2 Flow rate

### 1.2.1 Velocity and acceleration

In order to move sediment particles, a fluid must itself be moving. Intuitively, the flow of the fluid is a part of the key to understanding the sediment motion, so how should we describe and measure the rate of fluid flow?

One way is to measure the flow velocity, that is, the change in the position of a fluid particle in a moving fluid during a certain time interval. Figure 1.1 shows part of the path followed by a fluid particle within a moving fluid. At time $t_1$ the particle is at the position $x_1$, but at the later time of $t_2$ the same particle has reached $x_2$. The velocity $U$ is defined as the distance travelled by the particle divided by the time taken for the journey, that is, $(x_2 - x_1)/(t_2 - t_1)$. Under the International System of Units, the unit of time is the second, abbreviated to s, defined as the duration of a specified number of periods of a particular kind of atomic radiation (Table 1.1). Hence velocity is measured in metres per second, abbreviated to $m\ s^{-1}$ (Table 1.1). Now in Figure 1.1 the straight-line 'path' between $x_1$ and $x_2$ is noticeably shorter than the distance $s$ measured along the true particle path between the same two points. But as $x_2$ is moved closer to $x_1$, $(x_2 - x_1)$ approaches nearer and nearer to $s$ in value, and the straight-line path

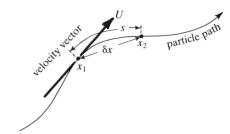

**Figure 1.1** Definition sketch for the velocity of a fluid particle.

becomes less and less distinguishable from the true particle path. Over a small increment of time, $\delta t$, the straight-line path becomes tangential to the true path at $x_1$, the particle will have traversed the small distance $\delta x$, and the velocity will be $\delta x/\delta t$. In the limit, then, the particle velocity is given by the differential coefficient $dx/dt$. The velocity is therefore a description of the particle motion at an instant of time. It has an interesting double property, for the motion has not only magnitude, called its speed, but also direction as shown by the arrow in Figure 1.1. This makes velocity a vector quantity.

Now imagine that point $x_1$ in Figure 1.1 is a point fixed in space through which a stream of fluid particles is passing. If the succession of particles through $x_1$ have identical velocity vectors, the fluid flow remains unchanged with time and is described as steady. The particle path shown in the diagram is then one of an infinity of streamlines which can be calculated and plotted to define the motion of the whole fluid. But what if the velocity vector changes for successive fluid particles passing through $x_1$? The paths followed by successive particles are not the same and we have what is called unsteady flow. Steady and unsteady flow are two important categories of fluid motion and they refer to the stability of the motion with respect to time. Should the velocity vector change for successive particles passing through $x_1$, then those particles are subject at our fixed point to a temporal acceleration, defined as the rate of change of velocity with time. Remembering how we derived velocity, the temporal acceleration is clearly $(\delta x/\delta t)/\delta t$ in terms of small increments, which becomes the differential coefficient $d^2x/dt^2$ in the limit. The units of temporal acceleration are therefore metres per second per second, abbreviated to $m\,s^{-2}$ (Table 1.1).

A fluid flow may experience another kind of acceleration. The flow at a tap filling a bath is fast but steady, whereas only sluggish currents are found in the large body of water already in the bath. Hence in merging with the larger volume, the flow from the tap experiences a spatial change in velocity, that is, a spatial acceleration. Consider the particle path in Figure 1.2. At $x_1$ the fluid particle has a velocity vector $U$ which at $x_2$, a small distance $\delta x$ along the path, has changed to $(U + \delta U)$, where $\delta U$ is another small increment. The velocity change is therefore $\delta U/\delta x$, which in the limit becomes the differential coefficient $dU/dx$. The spatial acceleration therefore has the units per second, abbreviated to $s^{-1}$ (Table 1.1). A fluid flow characterized by a non-zero spatial acceleration is called non-uniform. The term uniform flow is applied only when this acceleration is zero. Uniform and non-uniform flow are two further

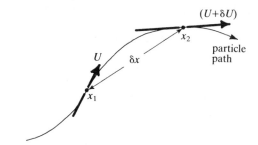

**Figure 1.2** Definition sketch for the acceleration of a fluid particle.

important categories of flow, and refer specifically to the stability of the motion with respect to distance. The path of a fluid particle is also a streamline only if the flow is either steady or uniform.

Hence we have four possible simple categories of flow as well as mixed types. The vast majority of real-world flows are both unsteady and non-uniform, that is, the velocity vector of fluid particles changes from point to point, and the velocity vector of a fluid particle passing any given point changes from instant to instant. In the laboratory, however, we can easily produce either a uniform steady flow, with no accelerations of either kind, or just a pure non-uniform or a pure unsteady motion, with an acceleration of a single kind.

### 1.2.2 Measures of discharge

The limitation on velocity and acceleration as measures of flow rate is that they refer to effectively massless particles, as you can see from their derivations and units of measurement. Other quantities are needed where we require to state the rate of transport of the matter involved in a fluid flow.

Stand at a place on the bank of a flowing river, and you will speedily realize that matter in the form of water is being carried past. Considering the whole river at that place, we can define this transport, or discharge, as the total mass of water flowing past in a certain period of time. The total mass discharge therefore has the units kilograms per second, abbreviated to $kg\,s^{-1}$ (Table 1.1) and may be represented by the symbol $Q_m$. But we concluded that water at the Earth's surface could be regarded as incompressible and of constant density $\varrho$. Making use of this result, we can convert the total mass discharge directly into the total volumetric discharge $Q_v = Q_m/\varrho$, which must therefore be measured in units of cubic metres per second, abbreviated to $m^3\,s^{-1}$ (Table 1.1). It is generally much more convenient to measure fluid discharges in volumetric terms rather than mass terms. Sediment discharges, however, are easiest to determine in terms of mass, and there are also sound physical reasons for preferring this particular measure (Ch. 4).

The total volumetric discharge tells us a good deal about what the river is doing at our fixed station, but reveals nothing about the local variations that might be expected across the flow. We now need to recognize that the flowing water possesses a certain cross-sectional shape (Fig. 1.3), which can be described using a vertical co-ordinate axis $y$ and a horizontal, flow-transverse co-ordinate axis $z$. The $x$-direction is taken as that of the flow itself. The local flow velocity is considered to vary over this cross section, that is $U(y, z)$, meaning '$U$ is a function of both $y$ and $z$'. Consider a unit area somewhere in the plane of the cross section. Remembering that density is defined as the mass per unit volume, the mass discharge through this local area is clearly $\varrho U$ kilograms per square metre per second, abbreviated to kg m$^{-2}$s$^{-1}$ (Table 1.1). Dividing this quantity by the fluid density, we duly obtain the local volumetric discharge, measured in cubic metres per square metre per second, abbreviated to m$^3$ m$^{-2}$s$^{-1}$ or m s$^{-1}$ (Table 1.1). Such a quantity describing transport through a surface is commonly called a flux. Let us suppose that the channel is $h$ metres deep at the site of our chosen unit area. If we were to sum up all the local values of the volumetric discharge over the depth $h$, we should obtain the volumetric discharge per unit width of the channel at that site, namely

$$q = \frac{1}{\varrho} \int^{h} \varrho U(y)\, dy \quad \text{m}^3\,\text{m}^{-1}\text{s}^{-1} \text{ or } \text{m}^2\,\text{s}^{-1} \quad (1.1)$$

where $q$ is the unit discharge (Table 1.1). An equivalent expression is

$$q = \frac{1}{\varrho}(\varrho h U_{\mathrm{m}}) \quad (1.2)$$

where $U_{\mathrm{m}}$ is the average of $U$ over the depth $h$, as may easily be proved by writing the integral for $U_{\mathrm{m}}$. Now we expect $q$ to vary with position across the channel. Therefore to obtain the total volumetric discharge, we can either sum $q$ over the channel width

$$Q_{\mathrm{v}} = \int^{w} q(z)\, dz \quad \text{m}^3\,\text{s}^{-1} \quad (1.3)$$

where $w$ is the width or, starting from a more fundamental position, write

$$Q_{\mathrm{v}} = \frac{1}{\varrho} \int^{h} \int^{w} \varrho U(y, z)\, dy\, dz \quad (1.4)$$

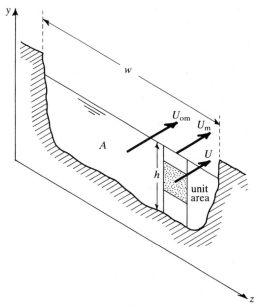

**Figure 1.3** Definition sketch for the various measures of fluid discharge.

As before, an equivalent expression to Equations 1.3 and 1.4 is

$$Q_{\mathrm{v}} = \frac{1}{\varrho}(\varrho A U_{\mathrm{om}}) \quad (1.5)$$

where $A$ is the area of the channel cross-section and $U_{\mathrm{om}}$ is the overall average of $U$. Equations 1.2 and 1.5 are fundamental relationships connecting discharge, flow depth and cross-sectional area, and mean velocity. The density terms were kept in these equations simply to demonstrate the origin of the units of discharge, and it is unnecessary in practice to retain them. Notice how these units change in steps of m$^{-1}$ as we go from the local, to the unit-width, to the total value.

## 1.3 Law of continuity (conservation of mass)

A basic axiom of classical physics is that mass can be neither created nor destroyed. As we have accepted that a fluid at the surface of the Earth is incompressible and of constant density, this axiom requires us to account in a fluid system for all volumetric quantities, that is, the discharge flowing in, the volume stored within, and the discharge flowing out must be balanced. This principle applies not only to fluids, but also to any sediment particles being transported and deposited (stored).

Imagine that you are filling a leaky tank with water, such that the volume $V$ already present is increased by the small increment $\delta V$ in a small increment of time $\delta t$ (Fig. 1.4). The rate of change of volume in the tank is therefore $\delta V/\delta t$ which, in the limit, becomes the differential coefficient $dV/dt$ m$^3$ s$^{-1}$. If $Q_{in}$ is the total volumetric discharge into the tank, and $Q_{out}$ the total of the discharge through the leaks, then the axiom requires that

$$Q_{in} - Q_{out} = \frac{dV}{dt} \quad \text{m}^3\,\text{s}^{-1} \tag{1.6}$$

This is called an equation of continuity, and it is the specific form appropriate to a fluid system with storage.

In many cases, storage is physically impossible, and the continuity equation takes the form

$$Q_{in} - Q_{out} = 0 \tag{1.7}$$

One such system is a pipe filled with flowing water (Fig. 1.5). Continuity requires that the discharge at section 2 equals that at section 1 further back along the flow. We can exploit this requirement, together with Equation 1.5, to calculate the overall mean velocity at section 2, provided that we know the corresponding velocity at section 1. Putting the argument in full

$$Q_1 = Q_2 \tag{1.8}$$

whence from Equation 1.5

$$A_1 U_1 = A_2 U_2 \tag{1.9}$$

and

$$U_2 = \frac{A_1}{A_2} U_1 \quad \text{m s}^{-1} \tag{1.10}$$

where $A_1$ and $A_2$ are the appropriate cross-sectional areas, and $U_1$ and $U_2$ the corresponding overall mean

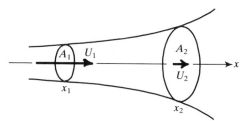

**Figure 1.5** Definition sketch for the law of conservation of mass (continuity) in a pipe of variable cross section.

velocities. Thus at a constant discharge, the mean velocity is inversely proportional to the cross-sectional area. This is a handy property of the continuity equation and is frequently exploited in analysing fluid motion.

## 1.4 Law of conservation of momentum

A particle of mass $m$ and velocity $U$ has a momentum of $mU$ newton seconds, abbreviated to N s (Table 1.1). It is more convenient in the case of a flowing fluid to take a unit volume rather than an arbitrary mass, in which case the momentum becomes equal to $\varrho U$, with the units newton seconds per cubic metre, abbreviated to N s m$^{-3}$. Now if we consider a fixed cross section in a flowing fluid, then the rate at which the fluid momentum is transported through unit area of that cross section equals $(\varrho U)U$ newton seconds per square metre per second, abbreviated to N s m$^{-2}$ s$^{-1}$ or N m$^{-2}$. This quantity, called the momentum flux, corresponds to the local bulk discharge (or flux) discussed under the continuity equation. However we choose to treat momentum, it is important to note that it is a vector quantity, as velocity appears in its definition.

Another axiom of classical physics is that the momentum of a system of moving masses may not be lost, although some of it may be changed into an impulsive force. Applied to the fluid flow in Figure 1.6 between points 1 and 2, we have

$$M_2 - M_1 = p \quad \text{N m}^{-2} \tag{1.11}$$

where $M_1$ and $M_2$ are the respective momentum fluxes and $p$ is the force. Notice that the impulsive force is a distributed one, as we are treating the flux of momen-

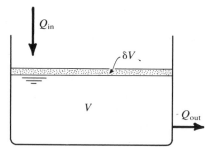

**Figure 1.4** Definition sketch for the law of conservation of mass (continuity).

**Figure 1.6** Definition sketch for the law of conservation of momentum.

tum, and that the units of the flux ($N\,s\,m^{-2}\,s^{-1}$) become equivalent to those of the stress ($N\,m^{-2}$) by cancelling out the seconds. As the momentum flux is a vector, the impulsive stress may act in a different direction from the momentum.

Now try using the momentum equation (Fig. 1.7). When a jet of water is directed normal to a rigid surface and does not rebound, the surface destroys the whole momentum of the jet, as measured in the direction of the jet. The momentum loss is equal to the impulsive force exerted by the surface on the water. Giving the water in the jet a uniform velocity $U$ and density $\varrho$, the impulsive stress becomes equal to $-\varrho U^2\,N\,m^{-2}$ and therefore acts in the opposite direction to the jet. The stress is equal in magnitude but opposite in direction to the force exerted by the water on the surface. This case is particularly simple and instructive because, at the surface itself, the velocity of the water parallel to the $x$-direction is zero, and the water therefore no longer has any momentum parallel to the original direction of the jet. The stress created at the surface is therefore easily seen to be equivalent to the destruction of momentum and is, in fact, equal to the momentum destroyed per second per unit area of flow. The same argument applies if we consider not the distributed force but the total force, equal to $-\varrho A U^2\,N$, where $A$ is the cross-sectional area of the jet.

How does momentum conservation work in the case of water following a right-angled channel bend, a situation with many close natural parallels? Imagine that the channel shown in Figure 1.8 is uniform in cross section, with depth $h$ and width $w$, so that the overall mean velocity $U_1$ in section 1 equals in magnitude the corresponding velocity $U_2$ in section 2. The velocity vectors are not parallel, whence they yield the resultant vector with magnitude $U_R$ shown in the diagram. Taking the whole flow cross section, the change in momentum is

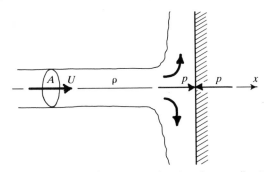

**Figure 1.7** Change of momentum in a jet of water directed normal to a plane surface.

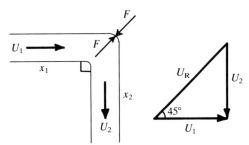

**Figure 1.8** Change in momentum of water flowing in a channel with a right-angled bend.

$(\varrho hwU_1)U_R$, so that the outer side of the channel must exert an impulsive force $F\,N$ on the water. As $U_1$ and $U_2$ are equal in magnitude, it follows from the geometry of the vector triangle that $F$ acts at $45°$ to the direction of the flow entering the bend. By the same geometry, the magnitude of the resultant vector is seen to be $U_1/\cos 45°$, whence the magnitude of $F$ becomes $(\varrho hwU_1)(U_1/\cos 45°)\,N$.

## 1.5 Law of conservation of energy

### 1.5.1 Work and power

Work must be done either to drive a certain amount of water along a channel or to blow up a bicycle tyre. Anyone who has pumped up a tyre will have come to appreciate that the work involved moving a force, the grip on the handle of the pump, through a certain distance, the number of pump strokes times the length of stroke. Work is in fact defined as force multiplied by the distance over which it acts, measured in the same direction as the force. As total force is measured in newtons, and distance in metres, the ordinary unit of work is the newton metre, otherwise known as the joule, abbreviated to J (Table 1.1).

The idea of power is closely allied to that of work, and relates to the work done per unit time, that is, the rate of doing work. Power is consequently defined as work divided by time. The ordinary unit of power is the joule per second, abbreviated to $J\,s^{-1}$, also called the watt, for which the abbreviation is W (Table 1.1). In problems relating to fluids, where it is more convenient to use distributed rather than total force, the appropriate measure of power is the watt per square metre, abbreviated to $W\,m^{-2}$ (Table 1.1), and represented by the Greek letter $\omega$ (omega). The two measures of power are similar in physical meaning. To perform a certain amount of work in a shorter period requires more power of either a machine or a fluid flow than to do the same amount of work in a longer period of time. Returning to one of our examples, the average bicyclist at the end of pumping up an empty tyre is no longer operating at the same power as at the start!

### 1.5.2 Energy and its kinds

We describe as energetic any person who is full of the capacity to get things done. In physics, capacity to do work is also called energy, which is therefore simply stored work. The ordinary unit of energy is consequently the joule, as in the case of work (Table 1.1). The more convenient measure of energy in the case of fluids, however, is the joule per cubic metre, abbreviated to $J m^{-3}$ (Table 1.1). Notice that its dimensions are the same as those of stress.

What kinds of energy typify moving fluids? The fact of the motion means that the fluid possesses kinetic energy. The kinetic energy per unit volume is equal to $\frac{1}{2}\varrho U^2$, where $\varrho$ is the fluid density and $U$ the flow velocity. But as the fluid is at or near the surface of the Earth, it is being attracted towards the centre, and will move there as soon as sufficient constraints are removed. Hence the fluid has energy of position. This is called potential energy, equal per unit volume to $\varrho g y$, where $g$ is the acceleration due to gravity and $y$ the height of the unit volume above a suitable datum. The acceleration due to gravity is (very nearly) a constant, equal to $9.81\ m s^{-2}$, which operates over the entire Earth and represents the attraction of bodies to the Earth. Multiplied by a mass $m$, it yields a weight force $mg\ N$, whence the $kg\,m\,s^{-2}$ in Table 1.1 is a true equivalent of the newton. A third energy form is pressure energy, given per unit volume by the normal stress (pressure) $p$. That this is a true form of fluid energy is easily proved by compressing the air in a bicycle pump. The handle springs back when released, in consequence of the work done on it by the compressed air.

Changes in heat energy invariably accompany the flow of any real-world fluid. This is because water and air and other natural fluids are viscous, being sticky and with capacity to resist change in shape. These thermal changes denote a loss of fluid energy (kinetic, potential, pressure), and manifest themselves as shear stresses, which we represent by the symbol $\tau$. Slide your hand briskly over a table top a few times to appreciate this point.

It frequently turns out that the changes in heat energy accompanying fluid flow are negligibly small in comparison with the other kinds. The fluid may then be treated as without viscosity (inviscid), a convenient simplification in the mathematical study of flow.

### 1.5.3 Bernoulli's equation

The third great axiom of classical physics is that energy cannot be lost from a mechanical system, although it may change from one form into another. In a flowing fluid, then, the total energy $E$ is a constant, namely

$$E = \tfrac{1}{2}\varrho U^2 + \varrho g y + p + E_{loss} = \text{constant} \quad J m^{-3} \quad (1.12)$$

summing up the four kinds of energy described. This statement of the law of conservation of energy is called Bernoulli's equation, after Daniel Bernoulli, its 18th century Swiss discoverer, or, alternatively, the energy equation. When $E_{loss}$ expressed as frictional heat is negligibly small, Equation 1.12 comprises just three terms. The total energy, a constant for each flow, is the sum of the kinetic energy per unit volume, the potential energy per unit volume, and the pressure energy.

Can we prove Bernoulli's statement? Imagine that in a steady flow of an incompressible fluid we select from the infinity of streamlines just those streamlines which define a tube (Fig. 1.9). This tube is called a stream tube. Clearly, if we go on reducing the number of contiguous streamlines that define it, our stream tube can be reduced to just one streamline. As the motion is steady, the streamlines defining the stream tube are also particle paths, whence no fluid particle may pass through the wall of the steam tube. The wall is frictionless, for it is simply the surrounding fluid. Hence the motion involves no frictional losses in the form of heat.

At section 1 in Figure 1.9, where the cross-sectional area of the stream tube is $A_1$ and the pressure $p_1$, a small quantity of fluid passes a small distance $\delta x_1$ down the tube in a small increment of time $\delta t$. The work done on this fluid by the flow pressing from behind is, by definition, the product of the pressure times the cross-sectional area times the distance moved. That is, the

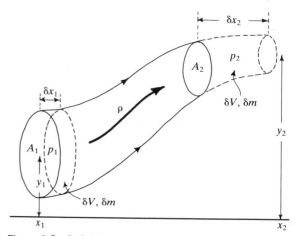

**Figure 1.9** Definition sketch for Bernoulli's equation.

8

work done equals $p_1(A_1 \delta x_1)$ J. Now $(A_1 \delta x_1)$ equals the small increment of volume $\delta V$, so the work done is $p_1 \delta V$. As the fluid is incompressible, an equal volume increment is discharged in the same time interval from the tube at section 2. This represents a loss of energy from the tube, equal to $p_2(A_2 \delta x_2) = p_2 \delta V$ J. Hence

$$\text{change in energy} = p_1 \delta V - p_2 \delta V = \delta V(p_1 - p_2) \quad \text{J} \tag{1.13}$$

This change should manifest itself as a change in the kinetic and/or potential energy.

As the fluid is incompressible and continuous, the fixed volume $\delta V$ by which the fluid advances down the tube is equivalent to the disappearance of a fixed increment of mass $\delta m$ from section 1 and its reappearance with different kinetic and potential energies at section 2. Hence we can also write

$$\text{change in energy} = \frac{1}{2} \delta m \left[ \left( \frac{\delta x_2}{\delta t} \right)^2 - \left( \frac{\delta x_1}{\delta t} \right)^2 \right]$$
$$+ \delta m\, g(y_2 - y_1) \quad \text{J} \tag{1.14}$$

using our previous definitions. Here the first term is the change in kinetic energy, where $\delta x_2/\delta t$ and $\delta x_1/\delta t$ are the flow velocities, and the second term is the change in potential energy, where $y_2$ and $y_1$ are elevations above datum.

We now have two expressions for the energy change. When equated and divided through by $\delta V$, they yield

$$p_1 - p_2 = \frac{1}{2} \frac{\delta m}{\delta V} \left[ \left( \frac{\delta x_2}{\delta t} \right)^2 - \left( \frac{\delta x_1}{\delta t} \right)^2 \right]$$
$$+ \frac{\delta m}{\delta V} g(y_2 - y_1) \quad \text{J m}^{-3} \tag{1.15}$$

On multiplying out, and bringing the terms for section 1 to the left and those for section 2 to the right, we obtain in the limit

$$p_1 + \tfrac{1}{2} \varrho U_1^2 + \varrho g y_1 = p_2 + \tfrac{1}{2} \varrho U_2^2 + \varrho g y_2 \tag{1.16}$$

where $U_1$ and $U_2$ are substituted respectively for $dx_1/dt$ and $dx_2/dt$, and $dm/dV = \varrho$, the fluid density. Hence for any stream tube or streamline in steady flow

$$p + \tfrac{1}{2} \varrho U^2 + \varrho g y = \text{constant} \tag{1.17}$$

which is the Bernoulli equation we set out to prove. At a constant potential energy, then, an increase in flow velocity should lead to a decrease in pressure, and vice versa.

This result is easily tested experimentally. Take two inflated balloons and, from a rail, suspend them by light threads at a small horizontal separation (Fig. 1.10a). What do you think happens when an air current is directed between the balloons? One's instinct is to say that the balloons move apart, but Bernoulli's equation insists otherwise. The current of air between the balloons lowers the pressure on the sides facing each other in comparison with the outer sides. Hence a net force acts from each balloon towards the other, and the bodies move together. Take a deep breath and blow gently between the balloons. They do indeed collide (Fig. 1.10b).

The Bernoulli equation is a powerful analytical tool with many applications. An interesting one concerns steady flow in open channels. Consider the free surface with the channel bottom as datum. Bernoulli's equation may be written

$$E - p = \tfrac{1}{2} \varrho U^2 + \varrho g h \tag{1.18}$$

(a)

(b)

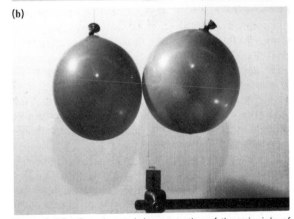

**Figure 1.10** Experimental demonstration of the principle of Bernoulli's equation. (a) Air-filled balloons hanging freely under gravity. (b) Balloons drawn together by the effect of a jet of air blown upwards between them.

9

where $E$ is the total energy (a constant), $p$ is the atmospheric pressure (a constant), $U$ the flow velocity considered uniform in the vertical, and $h$ the flow depth. Dividing through by $\varrho g$

$$\frac{E-p}{\varrho g} = \frac{1}{2}\frac{U^2}{g} + h \quad \text{m} \qquad (1.19)$$

where each term now has the dimensions of length and is called a head. The term to the left of the equals sign is a constant and may be called the total head $H$, an alternative way of expressing the total flow energy. Making use of Equation 1.2, we can replace $U$ by the unit discharge $q$ and depth $h$, whence Equation 1.19 becomes

$$H = \frac{1}{2}\frac{q^2}{gh^2} + h \qquad (1.20)$$

Equation 1.20 is a cubic in $h$, with a minimum value given by differentiating with respect to $h$ and setting $dH/dh = 0$. Thus

$$\frac{dH}{dh} = -\frac{2q^2}{2gh^3} + 1 = 0 \qquad (1.21)$$

$$\frac{q^2}{gh^3} = 1 \qquad (1.22)$$

Let us denote the depth and velocity at this minimum by respectively $h_{cr}$ and $U_{cr}$. Then as $U_{cr}^2 = q^2/h_{cr}^2$ by definition (Eqn 1.2),

$$\frac{U_{cr}}{(gh_{cr})^{1/2}} = 1 \qquad (1.23)$$

The group on the left in Equation 1.23 has no dimensions and is called the Froude number, denoted by $Fr$. It effectively measures the ratio of the inertial to the gravitational forces in the flow. When $Fr = 1$ the total flow energy is a minimum for a given discharge, and the flow is said to be critical. A subcritical flow is one for which $Fr < 1$. When $Fr > 1$ the flow is called supercritical. Thus unless $H$ is at its minimum value for each discharge, a flow may occur in either of two states, in one of which $h > h_{cr}$ and $U < U_{cr}$ (subcritical flow), and in the other of which $h < h_{cr}$ and $U > U_{cr}$ (supercritical flow). These states can often be seen to alternate along a fast-flowing river. Another demonstration is obtained by running a tap into a sink (Fig. 1.11). The radial expansion of the current causes an inner zone of super-

**Figure 1.11** Hydraulic jump formed round a vertical jet of water where it strikes and spreads out over a plane surface.

critical flow to become separated from an outer zone of much deeper, subcritical flow by a step-like feature called a hydraulic jump.

## 1.6 Energy losses during fluid flow

Bernoulli's equation for a viscous fluid, Equation 1.12, always includes an expression for frictional energy losses.

You can easily test this experimentally. Using identical T-pieces, attach a vertical glass tube to act as a manometer at each end of a long piece of rubber hose of uniform narrow bore. Run water steadily through the hose, arranged so that the lower ends of the manometers are at the same level (Fig. 1.12). You will see that the water stands at different heights $y$ in the manometer tubes, equivalent to a pressure difference $\Delta p = \varrho g \Delta y$ at the level of the T-pieces, where $\Delta y$ is the height difference. Now as the flow is steady, the hose of uniform bore, and the T-pieces of equal elevation, there is no change in the kinetic and potential energies between the ends of the hose. Therefore from Equation 1.12, $\Delta p = E_{loss}$. The pressure difference is thus the force driving the flow against friction at the walls of the hose. The total pressure force acting across the T-pieces is clearly $\Delta p(\pi r^2)$, where $r$ is the radius of the bore of the hose, and the term in brackets is the bore cross-sectional area. Correspondingly, the total resisting force is the product of the shear stress per unit area $\tau$ acting on the walls of the hose multiplied by the total wall area, i.e. $\tau(2\pi rL)$, where $L$ is the length of hose between the T-pieces. Equating the forces and writing out the resulting expression in terms of the resisting force gives

$$\tau = \Delta p \frac{(\pi r^2)}{2\pi rL} = \frac{\Delta pr}{2L} \, \mathrm{N\,m^{-2}} \qquad (1.24)$$

whence the stress increases linearly with the radius of the hose and inversely as its length.

We are here able to measure the shear stress directly, because we know the water density and the difference between the manometers. Direct measurement is impossible for most flows. However, notice that the kinetic energy is unchanged in the experiment. This suggests that we may write for a steady uniform flow

$$\tau = \text{coefficient of proportionality} \times (\tfrac{1}{2}\varrho U^2) \quad \mathrm{N\,m^{-2}} \qquad (1.25)$$

where the coefficient of proportionality is a pure number, varying from flow to flow. Equation 1.25 is called the quadratic stress law. The coefficient of proportionality is a friction coefficient, also called a 'drag' or 'resistance' coefficient. The law takes slightly

different forms depending on how the coefficient is defined. For a body in relative motion with an enclosing fluid, it is usually written

$$\tau = C_D(\tfrac{1}{2}\varrho U^2) \qquad (1.26)$$

where $C_D$ is called the drag coefficient. The law for a channelized flow is

$$\tau = \frac{f}{4}(\tfrac{1}{2}\varrho U^2) \qquad (1.27)$$

where $f$ is called the Darcy–Weisbach friction coefficient. Drag and friction coefficients have generally to be determined experimentally, but can for some flows be calculated theoretically. Try measuring $f$ for the hose, exploiting Equation 1.5 to obtain $U$.

Graphs, tables and empirical formulae for $C_D$ and $f$ are widely available, on account of the practical importance of these coefficients (e.g. Coulson & Richardson 1965, Guy et al. 1966, Clift et al. 1978).

## 1.7 Newton's laws of motion

### 1.7.1 First law

The conservation laws discussed above tell us what happens to mass, momentum and energy in a fluid system, and afford some insight into the origins of force. These insights are completed, and the relations of force to motion specified, by reference to Newton's three laws.

Newton's first law of motion states that a body will remain at rest, or will continue to move with constant velocity, unless an external force causes it to do otherwise. Alternatively, a change in the state of motion of a body is caused by a force.

The body shown in Figure 1.13 is at rest and acted on by four forces. As the body is stationary, the forces must balance, implying that $F_1$ is equal and opposite to $F_2$, and $F_3$ equal and opposite to $F_4$. Suppose that the body shown in Figure 1.14 moves at a constant velocity

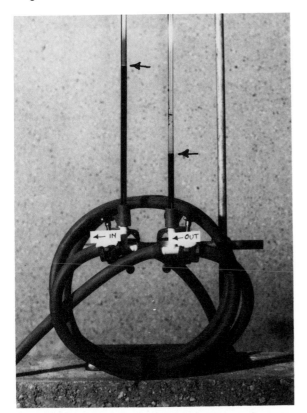

**Figure 1.12** Experiment to demonstrate energy loss as a consequence of fluid flow in a pipe. Note from the manometers (water darkened with ink) that the pressure at the entrance to the coil of piping exceeds that at the exit.

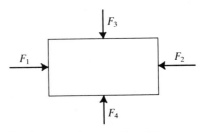

**Figure 1.13** Newton's first law of motion.

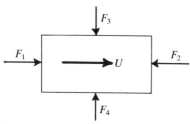

**Figure 1.14** Newton's first law of motion, in the case of a steadily moving body.

$U$. The law again requires that there is no resultant or net force acting on the body. Hence $F_1$ is again equal and opposite to $F_2$, and $F_3$ equal and opposite to $F_4$.

### 1.7.2 Second law

A resultant force acting on a body causes a change in the motion of the body from its previous state of either rest or constant velocity. This is Newton's second law, conveniently summarized as

$$F = ma \quad \text{N} \qquad (1.28)$$

where $F$ is the resultant force, $m$ the mass in kilograms of the body, and $a$ its acceleration in $\text{m s}^{-2}$. An expression of the general form of Equation 1.28 is called an equation of motion.

Let the body of mass 10 kg shown in Figure 1.15 be acted on by a force of 5 N to the right and another of 4 N to the left. The resultant force is therefore $(5 - 4) = 1$ N, affording the acceleration to the right of $0.1 \text{ m s}^{-2}$. Suppose that equal and opposite vertical forces also act on the body. It will continue to move to the right at the previous acceleration, for there is no resultant vertical force and therefore no vertical motion.

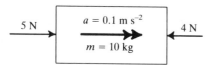

**Figure 1.15** Newton's second law of motion.

### 1.7.3 Third law

Newton's third law of motion states that action (a force) and reaction (another force) are equal and opposite. Consider a brick of mass 5 kg resting on a horizontal table (Fig. 1.16). The action is a force of $5g$ N acting vertically downwards from the brick to the table, and represents the attraction of the brick to the Earth. The reaction is consequently a force of $5g$ N acting vertically upwards from the table to the brick.

In Section 1.4 we saw that the momentum of a jet

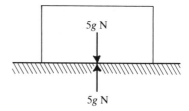

**Figure 1.16** Newton's third law of motion.

directed normal to a rigid surface is totally destroyed by the surface, with the result that an impulsive stress acting from the surface to the jet appears (Fig. 1.7). The reaction is an equal stress directed from the jet towards the surface. A reaction force also arises when the jet is inclined to the surface.

## 1.8 Fluid viscosity

Viscosity can be defined as the 'stickiness' of a fluid and its capacity to resist changes of shape. Practically everyone has experience of this particular fluid property. Compare the ease with which cups of water, lubricating oil and either liquid honey or sugar syrup at room temperature can be stirred. In terms of the force necessary to achieve a given stirring rate, you cannot avoid but rank the liquids in the same order as above.

How can we define viscosity more precisely? Suppose that a quantity of fluid is placed in the small gap of thickness $\delta y$ between two parallel plates, one of which is fixed and the other free to move in its own plane (Fig. 1.17). If a force of $F$ N is applied to the free plate, we should expect to see it move. Now the fluid molecules adjoining the plates stick to the plates, whence we should expect the motion of the free plate to be transmitted through the whole fluid layer, affording the velocity profile $U(y)$. We are now subjecting the fluid to the kind of deformation called simple shear. Let the force displace the free plate a small increment $\delta x$ in a small increment of time $\delta t$. The simplest proportionality we could expect between the rate of displacement $\delta x/\delta t$, the thickness of the fluid layer and the force is

$$\frac{\delta x}{\delta t} \propto \frac{F}{A} \delta y \qquad (1.29)$$

where $A$ is the wetted area of the moving plate. From our efforts to stir water, oil and syrup, we should expect $\delta x/\delta t$ to decline at a constant force as the fluid between plates a fixed distance apart became more viscous. Denoting the viscous quality by the Greek letter $\eta$ (eta),

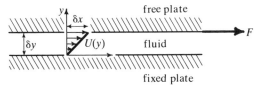

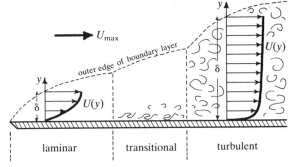

**Figure 1.17** Definition sketch for the viscosity of a fluid confined between parallel plates, one of which is moving in its own plane.

**Figure 1.18** Schematic representation in streamwise section of the boundary layer growing on a flat plate parallel to a fluid stream.

such that the value of $\eta$ increases with growing viscosity, Equation 1.29 can be rearranged and restated as the equality

$$\frac{\delta x/\delta t}{\delta y} = \frac{1}{\eta}\frac{F}{A} \quad \mathrm{s}^{-1} \tag{1.30}$$

In the limit, the left-hand term becomes the velocity gradient $dU/dy$, and noting that $F/A$ is the shear stress,

$$\tau = \eta\frac{dU}{dy} \quad \mathrm{N\,m}^{-2} \tag{1.31}$$

The units of viscosity are therefore newton seconds per square metre, abbreviated to $\mathrm{N\,s\,m}^{-2}$, also called the pascal second (Pa s) (Table 1.1). Water, for example, has a viscosity of $1.06 \times 10^{-3}\,\mathrm{N\,s\,m}^{-2}$ at $18°C$, whereas that of air is about $1.8 \times 10^{-5}\,\mathrm{N\,s\,m}^{-2}$. The viscosity of magmas is very variable but is orders of magnitude greater than that of water.

Equation 1.31 defines the coefficient of viscosity. Physically, viscosity represents the momentum that the flow loses in order to sustain a velocity gradient $dU/dy$, as you can see from the units of viscosity. A velocity gradient will be detectable, and there will be a shear stress, only when a fluid is moving relative to its boundaries, which may be either a rigid surface or another fluid with a different motion. There is no shear stress when the fluid is at rest relative to its boundaries. Viscosity causes shear stresses, and is the only cause of shear stresses in fluids.

## 1.9 Boundary layers

Where a steady fluid stream meets a stationary boundary, viscosity slows down the fluid in a thin, downstream-thickening zone adjoining the boundary (Fig. 1.18). In this zone, called the boundary layer, the streamwise velocity changes continuously normal to the boundary, that is, a viscosity-dependent shear stress is present. Outside the boundary layer, there is no velocity gradient and the fluid appears to be inviscid. The con-

cept of the boundary layer, due to the German aerodynamicist Ludwig Prandtl, therefore leads us to a useful 'two-layer' model of fluid flow.

The flat-plate boundary layer shown in Figure 1.18 is tripartite, with an initial laminar section. Where the motion is laminar, the temporal mean velocity at a fixed point and the instantaneous velocity at that point are exactly the same. The laminar motion also means that the streamlines do not intertwine, and that the momentum transfer implied by the velocity gradient proceeds only at a molecular scale. Experimentally, the theoretical expression

$$U = \frac{\tau}{\eta}\left(y - \frac{y^2}{2\delta}\right) \quad \mathrm{m\,s}^{-1} \tag{1.32}$$

describes rather well the velocity distribution within a laminar boundary layer on a flat plate, where $U$ is the streamwise velocity at height $y$ above the boundary, $\tau$ the shear stress at the boundary, $\eta$ the viscosity and $\delta$ the boundary-layer thickness. Equation 1.32 states that the velocity profile is a parabola. Substituting into the equation $y = \delta$ and $U = U_{max}$, where $U_{max}$ is the flow velocity outside the boundary layer, gives

$$\tau = \frac{2\eta U_{max}}{\delta} \quad \mathrm{N\,m}^{-2} \tag{1.33}$$

As the viscosities of air and water turn out to be very small, the shear stresses associated with naturally occurring laminar boundary layers are also very small.

At a certain distance along the plate (Fig. 1.18), the laminar boundary layer is so thick that the motion ceases to be stable and proceeds to become turbulent. By turbulent we mean that the flow comprises an assembly of more or less large parcels or eddies of fluid which, although travelling together, have a complex

13

internal motion and a certain independence of behaviour. The boundary layer is called transitional over the length of plate where this change is occurring. Excellent illustrations of instability waves, hairpin vortices and other organized motions accompanying transition to turbulence are given by Kegelman et al. (1983) and Taneda (1983).

Eddies are very successful in transferring high-momentum fluid from the outer to the inner parts of a turbulent boundary layer (Fig. 1.18). This effectively means that a turbulent fluid has a 'viscosity' that is some orders of magnitude greater than the true viscosity. The idea of a turbulent viscosity is helpful in solving many problems of turbulent flow, provided we remember that we are dealing with neither a true viscosity nor a constant quantity, even within a single boundary layer. Experimentally, the formula

$$\frac{U}{U_{\max}} = \left(\frac{y}{\delta}\right)^n \qquad (1.34)$$

describes rather well the velocity profile in a turbulent boundary layer, where $U$ is to be understood as the local time-averaged velocity. The exponent $n$ varies from about 1/5 near transition to about 1/7 much further down stream. Useful as is Equation 1.34, it does conceal the fact that a turbulent boundary layer is divisible into regions on the basis of the precise form of the velocity profile (Sec. 6.2.1). A particularly interesting and important one – the viscous sublayer – is thin and lies immediately adjacent to the wall. Within the viscous sublayer the flow is effectively laminar and the velocity increases linearly outwards from the boundary, as described by Equation 1.31. Nonetheless, a turbulent boundary layer exerts a much greater boundary shear stress than a laminar one, because of the large turbulent viscosity. Note the comparative steepness of the lower part of the profile in Figure 1.18. As the turbulent viscosity is not a constant, the stress has generally to be calculated with the help of empirical data.

The following experiment (Fig. 1.19) will give you an insight into turbulence. Take a glass tube with a bore of about 0.01 m and flare one end slightly in a flame. Attach to the other end a substantial length of rubber hose fitted with a screw clip, arranging the tube vertically in a large tank of still water to form a siphon. Draw out another glass tube into a fine capillary and fit it to a flexible bulb full of a solution of potassium permanganate. Clamp the capillary so that it points axially up the vertical tube. Fill the siphon and close the screw clip. Opening the clip induces a current in the tube. On simultaneously squeezing the bulb, the motion is seen as

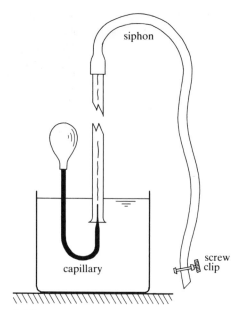

**Figure 1.19** Apparatus necessary for Reynolds's experiment on laminar and turbulent flow in a pipe.

the behaviour of a thread of coloured fluid. Open the clip slightly to obtain a gentle current. The coloured thread, effectively a streamline, remains straight and coherent along the flow (Fig. 1.20a). Now open the clip increasingly wide, to increase the flow velocity. At a certain velocity, and position in the tube, the thread loses coherence and colour invades the whole fluid (Fig. 1.20b). The motion is turbulent and the streamlines are intertwined. By blinking rapidly, you may be able to glimpse individual eddies.

Osborne Reynolds, a British 19th century physicist, realized from this experiment that the laminar–turbulent transition occurred when the inertial flow forces exceeded by a critical ratio the viscous ones. This ratio is expressed by the non-dimensional quantity called the Reynolds number, denoted by $Re$. For a pipe

$$Re = \frac{\text{inertial force}}{\text{viscous force}} = \frac{2\varrho r U_{\text{om}}}{\eta} \qquad (1.35)$$

where $\varrho$ is the fluid density, $r$ the radius, and $U_{\text{om}}$ the overall mean velocity. The critical Reynolds number for the laminar–turbulent transition is sensitive to experimental conditions, but is of the order of $10^4 - 10^5$. For a free-surface flow (or boundary layer)

$$Re = \frac{\varrho h U_{\text{om}}}{\eta} \qquad (1.36)$$

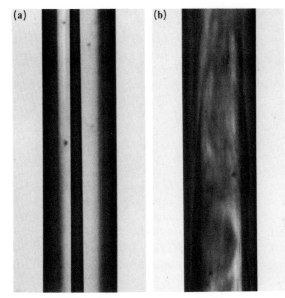

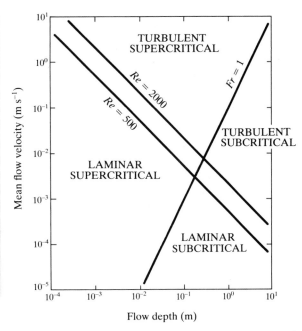

**Figure 1.20** Results from Reynolds's experiment. (a) Coherent thread of colour under laminar conditions. (b) Disturbed thread of colour under turbulent conditions. Flow is upwards.

**Figure 1.21** Regimes of free-surface flow, in terms of critical Reynolds and Froude numbers.

where $h$ is the flow depth (or boundary-layer thickness). A Reynolds number defined using downstream distance could instead be applied to the transition in Figure 1.18.

It is generally accepted that free-surface flows are invariably laminar at Reynolds numbers less than 500–2000. Combining the Reynolds number (Eqn. 1.36) with the Froude number (Eqn. 1.23), four possible regimes of channel or free-surface flow may be recognized (Fig. 1.21). Rivers are generally turbulent, with laminar motion restricted to sheet flows. The wind is invariably turbulent, as the atmospheric boundary layer is of the order of 1 km thick.

## 1.10 Flow separation

In both kinds of boundary layer, the near-wall fluid has little kinetic energy and, by Bernoulli's equation (Eqn. 1.12), would be arrested if it encountered even a slight positive (adverse) pressure gradient. The arrested fluid, forming a distinct and partly or wholly closed region within the fluid, would then displace outwards the still-flowing portion of the stream, separating it from the flow boundary. This is the phenomenon of flow separation, the region of recirculating arrested fluid forming what is called a separation bubble (Fig. 1.22). At S, called a separation point, two streamlines lying in the

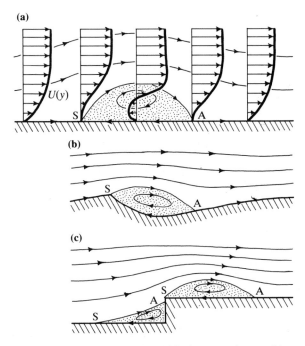

**Figure 1.22** Flow separation (a) in the general case, (b) at the crest of a ripple, dune, bar or forward-facing step, and (c) at a cliff or backward-facing step.

flow boundary join. The velocity profile here yields $dU/dy = 0$ at the boundary, in contrast to the large wall value in the boundary layer up stream. The streamline passing through the fluid, called the separation streamline, reattaches to the bed at A, known as an attachment point. Here again $dU/dy = 0$. Between the separation streamline and the bed lies the separation bubble, the flow outside the bubble being called the external or separated stream.

Flow separation is a widespread phenomenon of great practical importance (Chang 1970, Hunt *et al.* 1978). Sedimentologically, it occurs to leeward of such raised bed features as ripples and dunes (Fig. 1.22b), and up stream of bluff bodies such as stones on a beach and cliffs facing the wind (Fig. 1.22c).

## 1.11 Applying the concepts and rules

### 1.11.1 Posing a problem

Now that the basic concepts and rules of fluid mechanics have been sketched, how easily are they applied to sedimentological problems, and are the insights gained worth the trouble?

In an attempt to answer this question, we shall start an exploration of the problem of storm sedimentation, to be completed in the last chapter of this book (Ch. 13). Common experience teaches us that greatly increased wave activity accompanies storms at sea or on large lakes, but this is not the issue of immediate interest. We may reasonably suspect that a storm is also accompanied by currents and water-level changes due specifically to the storm. We may further suspect that these effects influence the erosion, transport and deposition of sediment during the storm, and that there are deposits in the rock record bearing the imprint of ancient storms. Our difficulty is that, at present, we know very little by direct observation about the sedimentological consequences of such events. Speculation and misunderstanding, on the other hand, are rife. Could a mathematical analysis afford an insight? The best insight will come when the problem has been well posed, that is, when we have (1) identified all its essential physical aspects, (2) made the minimum of assumptions, and (3) run calculations using the equations deduced.

### 1.11.2 Barometric effect

Storms normally accompany deep atmospheric depressions, that is, regions where the atmospheric pressure is significantly less than in adjoining anticyclones and ridges of high pressure. The level of a sea or lake should respond to these spatial pressure variations.

Consider the barometric effect accompanying a storm advancing across a large water body of uniform undisturbed depth (Fig. 1.23). As storms travel comparatively slowly, and water has little viscosity and so responds quickly to disturbing forces, it is reasonable to expect the response of the water surface to differ little from that under static conditions. Consider the forces on a horizontal plane below the water surface, under conditions of static equilibrium. The only forces relevant to the problem are therefore vertical ones. The total downward force acting on the plane at section 1 is the sum of the atmospheric hydrostatic pressure $p_{a,1}$ and the aqueous hydrostatic pressure $p_{w,1}$. An equal and opposite reaction acts upwards on the plane, by Newton's third law. At $x_2$ the corresponding total forces are the sum of $p_{a,2}$ and $p_{w,2}$. From our assumption

$$p_{a,1} + p_{w,1} = p_{a,2} + p_{w,2} \quad \text{N m}^{-2} \qquad (1.37)$$

since there would otherwise be motion in response to a force gradient. Bringing the terms for the water to the left, and those for the atmosphere to the right, we get

$$\Delta p_w = \Delta p_a \qquad (1.38)$$

Now $\Delta p_w = \varrho g \, \Delta y$, where $\varrho$ is the density of water, $g$ the acceleration due to gravity, and $\Delta y$ the difference in water level between the sections. Therefore

$$\Delta y = \frac{\Delta p_a}{\varrho g} \quad \text{m} \qquad (1.39)$$

The elevation difference is linearly proportional to the pressure difference; the water stands highest where the atmospheric pressure is least.

What does Equation 1.39 mean quantitatively? Normal atmospheric pressure, called the bar, is gener-

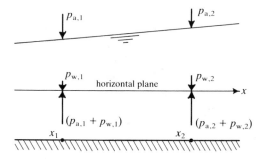

**Figure 1.23** Definition sketch for the barometric effect associated with a storm.

ally taken to equal the pressure due to a vertical column of mercury of density $1.36 \times 10^4 \text{ kg m}^{-3}$ measuring 0.76 m in length, that is, $1.014 \times 10^4 \text{ N m}^{-2}$. Atmospheric pressure in a deep depression is typically 0.05 bar less than in the surrounding areas, whence the expected water-level difference is approximately 0.5 m. If the water body is restricted, its actual rise above the undisturbed level will be less than 0.5 m, as a bulge in one place must by continuity be compensated by a drop in another. The calculated rise is not inconsiderable and, when accompanied by wave action, could contribute to the overtopping of coastal barriers and sea defences.

### 1.11.3  Wind-stress current

The wind blows during a storm and must therefore stress the water surface. Two distinct effects may be expected: (1) a drift of water in the same direction as the wind, and (2) a windwise tilting up of the water surface, in order by means of a pressure gradient to balance the wind shear. We can treat these effects only by separating them.

The shape and position of the water surface under barometric static equilibrium (Eqn. 1.37) is compatible with the existence of a steady wind-stress current if we imagine the surface to be replaced by a rigid lid moving in its own plane (Fig. 1.24). Thus we exclude the additional tilt due exclusively to the wind. As we can suppose that no other currents stir the water body, the simplest assumption about the internal motion is that it is laminar. The situation is like that in Figure 1.17, and for steady conditions we can write

$$\tau_a = \eta \frac{dU}{dy} \quad \text{N m}^{-2} \tag{1.40}$$

where $\tau_a$ is the shear stress exerted by the wind on the water surface, the wind blowing in the positive $x$-direction, $\eta$ is the viscosity of the water, and $U$ the velocity at height $y$. Integration of Equation 1.40 yields the velocity profile

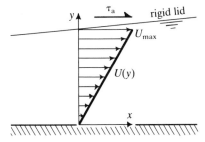

**Figure 1.24**  Definition sketch for the laminar wind-stress current associated with a storm.

$$U = \frac{\tau_a}{\eta} \int dy = \frac{\tau_a}{\eta} y + C \quad \text{m s}^{-1} \tag{1.41}$$

where $C$ is the integration constant. As the water molecules must stick to the bed, we have the boundary condition $U = 0$ at $y = 0$. Insertion of this condition into Equation 1.41 gives $C = 0$, whence the steady wind-stress current is distributed as

$$U = \frac{\tau_a}{\eta} y \tag{1.42}$$

and the velocity $U_{max}$ at the surface is

$$U_{max} = \frac{\tau_a}{\eta} h \tag{1.43}$$

where $h$ is the depth. Hence the wind-stress current flows with the wind, increases linearly upwards, and is a maximum at the water surface.

### 1.11.4  Gradient current

The other expected effect is the windwise upward tilt of the water surface, as a necessary consequence of the wind shear. In real cases, this tilt will be in addition to any slope related to barometric static equilibrium. In order to model the effect, however, let us suppose that the tilt shown in Figure 1.25 occurs under a uniform atmospheric pressure and represents a steady state.

With the wind blowing steadily in the positive $x$-direction, consider a windwise slice through the water body of length $\delta x$ and unit width. The pressure force acting on the vertical face at $x_1$ is

$$F_1 = \varrho g \int^h (h - y) \, dy = \tfrac{1}{2} \varrho g h^2 \quad \text{N m}^{-1} \tag{1.44}$$

and at $x_2$ is

$$F_2 = \varrho g \int^{h + \delta y} (h + \delta y - y) \, dy = \tfrac{1}{2} \varrho g [h^2 + 2h \, \delta y + (\delta y)^2] \tag{1.45}$$

where $\delta y$ is the small rise in water level over the small distance $\delta x$. Hence the net pressure force is

$$F_1 - F_2 = -\tfrac{1}{2} \varrho g [2h \, \delta y + (\delta y)^2] \tag{1.46}$$

As conditions are steady, this force by Newton's first law is balanced by a shear force $\tau_w \, \delta x$ exerted by the bed on the water body, implying the presence of another internal current. Equating the forces, the drag exerted by the water body on the bed is

$$\tau_w \, \delta x = -\tfrac{1}{2} \varrho g [2h \, \delta y + (\delta y)^2] \tag{1.47}$$

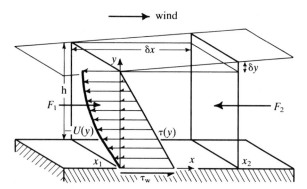

**Figure 1.25** Definition sketch for the laminar gradient current associated with a storm.

which in the limit, and neglecting squared terms, yields

$$\tau_w = -\varrho gh \frac{dy}{dx} \quad \text{N m}^{-2} \qquad (1.48)$$

where $dy/dx$ is the surface slope. The boundary shear stress therefore increases linearly with the water-surface slope, for uniform atmospheric pressure.

The driving force given by Equation 1.46 acts in the negative $x$-direction, implying that the tilt-related current flows also in this direction, and therefore contrary to the wind. Because of its origin, this flow is called the gradient current.

The gradient current may be modelled using the boundary conditions. At the bed, $y = 0$, the shear stress, Equation 1.48, is a maximum. At the free surface, however, $y = h$ and $dU/dy = 0$, where $U$ is the velocity at height $y$. Hence at the free surface there is no shear stress due to the gradient current. The simplest model for the variation of the shear stress $\tau$ within the water body is therefore the linear relationship

$$\tau = \tau_w \left(1 - \frac{y}{h}\right) \quad \text{N m}^{-2} \qquad (1.49)$$

Continuing to assume laminar flow

$$\tau = \eta \frac{dU}{dy} \qquad (1.50)$$

from Equation 1.31. Eliminating $\tau$ between Equations 1.49 and 1.50, and substituting for $\tau_w$ from Equation 1.48, gives

$$\frac{dU}{dy} = -\frac{\varrho gh(dy/dx)}{\eta} \left(1 - \frac{y}{h}\right) \quad \text{s}^{-1} \qquad (1.51)$$

The velocity profile for the gradient current is the integral of Equation 1.51

$$U = -\frac{\varrho gh(dy/dx)}{\eta} \left(y - \frac{y^2}{2h}\right) \quad \text{m s}^{-1} \qquad (1.52)$$

using the boundary condition $U = 0$ at $y = 0$ to determine the integration constant. The gradient current therefore has a parabolic profile, as stated for the laminar boundary layer (Eqn. 1.32), and opposes the wind.

### 1.11.5 Combined current and water-surface slope

Equation 1.42 gives us the wind-stress current and Equation 1.52 the current due to the wind-related surface tilt. The observed current is therefore their sum

$$U = \frac{\tau_a}{\eta} y - \frac{\varrho gh(dy/dx)}{\eta} \left(y - \frac{y^2}{2h}\right) \quad \text{m s}^{-1} \qquad (1.53)$$

The current cannot be calculated, however, without an expression for $dy/dx$.

This expression follows using continuity. The integral of Equation 1.53 over depth is the unit discharge of the combined currents. For a confined body of water like a lake, or for a shelf sea affected by an onshore wind, the unit discharge must be zero, that is, any flow in the shoreward direction will be precisely balanced by an offshore flow, since the shore is impenetrable. Thus

$$0 = \int_0^h \left[\frac{\tau_a}{\eta} y - \frac{\varrho gh(dy/dx)}{\eta} \left(y - \frac{y^2}{2h}\right)\right] dy \quad \text{m}^3\,\text{m}^{-1}\text{s}^{-1} \qquad (1.54)$$

which, evaluated for the tilt, affords

$$\frac{dy}{dx} = \frac{3\tau_a}{2\varrho gh} \qquad (1.55)$$

The tilt is linearly proportional to the wind shear and inversely proportional to the depth. Substituting from Equation 1.55 into Equation 1.53

$$U = \frac{\tau_a}{\eta} \left(\frac{3y^2}{4h} - \frac{y}{2}\right) \quad \text{m s}^{-1} \qquad (1.56)$$

Putting $y = h$ gives the surface current

$$U_{max} = \frac{\tau_a h}{4\eta} \qquad (1.57)$$

By differentiating Equation 1.56 and setting $y = 0$, it is easily seen using Equation 1.31 that $\tau_w = \frac{1}{2}\tau_a$.

Equation 1.56 is a parabola, indicating a windwise (shoreward) current in the upper one-third of the water body, and a gentler offshore flow in the two-thirds below (Fig. 1.26). As the bottom flow is off shore, quantities of sediment could be carried during a storm from the shallows and over the wave-stirred bottom into the deeps. As it stands, however, Equation 1.56 is strictly applicable only to conditions of laminar flow, and if used with the molecular viscosity of water (of the order of $1 \times 10^{-3}\,N\,s\,m^{-2}$) will be found to yield absurdly large currents for representative storm conditions and shelf depths. Conditions in fact are ordinarily turbulent at sea, so at the very least we must use a 'turbulent' viscosity for the water, likely to be orders of magnitude larger than the ordinary viscosity, and it may also be necessary to accept forms for the profiles of the wind-stress and gradient currents different from those in the laminar case. The principles which led to Equation 1.56 will nonetheless continue to apply in any analysis of the turbulent case. Equation 1.55 is not constrained in the above way and suggests that the wind stress accompanying a storm will create at the shore a significant superelevation of the water surface, depending on the shape of the water body. This superelevation, combined with any barometric changes, constitutes a storm surge (Heaps 1967, 1983). Hurricanes, for example, commonly force coastal water levels to surge several metres above normal, causing much erosion and flooding.

## Readings

Batchelor, G. K. 1968. *An introduction to fluid dynamics*. London: Cambridge University Press.

Bradshaw, P. 1971. *An introduction to turbulence and its measurement*. Oxford: Pergamon.

Chang, P. K. 1970. *Separation of flow*. Oxford: Pergamon.

Clift, R., J. R. Grace and M. E. Weber 1978. *Bubbles, drops and particles*. New York: Academic Press.

Coulson, J. M. and J. F. Richardson, 1965. *Chemical engineering*, Vol. 1. Oxford: Pergamon.

Feather, N. 1961. *Mass, length and time*. Harmondsworth: Penguin.

Francis, J. R. D. 1975. *Fluid mechanics for engineering students*, 4th edn. London: Arnold.

Guy, H. P., D. B. Simons and E. V. Richardson 1966. *Summary of alluvial channel data from flume experiments, 1956–61*. Prof. Pap. US Geol. Surv., no 462-I.

Heaps, N. S. 1967. Storm surges. In *Oceanography and marine biology*, H. Barnes (ed.), Vol. 5, 11–47. London: Allen & Unwin.

Heaps, N. S. 1983. Storm surges, 1967–82. *Geophys. J. R. Astron. Soc.* **74**, 331–76.

Hunt, J. C. R., C. J. Abell, J. A. Peterka, and H. Woo, 1978. Kinematical studies of the flows around free or surface-mounted obstacles; applying topology to flow visualization. *J. Fluid Mech.* **86**, 179–200.

Kegelman, J. T., R. C. Nelson and T. J. Mueller 1983. The boundary layer on an axisymmetric body with and without spin. *J. Am. Inst. Aeronaut. Astronaut.* **21**, 1485–91.

Milne-Thomson, L. M. 1962. *Theoretical hydrodynamics*, 4th edn. London: Macmillan.

Sadler, A. J. and D. W. S. Thorning, 1983. *Understanding mechanics*. Oxford: Oxford University Press.

Scorer, R S. 1978. *Environmental aerodynamics*. Chichester: Horwood.

Sellin, R. H. J. 1969. *Flow in channels*. London: Macmillan.

Taneda, S. 1983. Visual observations on the amplification of artificial disturbances in turbulent shear flows. *Phys. Fluids* **26**, 2801–6.

Tritton, D. J. 1977. *Physical fluid dynamics*. Wokingham: Van Nostrand.

Turner, J. S. 1973. *Buoyancy effects in fluids*. Cambridge: Cambridge University Press.

Van Dyke, M. 1982. *An album of fluid motion*. Stanford: Parabolic Press.

Yih, C. S. 1980. *Stratified flows*, 2nd edn. New York: Academic Press.

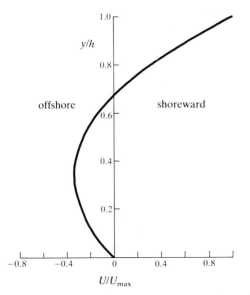

**Figure 1.26** The combined laminar current (gradient plus wind-stress currents) due to a storm, normalized by the velocity at the water surface.

# 2 Pressed down and running over

Composition and density of sediment particles – particle size and its description – particle form – particle packings – interparticle voids – controls on particle packing – angles of repose – stability of slopes on sand.

## 2.1 Introduction

Discrete mineral particles are the other essential ingredient of the two-phase fluid–particle system we are exploring. Many of these detrital particles will have become detached as the result of weathering from rocks exposed on the Earth's surface. Others may have been produced by organisms manufacturing hard tissues. Some particles will have been formed through the precipitation of salts from natural waters.

A walk along the seashore or a river bank highlights many of the individual and bulk properties of detrital particles. They range enormously in form and size. Here are discoidal stones so large that they cannot easily be picked up, but there is a sand bed composed of round grains just a millimetre or so across, and there a patch of mud composed of platy or needle-shaped particles so small that most are invisible beneath a powerful lens. The composition of detrital particles is equally varied. Whereas most sand grains each consist of a single mineral, stones are likely to comprise a mineral aggregate and to be referable to some definite rock type, for example, granite or greywacke. Dominating the mud are mineral species belonging to the large and diverse clay mineral group. The kinds of particles mentioned do not occur alone, but in the sediment are packed to form a more or less porous and permeable aggregate. The bulk properties of these packings are affected by particle size, because in detrital sediments there is a correlation between size and mineral composition. Dry particles of sand and stones, and particles of these sizes when water-saturated, do not adhere together because there are no significant interparticle surface forces. Sediments dominated by such particles may therefore be described as cohesionless: dry sand, for example, can be freely poured from one container to another. But mud displays cohesion, that is, it sticks together, resisting disruption, because the tiny clay mineral grains which predominate in its composition are bound together by powerful surface forces. Hence muddy sediments have quite different bulk properties and behaviour from cohesionless ones, and are conveniently reserved to a later chapter.

## 2.2 Particle composition and density

Table 2.1 lists the chemical composition and density of the chief detrital minerals present in cohesionless sediments. Of these minerals, the commonest are quartz, the feldspars (orthoclase, microcline, plagioclase series), calcite and aragonite. Their density is roughly 2.5–3 times that of water, but about $2 \times 10^3$ times the density of air at normal temperature and pressure. The so-called accessory minerals normally total no more than a small fraction of 1% by weight, but under some circumstances can form a substantial or even predominant part of the sediment. These minerals are significantly more dense than the common species, and for this reason are often described as heavy minerals.

The chief clay minerals appear in Table 2.1 for comparison, although they are rare to absent in cohesionless sediments. In terms of solids density, clay minerals compare with quartz, feldspar and the carbonates, but in turbid waters and liquid muds typically occur as large

**Table 2.1** Density of common minerals (solids density) and rocks (bulk density).

| Name | Formula | Density ($kg\ m^{-3}$) |
|---|---|---|
| **Common minerals** | | |
| aragonite | $CaCO_3$ | 2930 |
| biotite | $K(Mg,Fe)_3(AlSi_3O_{10})(OH)_2$ | 2800–3400 |
| calcite | $CaCO_3$ | 2710 |
| dolomite | $CaMg(CO_3)_2$ | 2870 |
| gypsum | $CaSO_4 \cdot 2H_2O$ | 2320 |
| microcline | $KAlSi_3O_8$ | 2560 |
| muscovite | $KAl_2(AlSi_3O_{10})(OH)_2$ | 2800–2900 |
| orthoclase | $KAlSi_3O_8$ | 2550 |
| plagioclase series | $(Na,Ca)(Al,Si)AlSiO_2O_8$ | 2620–2760 |
| quartz | $SiO_2$ | 2650 |
| **Accessory minerals** | | |
| apatite | $Ca_5(PO_4)_3(F,Cl,OH)$ | 3100–3250 |
| corundum | $Al_2O_3$ | 3990 |
| epidote | $Ca_2(Al,Fe)_3Si_3O_{12}(OH)$ | 3250–3500 |
| fluorite | $CaF_2$ | 3180 |
| garnet group | $(Fe,Al,Mg,Mn,Ca,Cr)_5(SiO_4)_3$ | 3560–4320 |
| haematite | $Fe_2O_3$ | 5270 |
| hornblende | $NaCa_2(Mg,Fe,Al)_5(Si,Al)_8O_{22}(OH)_2$ | 3000–3470 |
| ilmenite | $FeTiO_3$ | 4790 |
| kyanite | $Al_2SiO_5$ | 3670 |
| monazite | $(Ce,La,Y,Th)PO_4$ | 5270 |
| olivine series | $(Mg,Fe)_2SiO_4$ | 3210–4390 |
| pyrite | $FeS_2$ | 5020 |
| pyroxene group | $(Ca,Mg,Fe)_2(Si,Al)_2O_6$ | 3200–3550 |
| rutile | $TiO_2$ | 4250 |
| staurolite | $FeAl_4Si_2O_{10}(OH)_2$ | 3700 |
| tourmaline | $Na(Mg,Fe)_3Al_6(BO_3)_3(Si_6O_{18})(OH)_4$ | 3030–3100 |
| zircon | $ZrSiO_4$ | 4670 |
| **Clay minerals** | | |
| kaolinite | $Al_4Si_2O_5(OH)_4$ | 2600 |
| illite | $KAl_2(AlSi_3O_{10})(OH)_2$ | 2800–2900 |
| chlorite series | $(Mg,Fe,Al)_6(Al,Si)_4O_{10}(OH)_8$ | 2600–3300 |
| smectite | $Al_2Si_4O_{10}(OH)_2 \cdot nH_2O$ | 2000–2700 |
| **Rock types** | | |
| diorite | | 2720–2960 |
| gabbro | | 2850–3120 |
| gneiss | | 2660–2730 |
| granite | | 2520–2810 |
| granodiorite | | 2670–2790 |
| sandstone | | 2170–2700 |
| schist | | 2700–3000 |
| slate | | 2720–2840 |

aggregate grains called floccules. These can reach several millimetres in size and consist of an open mesh of much smaller clay mineral crystals. Because of the relatively large amounts of fluid trapped within the mesh, the bulk or overall density of these floccules exceeds by only a little that of water. Some biogenic carbonate grains also have a bulk density substantially smaller than that of the mineral involved. This derives from the presence of empty pores within the organic hard tissue, as in the calcite plates secreted by sea urchins. The bulk density of such biogenic particles as snail shells and the tests of foraminifera is also low, on account of the enclosed empty spaces.

The bulk density of the common rocks, which as such contribute significantly to many sediments, is similar to that of the common rock-forming minerals, an obvious reflection of their mineralogical composition. The sandstones listed in Table 2.1 are variable in bulk density because some examples contain more empty pores than others. The density range for such as granite reflects the permissible variation in composition of this rock type.

## 2.3 How big is a particle?

Carefully examine a pebble drawn at random from a heap. What is the pebble's relative size or bigness? The question is not easily answered because the particle lacks a regular shape. What the question asks, however, is how much space the pebble occupies or, so to speak, defines by its presence. Our answer can vary depending on whether we choose to consider the pebble in one, two or three dimensions.

The first thing to notice is that the pebble can be described by measuring with a caliper gauge the long $a$, intermediate $b$, and short $c$ dimensions on orthogonal axes (Fig. 2.1). Any one of these measurements $a > b > c$ could be used to characterize the pebble, but such a representation would neglect its solidity.

An improved choice would be the mean of any two of the axes. The quantity $(a + b)/2$ would be best, since this represents the pebble in its most stable horizontal position. Even so, only two intersections of the pebble have

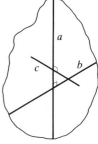

**Figure 2.1** The three dimensions of an irregular sedimentary particle.

actually been measured. Alternatively, we could measure the area of the pebble as seen in the stable horizontal position, and then calculate the diameter of the circle of the same area, that is, the quantity $D_A = (4A/\pi)^{1/2}$, where $D_A$ is the equivalent-area diameter and $A$ the pebble projected area.

A better choice still would be the quantity $(a + b + c)/3$, that is, the mean of the three axial lengths. But a previous criticism can again be raised, for we have merely used three intersections and neglected the remaining substance. The only way to involve the whole pebble in the determination of a characteristic dimension is to measure its volume, say, by dropping the water-saturated particle into a measuring cylinder of water. If $V$ is the volume displaced, equal to the overall pebble volume, then $D_V = (6V/\pi)^{1/3}$ is the equivalent-volume diameter of the particle.

In each of these cases the relative size or bigness of the pebble has been specified in terms of a single length, that is, one-dimensionally. In one case, however, this length is proportional to a measured area, and in another to a measured volume. The description of a particle in areal or volumetric terms is just as valid as a linear characterization, and in many dynamical problems is more appropriate.

The available methods of measurement (Allen 1968) specify particle size in a variety of ways. It is customary to measure only the long axes of sand grains mounted for microscopical study. Not only are the particles in this instance characterized one-dimensionally, but the basic measurements are restricted to one dimension. Similarly, whether or not a sand grain passes through a sieve depends on its intermediate dimensions relative to the mesh size. In contrast, the so-called hydraulic methods, modern as well as classical, describe the particle volumetrically. The pipette and hydrometer methods, and the sedimentation tower, all depend on the fall velocity of a particle, a function partly of the volume. In the Coulter counter, size is assessed as the extent to which the grain reduces the volume of an electrolyte-filled space carrying an electric current.

Having obtained a measurement of particle size, we next require a size classification for the purpose of description and communication. Size classifications or grade scales are simply arbitrary divisions of a continuous range of particle sizes, such that each scale unit or grade is of a convenient magnitude. One problem immediately presents itself, in view of the huge range of sizes of particles apparent in natural sediments, from clay mineral crystals a few micrometres ($\mu$m) long to blocks of rock tens of metres across. Division of such a large range into equal arithmetic units will clearly not

suffice. A scale of ratios, or geometric scale, seems preferable. Wentworth's (1922) ratio scale, with a class interval of 2 (or $2^{1/2}$, $2^{1/4}$ if finer subdivisions are required), is the one most favoured by sedimentologists (Table 2.2). Now a geometric scale is also logarithmic, since the logarithm $y$ of any number $x$ is defined as $x = n^y$, where $n$ is a number convenient for the base. This means that we can transform the geometrically increasing linear dimensions of Table 2.2 into an arithmetically increasing series of pure numbers, which are much more convenient for purposes of calculation, although not necessarily as telling physically. The transformed Wentworth scale, called the phi scale, is defined as $\phi = -\log_2 D$, where the base of the logarithms is 2 and $D$ is the particle diameter in millimetres ($10^{-3}$ m). Notice that the phi values equivalent to Wentworth's class limits vary inversely as the physical particle size.

**Table 2.2** Scale of grade.

| Name | Grade limits stated as particle diameters (mm) | Corresponding phi values |
|---|---|---|
| Gravel | | |
| boulder | >256 | < −8 |
| cobble | 64–256 | −6 to −8 |
| pebble | 4–64 | −2 to −6 |
| granule | 2–4 | −1 to −2 |
| Sand | | |
| very coarse sand | 1–2 | 0 to −1 |
| coarse sand | 0.5–1 | 1 to 0 |
| medium sand | 0.25–0.5 | 2 to 1 |
| fine sand | 0.125–0.25 | 3 to 2 |
| very fine sand | 0.0625–0.125 | 4 to 3 |
| Silt | 0.001 95–0.0625 | 9 to 4 |
| Clay | <0.001 95 | >9 |

Note: 1 mm = 1000 $\mu$m.

Each natural sediment is an aggregate of particles which vary continuously over a certain range, as can be proved by examining sand from a beach or river bed using a hand lens. A simple way to measure the size distribution of such a deposit is to shake a sample in a nest of sieves of upward-increasing apertures, so that the grains become separated into their respective grades. The weight percentage of the total sample retained on each of the sieves measures the frequency in the sample of the grains ranging in size between the aperture of that

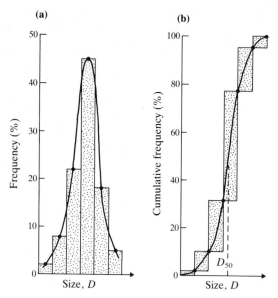

**Figure 2.2** Graphical representations of the distribution of particle size in a sediment: (a) histogram and smooth frequency, and (b) cumulative frequency curve (per cent less than stated size) and its relationship to the histogram.

sieve and the next coarsest. The block graph of these frequencies, called a histogram (Fig. 2.2a), is one way of illustrating the grain size distribution. A smooth frequency curve can be obtained from the histogram by drawing a continuous line through the frequency values at the midpoints of the grades (Fig. 2.2a). This curve is the differential form $dp/dD$ of the function $p(D)$ describing the probability $p$ of finding particles in the sample of a size larger (or smaller) than the stated diameter $D$. The cumulative or integral function $p(D)$ follows by summing in order the blocks which form the histogram (Fig. 2.2b).

Frequency and cumulative distributions such as appear in Figure 2.2 can be analysed statistically in various ways to give a series of parameters descriptive of a sediment. The most useful are the mean size $D_m$, the median size $D_{50}$ and the standard deviation. The median size is the particle diameter corresponding to the 50th percentile in a cumulative distribution (Fig. 2.2b) and divides the size distribution into equally weighted halves. The standard deviation describes the spread of grain size values, that is, the quality of a sediment's sorting. An ill-sorted sediment, for example, has a large standard deviation relative to the mean; well sorted deposits present only a relatively narrow range of grain sizes. Why are particles of a range of sizes, and not just one size, to be found in natural sediments? Bridge (1981) argues

that the size range reflects the short-term random variation of the fluid forces involved in sediment transport (see also Middleton 1977).

## 2.4 What form has a particle?

One has only to examine a few sands to realize that the shape of the particles involved is very variable. Ooids formed by the precipitation of calcium carbonate in sea water are almost perfectly spherical (Fig. 2.3a), whereas quartz grains correspond less satisfactorily to this regular shape, even after long stirring and abrasion by tidal currents (Fig. 2.3b). Many beach and river pebbles are either discoidal, resembling oblate spheroids, or roller shaped, and therefore like a prolate spheroid. Volcanic pumice grains vary from equant to elongate (Fig. 2.3c), and have surfaces deeply pitted by ruptured gas bubbles. Perhaps the most extreme shape variation is to be found amongst biogenic particles. Some organisms, like the carbonate-secreting alga *Halimeda* (Fig. 2.3d), consist of plate-like particles, whereas others, like the bivalves, yield such grains only when their shells are crushed (Fig. 2.3e). Cylindrical particles result from the breakage of branching corals, bryozoans and certain other carbonate-secreting algae (Fig. 2.3f). Gastropods on death yield conical particles (Fig. 2.3g). Rock parted by several sets of joints commonly affords approximately tetrahedral fragments on being weathered. Particles resembling spheroidal shells result by the death of bivalve molluscs and the separation of their valves (Fig. 2.3h).

Implicit in the preceding survey is the idea that a particle can be characterized as regards shape by comparing it to regular geometrical solids. This device is helpful qualitatively, but it is of little use in the quantitative description and comparison of particles. One of the most valuable schemes for quantitative comparison is Zingg's (1935) shape classification (Fig. 2.4). This has few shape categories and allows particle suites to be compared visually using sample plots of the ratios of the axes $a > b > c$.

The sphere is widely used as a shape model against which to compare other particles, and from this has arisen the concept of particle sphericity. In its practical form, sphericity is defined as $D_A/D_c$, where $D_A$ is the equivalent-area diameter of the particle in its most stable horizontal position, and $D_c$ the diameter of the smallest circumscribing circle (Fig. 2.5a). Rittenhouse's (1943) visual comparison chart permits the rapid assessment of projection sphericity, with ranges between 0 and 1 (sphere). Regardless of long-established practice,

**Figure 2.3** Examples of sediment particles (scale bars 0.01 m): (a) oolite sand; (b) quartz sand; (c) lightly abraded pumice; (d) *Halimeda* fragments; (e) abraded fragments of cockle shells; (f) *Lithothamnium* sand; (g) sand largely composed of broken gastropod shells; (h) separated valves of the common cockle.

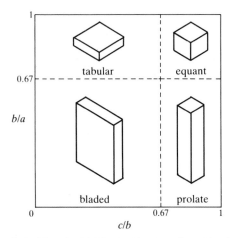

**Figure 2.4** Zingg's (1935) classification of particle shape.

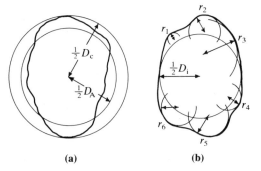

**Figure 2.5** Definition sketches for particle (a) two-dimensional sphericity and (b) two-dimensional roundness.

however, there are no compelling physical reasons why sedimentary particles should be like spheres, either in terms of the factors governing their primary shapes, or those controlling their subsequent modification by abrasion and breakage. The view is taken here that, except where there has been breakage subsequent to release, sedimentary particles retain their primary form in a substantial and recognizable degree even after more than one cycle of transport. That underlying shape seems to be predetermined by such factors as the disposition of crystal boundaries (e.g. grains released from igneous and metamorphic rocks), and the attitude and spacing in the rock of planar features including bedding, lamination, joints and cleavage (e.g. many gravel-sized particles).

The sphere, which is perfectly rounded, is also widely used to model the rounding of particle edges and corners, the so-called roundness. Again the practical measure of roundness is based on the stable particle pro-

jection, and is defined as $D_{mc}/D_i$, where $D_{mc}$ is twice the average of the radius of curvature of the visible corners, and $D_i$ the diameter of the largest inscribed circle (Fig. 2.5b). Like sphericity, roundness ranges between 0 and 1 (sphere). Krumbein (1941a) gives a visual comparison chart suitable for the rapid measurement of roundness. Contrary to general belief, roundness and sphericity as defined above are not fundamentally independent, but necessarily vary together (Flemming 1965). An ellipsoid is on geometrical grounds a much better particle model and, so Kuenen (1956) found from his experiments, is the shape to which debris tends as a consequence of transport abrasion.

Whereas the axial dimensions of a particle tend to be retained during its transport history as a record of original form, the roundness changes rapidly and is a sensitive indicator of abrasion. The main controls on particle roundness have emerged from careful field observations (Plumley 1948, Bluck 1969) and controlled laboratory experiments (Krumbein 1941b, Kuenen 1956). Roundness increases with (1) transport distance and (2) particle size, but declines with (3) hardness of the particle. Figure 2.6 shows Krumbein's results for 27 limestone fragments 0.045–0.054 m in diameter rolled in a tumbling barrel for increasing periods of time, equivalent to increasing increments of transport distance. The average particle mass decreases at a slowly falling rate with increasing transport. Roundness shows an initially rapid followed by a more gradual decline.

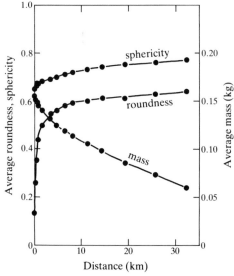

**Figure 2.6** Change in the average mass, roundness and sphericity of limestone fragments with increasing distance of transport in a tumbling barrel. Data of Krumbein (1941b).

The sphericity, measured on a volumetric basis, declines very gradually. Beyond about 7 km, roundness and sphericity change at about the same rate, pointing over these longer distances to a close interrelationship between the two. As illustrated by Krumbein's experiments, transport distance is sedimentologically the most interesting control on roundness. Other factors influencing roundness are (4) the transport medium, (5) particle form and (6) the associated debris. Sand acts as an abrasive when transported with gravel particles, causing them to round quickly, but pebbles can be broken into more angular shapes as the result of crushing between larger stones.

## 2.5 How close is a packing?

A bed of sand or gravel is an aggregation or packing of particles, the individual elements of which support each other in the gravity field and define interparticle voids. How dense or close is a packing?

Several measures are available. The bulk or overall density of a packing is

$$\gamma = \frac{\text{total mass of grains}}{\text{overall volume of packing (grains plus voids)}} \ \text{kg m}^{-3} \tag{2.1}$$

where $\gamma$ is the Greek letter gamma. This dimensional quantity varies with the solids density of the minerals involved, as well as with the fraction of void space. It is the correct measure in problems involving the forces acting on cohesionless sediments, but tells us little about the volumetric density of a packing. More appropriate is the fractional volume concentration

$$C = \frac{\text{total volume of grains}}{\text{overall volume of packing (grains plus voids)}} \tag{2.2}$$

which is non-dimensional. The fractional porosity $P = (1 - C)$ is defined as

$$P = \frac{\text{total volume of voids}}{\text{overall volume of packing (grains plus voids)}} \tag{2.3}$$

which is also non-dimensional. Because of its practical significance, the porosity is quoted much more often than concentration, but is the less fundamental quantity, for it is grains that define voids. A third property expressing closeness of packing is the co-ordination $N$,

that is, the average number of grains with which each particle is in contact. This property is not easily measured, but one of the simplest ways in the case of large particles is to dissect an aggregate through which paint has been poured and allowed to dry, counting the menisci on each particle. Co-ordination is significant particularly in connection with the strength of cohesionless sediments. Co-ordination increases with concentration, but the two are not uniquely related.

## 2.6 Kinds of packing

### 2.6.1 Haphazard packings or heaps
Pour a quantity of rice into a bowl. The grains descend under the influence of a force, in this case gravity, and jostle each other as they lose energy and pack to form a static bed. What exists in the bowl is a haphazard packing or heap of grains. The grain long axes lie at many angles, and the voids between the particles appear to be of no fixed shape or size. The arrangement of the particles apparently is markedly random. One would be mistaken, however, in supposing that the packing was entirely structureless, for the grains were added from one direction, and under the influence of a strong directed force. It is quite probable that the grains have a statistically preferred long-axis orientation and that other bulk properties correspondingly differ vertically and horizontally. All naturally occurring cohesionless sediments are haphazard packings, formed under either the influence of gravity alone, as in an avalanche down a talus slope, or gravity combined with a fluid force, as on a beach or river bed.

### 2.6.2 Random packings
These totally disordered packings are theoretically important for several reasons, but can be created only with the help of powerful computers. Essentially, a random packing is one produced independently of a directed force. For example, we can generate a 'cloud' of particles by bringing grains together randomly one at a time from all possible directions. Alternatively, particles can be introduced randomly into a specified large volume, until all spaces capable of holding a complete particle have been used up.

### 2.6.3 Regular packings
A regular packing is one composed of particles of a uniform size and shape arranged in touching contact according to a repeated regular geometrical configuration conforming to the rules of symmetry. There is no restriction on shape but, if non-spherical particles are

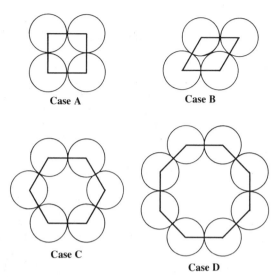

**Figure 2.7** Four regular arrangements of equal spheres touching in a plane.

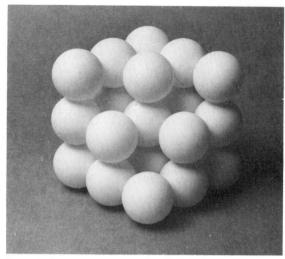

**Figure 2.8** Cubical packing of equal spheres.

used, a constraint on their orientation is implied. At first sight, such highly structured packings have no practical significance, and are merely amusing geometrical puzzles, yet they furnish many valuable insights into naturally occurring aggregates. It is worth modelling some representative examples by gluing ping-pong balls together.

Figure 2.7 shows four possible ways in which equal spheres may be regularly arranged in a plane so as to touch but not overlap. Each arrangement contains the minimum of particles necessary to define the fundamental unit of the pattern, which may therefore be indefinitely extended away by simply adding more spheres to the plane in an identical manner. In case A four spheres lie with their centres at the corners of a square of side $D$ equal to the sphere diameter. Case B shows four touching spheres at the corners of a rhombus. This is also of side $D$, but the perpendicular distance between any two rows of sphere centres is now $D\sqrt{(3/4)}$, from the geometry of the arrangement. Comparing these cases, A can be changed into B by translating one row of spheres parallel with the other. Case C involves six spheres arranged at the corners of an equiangular hexagon. The perpendicular distance between opposite rows of spheres is $2D\sqrt{(3/4)}$ and the area of the hexagon $3D^2\sqrt{(3/4)}$. In case D eight spheres lie along the sides of a square of side $D(1 + \sqrt{2})$, so as to form an equiangular octagon.

Each of these arrangements can be made into a three-dimensional packing by adding another layer. By adding to A a second square layer at a spacing of $D$ we obtain a packing represented by a unit cell in which eight touching spheres lie at the corners of a cube. This cube, of side $D$, encloses what amounts to one whole sphere of volume $\frac{4}{3}\pi(D/2)^3$, making the packing concentration $C = \frac{4}{3}\pi(D/2)^3/D^3 = \pi/6 \simeq 0.524$, independently of sphere size. The co-ordination is $N = 6$ in an extensive array of spheres packed in this cubical mannner (Fig. 2.8). When another rhombic layer is added to B so that the spheres in one layer fit the hoppers in the layer beneath, that is, the layer spacing is $D\sqrt{(2/3)}$, we obtain a rhombohedral packing of $C = (\pi/6)\sqrt{2} \simeq 0.740$ and $N = 12$ (Fig. 2.9). This packing can be derived from the cubical packing by a double translation, first of one layer of spheres parallel to the rows in the other, and then in the same plane but at right angles to the first translation. Whereas the concentration changes continuously during this process, note that the co-ordination increases in steps. There also exists a series

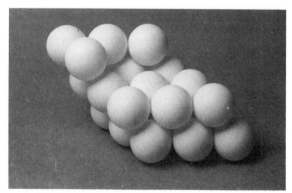

**Figure 2.9** Rhombohedral packing of equal spheres.

**Figure 2.10** Hexagonal packing of equal spheres with the particles in one plane directly above those in the plane below.

**Figure 2.11** Hexagonal packing of equal spheres with the particles in one plane displaced relative to those in the plane below.

tetragonal packing formed by adding a second layer to case D at the maximum layer spacing of $D$ (Fig. 2.12). This packing yields $C = 2\pi/3(1 + \sqrt{2})^2 \simeq 0.359$ and $N = 5$, the co-ordination being the same as in the corresponding hexagonal packing (Fig. 2.0).

These packings are just a few of the many possible regular arrangements of equal spheres; in every case the concentration is independent of the particle size. Graton and Fraser (1935) are amongst several authors who have exhaustively explored the properties of packings in the cubical–rhombohedral series, which includes, as

**Figure 2.12** Tetragonal packing of equal spheres.

of possible packings based on the hexagonal arrangement of case C. Putting two layers together at the maximum possible layer spacing of $D$, we obtain a packing with $C = (\pi/9)\sqrt{(4/3)} \simeq 0.403$ and $N = 5$ (Fig. 2.10). The other end member (Fig. 2.11), marked by $C = (\pi/9)\sqrt{2} \simeq 0.494$ and $N = 9$, is made by translating one hexagonal layer of spheres over the other in a direction parallel with a line drawn through the midpoints of opposite sides of the hexagon. Whereas the cubical and rhombohedral unit cells cut off segments totalling one whole sphere, those representing hexagonal packings enclose two whole spheres. Segments representing four whole spheres are cut off by the unit cell of the

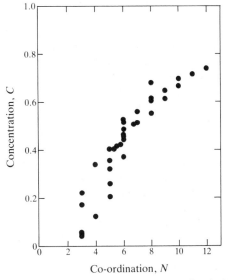

**Figure 2.13** Co-ordination versus concentration in known regular packings of equal spheres.

represented by the rhombohedral case, the closest possible regular packing of equal spheres. Although the tetragonal arrangement of Figure 2.12 has quite a low density, several regular packings even more open are known (Melmore 1942a,b), with co-ordinations as low as either $N = 3$ or $N = 4$. Figure 2.13 shows how co-ordination and concentration vary over a large number of regular sphere packings. The two measures of density increase together, but are not uniquely related. The reason for this is evident from the way in which, in the cubical–rhombohedral packing series, for example, the concentration changes continuously while the co-ordination varies stepwise. A similar increase of co-ordination with concentration may be expected to mark haphazard packings, but with both properties changing continuously because of the irregular particle arrangement.

## 2.7  Voids

The shape, size, disposition and connectedness of the voids in a packing are of great practical importance, as they govern the ease with which fluids such as water, petroleum and natural gas can pass through and be trapped within the particle aggregate.

Regular packings can be attacked geometrically. Consider the cubical arrangement of Figure 2.8. Eight spheres define a chamber with the shape of a concave octahedron, having eight curved faces and six cross-shaped entrances or throats, as shown by a plastic-filled packing from which the particles have been removed (Fig. 2.14a). From the geometry of four touching spheres at the corner of a square, each throat is of maximum diameter $D$ and minimum $D(\sqrt{2} - 1)$. Now consider the spheres in a slice between diagonally opposite edges of the cubical arrangement, that is, in the plane containing the centre of the unit void. In this slice the chamber will be found to have a maximum diameter of $D\sqrt{2}$ and a minimum of $D(\sqrt{3} - 1)$. The unit void in the rhombohedral packing (Fig. 2.9) is more complex, consisting of a concave octahedron lying between two smaller concave tetrahedra. The voids are shown in Figure 2.14b, illustrating a plastic-filled rhombohedral packing with the particles removed.

It is possible to give only a statistical description of the shape and throat characteristics of voids in haphazard packings, whether of particles of single or mixed sizes, and no wholly satisfactory method of doing this has yet been devised. As might be expected, plastic-filled haphazard packings of uniform spheres as well as natural grains reveal voids of bewilderingly many sizes and shapes (Figs 2.14c & d).

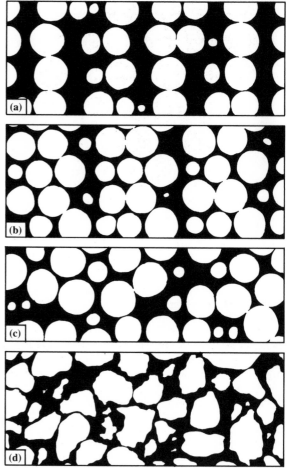

**Figure 2.14** Two-dimensional void shapes (black) in artificial granular packings: (a) equal spheres cubically packed; (b) equal spheres rhombohedrally packed; (c) equal spheres haphazardly packed; (d) quartz sand. The particles used to make (a)–(c) are commercially available glass balls and are only nominally spheres.

## 2.8  Controls on packing

### 2.8.1  Edge effects

All real packings are finite, so what happens at their edges, where the particles encounter a constraint? This question is important for two reasons: (1) packings must be sampled in order for their properties to be estimated, and (2) where particles of disparate sizes are mixed together, the larger ones may act as the effective boundaries to packings of the smaller.

Figure 2.15 shows the effect of packing a single layer of uniform spheres into a container of complex shape.

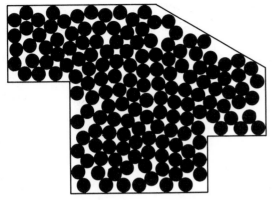

**Figure 2.15** Edge effects in the packing of equal spheres in a plane.

**Figure 2.16** 'Cubical' packing of equal prolate spheroids.

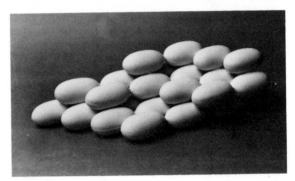

**Figure 2.17** 'Rhombohedral' packing of equal prolate spheroids.

The voids are invariably larger near the edges of the packing than in the interior, regardless of whether the interior arrangement is regular or haphazard. Measurements show that the effect of the walls in reducing the particle concentration, that is, increasing the porosity, persists inwards over a distance of about five sphere diameters (Ridgway & Tarbuck 1966). A container at least a few hundred particle diameters across should therefore be used to sample an unknown packing. For the same reason, the porosity of small particles filling the voids between significantly larger ones should be larger than when the small ones are packed alone.

### 2.8.2 Particle shape and orientation

What happens when non-spherical particles are used to make packings? Regular arrangements of uniform prolate spheroids afford a useful insight. A prolate spheroid is the solid shape defined by rotating an ellipse of major semi-axis $a$ and minor semi-axis $b$ about the major axis. Since the axes are unequal, such a particle can be given an orientation, and in making regular packings from it, we shall have to specify that orientation as well as the relative axial dimensions.

First constrain the packing by keeping the major axes parallel. One possible arrangement is to put the centres of touching prolate spheroids at the corners of an orthogonal parallelepiped. Figure 2.16 shows this arrangement, which evidently corresponds, since a spheroid is merely a stretched sphere, to the cubical packing of equal spheres (Fig. 2.8). The volume of the unit cell of the spheroid packing is seen to be $(2a)(2b)(2b) = 8ab^2$. Now as the unit cell cuts off segments totalling exactly one spheroid, and the spheroid is of volume $\frac{4}{3}\pi ab^2$, the concentration is $C = \pi/6$ and the co-ordination is $N = 6$, as in the cubical packing of spheres. Again retaining parallelism amongst the major axes, we can arrange the spheroids in a regular packing (Fig. 2.17) corresponding to the rhombohedral sphere arrangement (Fig. 2.9). The concentration as before is $C = (\pi\sqrt{2})/6$ and the co-ordination is $N = 12$. As in the case of spheres, the 'cubical' and 'rhombohedral' arrangements of equal prolate spheroids are end-member packings.

Suppose that the spheroids are allowed more than one orientation, rather than the strict parallelism of Figures 2.16 and 2.17? A 'cardhouse' arrangement, with the spheroid centres at the corners of a square, is shown in Figure 2.18. The co-ordination is $N = 6$ but the concentration is $C = 2\pi ab/3(a + b)^2$, decreasing as $b/a$ becomes smaller. Another possible arrangement appears in Figure 2.19, the centres of the spheroids in a layer now lying at the midpoints of the sides of a square. This arrangement also has a co-ordination $N = 6$ but the concentration is now $C = \pi ab/(a^2 + b^2)$. Notice that the loss of parallelism amongst the spheroids in these two packings is restricted to a single plane, namely that

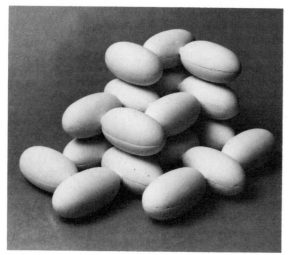

**Figure 2.18** 'Cardhouse' packing of equal prolate spheroids.

**Figure 2.20** Another open packing of equal prolate spheroids.

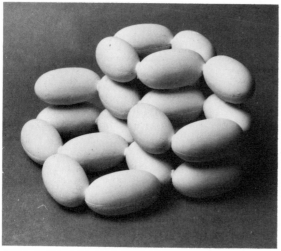

**Figure 2.19** An open packing of equal prolate spheroids.

parallel to the square layers. Full anisotropy can be achieved by arranging the spheroids parallel to the edges of a cube (Fig. 2.20). The co-ordination is increased to $N = 8$ and the concentration can be shown to be $C = \pi ab^2/(a^2 + b^2)^{3/2}$.

Thus by removing the constraint of strict dimensional parallelism we can obtain regular packings whose concentration is a function of particle shape (axial ratio) as well as geometry of arrangement. For each arrangement discussed, the concentration falls with decreasing ratio $b/a$. This suggests that the concentration of haphazard packings of real particles should vary with the strength

of the preferred orientation of the particles. It also suggests that high porosities could typify packings made from stick-shaped, lath-like, shell-shaped or platy grains. There is experimental evidence to support this conclusion (Allen 1970a, 1974).

### 2.8.3 Sorting

How is packing density affected by mixing particles of more than one size? The regular sphere packings discussed above can give us some insight into the question. The remainder must come from looking at natural or synthetic haphazard packings.

Consider the cubical arrangement of equal spheres (Fig. 2.8). Its concave ocatahedral unit void has a least diameter of $D(\sqrt{3} - 1)$. Obviously this is the diameter of the largest secondary sphere that can be associated with the eight primary ones, without upsetting their cubical arrangement and touching. The presence of this secondary sphere increases the packing concentration to $C = (\pi/6)[1 + (\sqrt{3} - 1)^3] \simeq 0.729$ and raises the co-ordination to $N = 10$, each a significant change. Turning to the rhombohedral packing (Fig. 2.9), it will be remembered that the unit void is a concave octahedron flanked by two smaller concave tetrahedra. Inspection of the ping-pong ball model reveals that the diameter of the largest secondary sphere fitting the unit void is $D(\sqrt{2} - 1)$. Table 2.3 shows that the effect of the secondary spheres is to increase the fractional concentration by about 0.05. The table also lists the sizes and effects of the tertiary, quaternary and quinary spheres that can be fitted in turn into the rhombohedral packing (White & Walton 1937). The progressive introduction of these

**Table 2.3** Effect of adding void-filling spheres to a rhombohedral packing of equal spheres.

| Sphere(s) | Diameter | Fractional concentration | Total spheres in packing |
|---|---|---|---|
| primary | $D$ | 0.740 | 8 |
| secondary | $D\sqrt{2}$ | 0.793 | 16 |
| tertiary | $D(\sqrt{2}-1)$ | 0.810 | 32 |
| quaternary | $0.225D$ | 0.843 | 96 |
| quinary | $0.155D$ | 0.852 | 160 |

spheres is tantamount to worsening the sorting of what has become a mixture. Hence natural mixtures should increase in concentration, and decrease in porosity, as their sorting becomes poorer. The model has obvious limitations, however, if only because it is supposed that further spheres can be inserted into the primary packing without disturbing that configuration.

The model is nonetheless obeyed qualitatively by artificial mixtures continuously graded according to prescribed size distribution patterns. Synthetic sands were studied experimentally by Sohn and Moreland (1968), who chose a Gaussian (normal) size distribution, and by Rogers and Head (1961), who took the log-normal distribution pattern. Wakeman (1975) explored mixtures of spherical glass beads graded log-normally. In all three cases the concentration generally increased as the sorting of the mixtures worsened. Figure 2.21 is based on Sohn and Moreland's work and shows the concentration for sands obtained from two sources as a

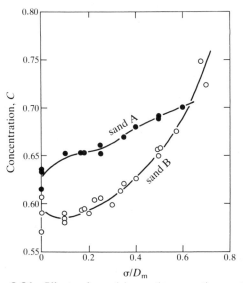

**Figure 2.21** Effect of particle sorting on the particle concentration of two synthetic quartz sands. Data of Sohn and Moreland (1968).

function of the relative standard deviation, $\sigma/D_m$, where $\sigma$ is the deviation of the size distribution and $D_m$ its arithmetic mean diameter. At zero value of this ratio, the sediment is perfectly sorted, and consists of particles effectively of a single size. It is only over a narrow range of very small values of $\sigma/D_m$, such as is seldom found amongst natural sediments, that the concentration declines with worsening sorting.

### 2.8.4 Surface properties
The particles that form natural packings are not smooth but possess surface irregularities and asperities which can affect packing density. There are incidental experimental data to suggest that rough-surfaced particles pack a little less densely than smooth ones of a similar shape and size, but the question has yet to be investigated systematically.

### 2.8.5 Deposition conditions
Rice or granulated sugar stored in a jar constitutes a haphazard packing. Every cook knows that it is necessary to tap the jar in order to maximize the amount of sugar or rice that can be stowed away. The tapping supplies energy to the packing, in consequence of which the grains jostle each other, so assuming a denser arrangement. It must therefore be concluded that packing density is affected by the conditions of deposition of the grains.

Some instructive experiments can be made with rice kept in a jar. Place the empty jar on a table and try dumping the rice into it, as far as possible in a single action, afterwards carefully marking on the outside the shape of the free surface formed by the grains. Repeat this several times using the same quantity of rice. The heights of the surface will be very similar, suggesting that dumping – the rapid deposition of densely arrayed grains – leads to a reproducible packing. This particular arrangement, its properties varying from one material to another, is called loose haphazard packing. Dump the rice again, but afterwards lightly tap the jar. The height of the packing will be seen to fall, rapidly with the first few taps, but then increasingly slowly, until eventually there is no further change. Mark the height of the packing. Repeat this experiment several times without varying the amount of rice. Another reproducible packing will have been formed. This packing, closer than the first, is called dense haphazard packing. Its properties again vary with the material forming the packing.

The first four columns of Table 2.4 list the concentrations in loose and dense haphazard packing that have been measured from a number of synthetic and natural

**Table 2.4** Packing concentrations and slope angles attained by selected granular materials (Allen 1970a).

| Material | Medium | Fractional concentration | | Slope angle (deg) | | |
|---|---|---|---|---|---|---|
| | | Loose packing | Dense packing | $\phi_r$ | $\phi_{i,max}$ | $\Delta\phi_{max}$ |
| quartz sand ($D_m = 270\ \mu$m) | air | 0.560 | 0.640 | 31.5 | 45.1 | 13.6 |
| | water | 0.558 | 0.651 | 31.4 | 44.6 | 13.2 |
| quartz–slate sand ($D_m = 1040\ \mu$m) | air | 0.543 | 0.628 | 33.3 | 46.4 | 13.1 |
| | water | 0.541 | 0.628 | 33.4 | 46.9 | 13.5 |
| *Lithothamnium* sand ($D_m = 3200\ \mu$m) | air | 0.584 | 0.700 | 36.2 | 49.9 | 13.7 |
| | water | 0.587 | 0.703 | 36.5 | 53.1 | 16.6 |
| glass beads ($D_m = 470\ \mu$m) | air | 0.587 | 0.645 | 23.9 | 32.6 | 8.7 |
| long-grain rice ($D_m = 4300\ \mu$m) | air | 0.532 | 0.602 | 36.4 | 44.4 | 8.0 |
| chopped spaghetti ($D_m = 7400\ \mu$m) | air | 0.446 | 0.556 | 31.4 | 45.2 | 13.8 |

granular materials (Allen 1970a). Rather precise values of 0.601 in loose and 0.637 in dense packing were obtained by Scott (1960) for polished ball-bearings nominally 0.00318 m in diameter. Scott and Kilgour (1969) and Finney (1970) obtained the even more precise values of respectively $0.6366 \pm 0.0005$ and $0.6366 \pm 0.0004$ for densely packed ball-bearings. Synthetic and natural grains, neither as spherical nor as perfectly graded as ball-bearings, yield packing densities similar to uniform spheres but ranging more widely between the loose and dense states. This is most noticeable with markedly elongated particles, such as *Lithothamnium* sand (Fig. 2.3f) and spaghetti chopped into longish lengths. One is reminded of the large fabric-related differences in the concentration of regular packings of prolate spheroids. The fluid medium has little influence on the two packing states.

Generally speaking, natural cohesionless sediments become packed neither through being dumped nor through being vibrated in a container, but it does seem likely that the loose and dense haphazard packings described above represent limits on the concentration of natural mixtures.

Laboratory experiments show that, within the bounds represented by these limiting packings, conditions of deposition have a strong effect on the concentration of natural sediments and synthetic materials. Steinour (1944) settled tapioca in lubricating oil, and Kolbuszewski (1948a,b, 1950) and Walker and Whitaker (1967) allowed quartz sand to sediment in water or air. Macrae and Gray (1961) studied the settling of glass

spheres (amongst other materials) in air. Figure 2.22 summarizes the packing concentrations measured by these workers, the 'conditions of deposition' being expressed in terms of the dry mass of particles falling

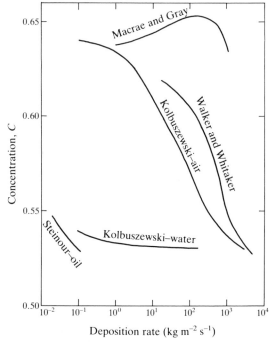

**Figure 2.22** Summary of experimental results relating to the effect of depositional conditions on particle concentration in granular sediments.

over unit bed area in unit time ($kg\,m^{-2}\,s^{-1}$). The graph suggests that the concentration varies from a constant value, comparable with that of dense haphazard packing, at a small intensity of deposition, to a lower constant value comparable with that of loose haphazard packing, provided that the intensity of deposition is high.

What happens when particles are deposited to form a static bed? Kolbuszewski (1950) envisaged two possible effects, one related to the energy of the arriving particles, and the other dependent on their concentration just above the bed. Every falling particle carries to the bed a certain amount of energy, which can be dissipated only through frictional and inelastic collisions amongst the grains already emplaced, so bringing them into a closer packing. Now the energy of a particle increases with its size and excess of density over the fluid medium. It is presumably for this reason that, for the same intensity of deposition, sand deposited in air has a higher concentration than sand settled in water (Fig. 2.22). On the other hand, if the particles are densely arrayed near the bed, so that they arrive relatively close together, the movement of grains that have just touched the bed, as well as of those recently accumulated, is inhibited and the packing becomes locked. A large concentration of sedimenting grains would therefore seem to favour a loose arrangement.

Two natural agents deposit sediment at a high intensity. These are turbidity currents (Ch. 12), carrying sand far out over ocean and lake floors, and the avalanches that sweep talus slopes and the leeward sides of desert dunes and dunes shaped by river and tidal currents (Ch. 5). Avalanche deposits are known to be relatively loosely packed (Denekamp and Tsur-Lavie 1981), a result with important implications for their load-bearing capacity.

## 2.9 How steep is a heap?

This question is not just academic, but bears on the limiting steepness of such widespread and important natural features as cliff talus and the leeward slopes of desert and other dunes.

Everyone at one time or another has tried heaping up sugar crystals in a bowl, only to discover that they cannot be made to form a slope steeper than a certain limiting angle. To put the experiment on a more serious footing, half-fill a cylindrical, screw-topped glass jar with dry sand (Fig. 2.23). Laying it sideways on a table covered with blank paper, slowly roll the jar through a few complete turns, so as thoroughly to mix the con-

**Figure 2.23** A jar half-filled with dry sand tilted at an angle steep enough for avalanching.

tents. Turn the jar back so that the surface of the sand becomes horizontal, as shown by a spirit level, and mark on the paper the place where jar and table touch. Now begin very slowly to roll the jar, stopping as soon as an avalanche of grains is initiated. Again mark where the jar touches the paper. Continue turning the jar, plotting its position every time an avalanche is released.

How should these results be interpreted? You will have confirmed that there is indeed a limiting steepness to a heap of loose grains, and also discovered that avalanching reduces this slope by a fixed and reproducible amount. The greatest angle through which the sand surface was tilted, proportional to the distance between the first and second marks on the paper divided by the circumference of the jar, is called the angle of initial yield $\phi_i$. The change in angle due to avalanching, known as the dilatation angle $\Delta\phi$, is given by the distance relative to the circumference between any two successive marks on the paper, beginning with the second. Finally, the value of the slope immediately after avalanching, termed the residual angle after shearing $\phi_r$, is simply $\phi_i$ less the average of $\Delta\phi$.

More sophisticated experiments of the same sort may be made by tilting granular materials in a large motorized drum with transparent sides (Allen 1970a). Some measurements made in this way appear in the last three columns of Table 2.4. The residual angle after shearing of each material is sensibly constant and can be identified with the condition of loose haphazard packing. The angle of initial yield increases above $\phi_r$ as the grain concentration increases above the loose haphazard state, to reach a maximum value $\phi_{i,max}$ in dense haphazard packing. The dilatation angle $\Delta\phi_{max}$ based on the two states ranges from about $8°$ for smooth glass beads (and also polished ball-bearings) to approxi-

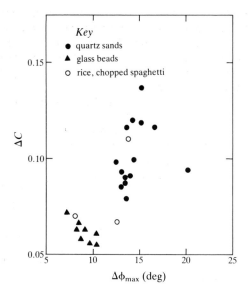

**Figure 2.24** Experimental results on the differences in angle of initial yield and particle concentration for various natural sands, glass beads and cereal grains in loose and dense haphazard packing. Data of Allen (1970a).

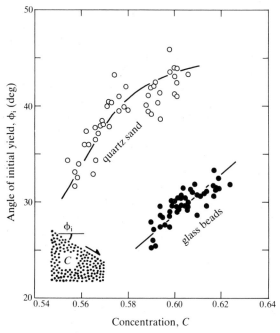

**Figure 2.25** Experimental variation of angle of initial yield with particle concentration for a quartz sand and one size of glass beads. Data of Allen (1970b).

mately 14° for rough-textured natural grains. As with packing concentration, there is little effect from the medium. Inspection of Table 2.4 suggests that over the wide range of materials studied the 'improvement' in slope steepness represented in $\Delta\phi_{max}$ increases with the 'improvement' in packing density, as measured by the difference $\Delta C$ in concentration between dense and loose haphazard packing. This is roughly confirmed by the more extensive plot of data given in Figure 2.24. Allen (1970b) also measured on a number of materials the variation of $\phi_i$ with grain concentration in the range between the limiting packing states. The results for closely graded spherical glass beads (diameter 210–297 $\mu$m) and a well sorted very fine-grained quartz sand are compared in Figure 2.25. As would have been expected from the variation of $\Delta\phi_{max}$ with $\Delta C$ (Fig. 2.24), the angle of initial yield increases with the concentration.

## 2.10   Building houses on sand

We are advisedly warned against building houses on sand, but it is untrue that cohesionless granular materials are totally lacking in strength, although there are circumstances under which even very small forces promote failure. Indeed, many civil engineers are daily engaged in successfully designing and erecting structures on such uncertain substrates. If further proof of the strength of loose granular masses were needed, consider the fact that one can walk safely over most sandy beaches and river bars, as well as the fact that the sand in the jar failed only after being tilted beyond a critical degree. But what are the origins of the strength of a material which, given the right circumstances, can be freely poured from one container to another?

Consider in Figure 2.26 a plane ABCD lying at a vertical depth $h$ below an infinite slope of inclination $\beta$ formed on a uniform cohesionless granular material of bulk density $\gamma$. The weight of material acting on the plane is

$$F_g = (abh)\gamma g \quad \text{N} \tag{2.4}$$

where $a$ is the horizontal length of the plane, $b$ its width, and $g$ the acceleration due to gravity. Hence the upslope-directed tangential resisting stress acting over unit area of the plane is

$$\tau = \frac{(abh)\gamma g \sin\beta}{(ab/\cos\beta)} \quad \text{N m}^{-2} \tag{2.5}$$

36

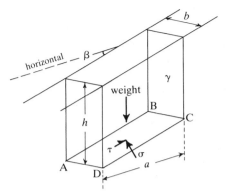

**Figure 2.26** Definition sketch for the avalanching of a loose sand.

where $(ab/\cos\beta)$ is the area of ABCD. The corresponding normal stress is

$$\sigma = \frac{(abh)\gamma g \cos\beta}{(ab/\cos\beta)} \quad \mathrm{N\,m^{-2}} \qquad (2.6)$$

It follows that

$$\frac{\tau}{\sigma} = \tan\beta \qquad (2.7)$$

Now the jar experiment suggests that the material above ABCD will remain in place only for so long as $\beta$ is sufficiently small. If the slope is steepened to $\beta = \phi_i$, where $\phi_i$ is the angle of initial yield appropriate to the packing of the material, the internal strength of the material no longer exceeds the tangential stress (Eqn. 2.5). A failure therefore occurs over ABCD, and grains avalanche away. Then

$$\frac{\tau}{\sigma} = \tan\phi_i \qquad (2.8)$$

whence the angle of initial yield is equivalent to an angle of friction characteristic of the material, and its tangent is a characteristic friction coefficient. The angle of initial yield therefore describes the ability of the granular material to resist an applied force without totally failing. The measurement and understanding of the friction angles characteristic of granular materials has been one of the most important tasks of workers in the field of soil mechanics (Rowe 1962).

One control on the angle of initial yield is already identifiable as the grain concentration (Figs 2.24 & 2.25), but wherein does the control lie? Consider the regular packings of the cubical–rhombohedral series (Figs 2.8 & 2.9) in which the particles become increasingly interlocked with ascending concentration. Spheres cubically packed lie directly above each other, so that in order to shear one layer of particles over another we do not need to dilate or expand the whole mass. Indeed, a heap packed in this way at once collapses and sharply declines in volume when shearing starts, as the spheres in one layer move down into the hoppers defined by the particles below. In a rhombohedral packing, however, the spheres already occupy the hoppers shaped by the lower particles (Fig. 2.9). In shearing such a packing, each sphere must be raised against gravity a distance of either $D[1 - \sqrt{(2/3)}]$ or $D[\sqrt{(3/4)} - \sqrt{(2/3)}]$, with an accompanying expansion of the bed, depending on whether the particle is carried directly over a sphere below or between two spheres. Work must be done on the packing in order to effect the expansion. Hence a packing with interlocked particles is stronger than one with no interlocking, the strength, partly depending on the degree of interlocking, increasing with the concentration.

The other likely contributor to the angle of initial yield is interparticle friction. A repetition of the jar experiment using one-sized glass beads illustrates the importance of this factor. These particles are fire-polished, practically uniform in size and almost perfectly spherical. Typically, the angle of initial yield for such beads is $30-34°$. By etching the beads in dilute hydrofluoric acid (do this either out of doors or in a properly equipped fume cupboard), the angle can be increased by $5-10°$. As the etching has no sensible effect on either their size, shape or mode of packing, it must be supposed that the increase in the angle is due to the visible destruction of the polish by the acid vapour. The same effect can be detected – but much more expensively – using steel ball-bearings, whose angle of initial yield when polished and grease-free is about $25°$. By allowing them to rust overnight out of doors, however, the angle can be increased to as high as $34°$.

Thus two effects contribute to the angle of initial yield and therefore to the strength of a cohesionless granular material. One is the packing density, particularly its expression in terms of particle interlocking, and the other is interparticle friction, which depends on the particle surface roughness. The two effects contribute broadly equally to the strength of natural cohesionless sediments (Rowe 1962).

## Readings

Allen, J. R. L. 1970a. The avalanching of granular solids on dune and similar slopes. *J. Geol.* **78**, 326–51.

Allen, J. R. L. 1970b. The angle of initial yield of haphazard assemblages of equal spheres, in bulk. *Geol. Mijnb.* **49**, 13–22.

Allen, J. R. L. 1974. Packing and resistance to compaction of shells. *Sedimentology*. 21, 71–86.

Allen, T. 1968. *Particle size measurement*. London: Chapman and Hall.

Bluck, B. J. 1969. Particle rounding in beach gravels. *Geol. Mag.* 106, 1–14.

Bridge, J. S. 1981. Hydraulic interpretation of grain-size distributions using a physical model for bedload transport. *J. Sed. Petrol.* 51, 1109–24.

Denekamp, S. A. and Y. Tsur-Lavie, 1981. The study of relative density in some dune and beach sands. *Engng Geol.* 17, 159–73.

Finney, J. L. 1970. Random packings and the structure of liquids. I. The geometry of close packing. *Proc. R. Soc. Lond.* A 319, 479–93.

Flemming, N. C. 1965. Form and function of sedimentary particles. *J. Sed. Petrol.* 35, 381–90.

Graton, L. C. and H. J. Fraser 1935. Systematic packing of spheres – with particular reference to porosity and permeability. *J. Geol.* 43, 785–909.

Kolbuszewski, J. 1948a. An experimental study of the maximum and minimum porosities of sands. *Proc. 2nd Int. Congr. Soil Mechanics*, Vol. 1, 158–65.

Kolbuszewski, J. 1948b. General investigation of the fundamental factors controlling loose packing of sand. *Proc. 2nd Int. Congr. Soil Mechanics*, Vol. 7, 47–9.

Kolbuszewski, J. 1950. Notes on deposition of sands. *Research* 3, 478–83.

Krumbein, W. C. 1941a. Measurement and geological significance of shape and roundness of sedimentary particles. *J. Sed. Petrol.* 11, 64–72.

Krumbein, W. C. 1941b. The effects of abrasion on the size, shape and roundness of rock fragments. *J. Geol.* 49, 482–520.

Kuenen, P. H. 1956. Experimental abrasion of pebbles, 2: rolling by current. *J. Geol.* 64, 336–68.

Macrae, J. C. and W. A. Gray 1961. Significance of the properties of materials in the packing of real spherical particles.

*Br. J. Appl. Phys.* 12, 164–72.

Melmore, S. 1942a. Open packing of spheres. *Nature* 149, 412.

Melmore, S. 1942b. Open packing of spheres. *Nature* 149, 669.

Middleton, G. V. 1977. Hydraulic interpretation of sand size distributions. *J. Geol.* 84, 405–26.

Plumley, W. J. 1948. Black Hills terrace gravels: a study in sediment transport. *J. Geol.* 56, 526–77.

Ridgway, K. and K. J. Tarbuck 1966. Radial voidage variation in randomly-packed beds of spheres of different sizes. *J. Pharm. Pharmacol.* 18 (Suppl.), 168–75S.

Rittenhouse, M. G. 1943. A visual method of estimating two-dimensional sphericity. *J. Sed. Petrol.* 13, 79–81.

Rogers, J. J. W. and W. B. Head 1961. Relationship between porosity, median size, and sorting coefficients of a synthetic sand. *J. Sed. Petrol.* 31, 467–70.

Rowe, P. W. 1962. The stress–dilatancy relation for static equilibrium of an assembly of particles in contact. *Proc. R. Soc. Lond.* A 269, 500–27.

Scott, G. D. 1960. Packing of spheres. *Nature* 188, 908–9.

Scott, G. D. and D. M. Kilgour, 1969. The density of random close packing of spheres. *Br. J. Appl. Phys.* 2, 863–6.

Sohn, H. Y. and C. Moreland 1968. The effect of particle size distribution on packing density. *Can. J. Chem. Engng* 46, 162–7.

Steinour, H. H. 1944. Rates of sedimentation. *Ind. Engng Chem.* 36, 618–24, 840–7, 901–7.

Wakeman, R. J. 1975. Packing density of particles with log-normal size distributions. *Powder Technol.* 11, 297–9.

Walker, B. P. and T. Whitaker 1967. An apparatus for forming uniform beds of sand for model foundation tests. *Geotechnique* 17, 161–7.

Wentworth, C. K. 1922. A scale of grade and class terms for clastic sediments. *J. Geol.* 30, 377–92.

White, H. E. and S. F. Walton 1937. Particle packing and particle shape. *J. Am. Ceram. Soc.* 20, 155–6.

Zingg, T. 1935. Beitrag zur Schotteranalyse. *Schweiz. Mineral. Petrog. Mitt.* 15, 39–140.

# 3 Sink or swim?

Settling of an array of grains – fluidization – permeability – Stokes's and impact laws of settling – settling of non-spherical particles

## 3.1 Two introductory experiments

What physical situation arises when either a single grain or an array of sediment particles is released in a fluid of contrasted density? This question is sedimentologically interesting for at least two reasons. It relates to the ability of the wind, rivers, tidal currents and turbidity currents to transport sediment in turbulent suspension, and to the attitudes assumed by particles on deposition. Some valuable insights will come from the following experiments.

The first experiment will show how the velocity and mode of fall of sediment particles depend on the fluid and on grain size, density and shape. We require (Fig. 3.1) a 2 litre glass measuring cylinder, clean water and glycerol sufficient in each case to fill the cylinder, a quantity of hand-picked quartz sand grains graduated in diameter between about one-quarter and several millimetres, several small bivalve mollusc shells (tellins or juvenile common cockles are ideal), a number of flat circular discs ranging in diameter between about 10 and 40 millimetres cut from aluminium kitchen foil coloured on one side with a spirit stain, a stopwatch, a metre stick and a pair of forceps. Slip a rubber band around the cylinder to act as an adjustable mark when measuring the fall rate of the particles, first placing the band about 0.1 m from the top. With the forceps, release the sand grains one by one, starting with the smallest, first into water and then glycerol. Time the fall between the mark and the bottom of the cylinder with the stopwatch. Repeat with the shells, releasing some with the convex side upwards and others with the concave side upwards, and finally with the smallest of the aluminium circles.

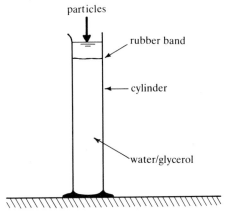

**Figure 3.1** Apparatus to demonstrate some of the controls on particle settling.

Note the behaviour of each particle as it falls. Measure the height of the liquid column below the mark for each set of runs. Next, by sliding the rubber band to various positions along the cylinder, determine the fall time of one of the sand grains in glycerol over a range of distances of fall. Finally, release the shells and metal circles into the still air of a room (a newspaper on the floor will break their fall), timing the descent and noting the behaviour of each. Measure the distance of fall. From these data, and with the aid of densities looked up in physical tables, tabulate the influence of fluid medium and particle size, shape and density on the velocity and mode of particle fall.

Five generalizations should emerge: (1) a particle sinks when its density exceeds that of the fluid; (2) a particle released from rest attains a steady terminal

39

fall velocity after an initial period of acceleration; (3) the terminal fall velocity increases with particle size; (4) the terminal fall velocity decreases with increasing fluid viscosity $\eta$ (glycerol $\eta \simeq 2.88\,\mathrm{N\,s\,m^{-2}}$, water $\eta \simeq 1.15 \times 10^{-3}\,\mathrm{N\,s\,m^{-2}}$, air $\eta \simeq 1.78 \times 10^{-5}\,\mathrm{N\,s\,m^{-2}}$); and (5) the behaviour during fall depends on both particle size and shape and on the fluid medium. Concerning the last generalization, the metal circles fall obliquely and tumble in the air, their unpainted sides flashing regularly as the particles turn over and over, but in water merely slide from side to side while slowly sinking broadside-on. Again, the shells sink broadside-on and concave-up in water and glycerol, rapidly turning to this position if released in some other attitude.

What forces govern the motion just studied? Since the rate of fall becomes steady after an initial period of acceleration, there must be a fixed driving force. The first and second generalizations imply that this force is proportional to the particle mass, but that the mass is in some way modified by the presence of the fluid. Hence the driving force must be a body force equal to the immersed (or effective) particle weight. Again, since a steady terminal fall velocity was attained, there must by Newton's first law of motion be a resisting force equal and opposite to the driving force. Our fourth generalization suggests that the resisting force arises from the fluid as it is distorted and disturbed by the particle moving through it, and that the resistance increases with fluid viscosity. The resisting or drag force will in fact comprise two elements, one expressing friction between the fluid and the particle surface (skin friction), and the other deriving from the uneven distribution of fluid pressure over the particle shape (form drag).

Do we encounter a fundamentally different physical situation when, instead of the grain moving through the stagnant fluid, as in the first experiment, the fluid is moving in such a way that the particle remains stationary relative to the ground? A slightly more difficult experiment will provide an answer. Procure a glass tube about 1 m long and with a bore of 5–10 mm. Clamp the tube vertically in a sink and, using a rubber tube fitted with a screw clip, connect the lower end to the water supply. Flood the tube with water and then close the screw clip tightly. Now drop into the tube the quartz grain whose fall velocity was earlier measured over a range of distances. By very cautiously opening and closing the screw clip, the grain can be made to hover within the tube, without either being flushed out or sinking out of sight. Next bring the tube to the horizontal and measure the discharge of water through it by timing a known volume into a measuring cylinder. Measuring

the bore of the tube, and using Equation 1.5, calculate the average flow velocity in the tube.

This experiment shows that the terminal velocity of a particle falling through a stagnant fluid is not sensibly different from the vertically upward steady velocity of the same fluid necessary to maintain that particle stationary relative to the ground. Therefore there is no fundamental difference between the two situations mentioned; it is a question merely of the frame of reference chosen. Whether we decide to move with the fluid or with the particle, and when evaluating the forces in action, we need only consider the *relative* motion of fluid and particle. Of course, there can be important practical differences between the two cases. A particle settling in a stagnant fluid is in a closed system: a continuing supply of fluid is unnecessary for the motion, but the particle will eventually be arrested by a flow boundary. By contrast, the state of rest of a particle hovering in an upward stream of fluid requires a continuing fluid supply, whence the system must be an open one constrained by the duration of the flow. In addition, the fluid may have acquired some additional properties, for example, turbulence absent from the corresponding closed system.

## 3.2 Settling of spherical particles arrayed in a stagnant fluid

### 3.2.1 Posing the problem

Let us pose the problem as generally as possible. Consider the settling towards the horizontal boundary of an infinite half-space of an infinite array of particles distributed throughout an infinite homogeneous stagnant fluid (Fig. 3.2). The particles are smooth solid spheres of uniform diameter $D$ and density $\sigma$, arrayed in a statistically uniform manner at an average fractional volume concentration $C$. They are falling at a steady terminal velocity $V$ relative to the ground. But as the particles displace fluid upwards during descent, their velocity relative to the interparticle fluid is $V_{\mathrm{rel}} > V$. The fluid is assumed to be of uniform density $\varrho$ and viscosity $\eta$.

An array of particles is chosen because this condition most closely approximates to natural situations. The selection of an infinite array, and a statistically uniform distribution, allows us to disregard the special modes of behaviour exhibited by finite clouds of particles. By choosing an infinite and stagnant fluid medium, we can ignore effects due to turbulence and vertical flow boundaries.

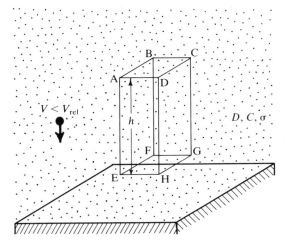

**Figure 3.2** Definition sketch for the settling of particles in an infinite cloud contained in an infinite fluid half-space.

### 3.2.2 Body forces

Our first experiment revealed that the motion of a sinking particle was driven by body forces. Consider in Figure 3.2 the column of fluid and particles of vertical height $h$ and unit basal area contained between the horizontal planes ABCD and EFGH moving downwards with the particles at their terminal velocity $V$. Neglecting the fluid for the moment, the force exerted on the base of the column by the weight of particles contained in the column is equal to $\sigma Chg\,\mathrm{N\,m^{-2}}$, where $g$ is the acceleration due to gravity. Notice that this force has the quality and dimensions of pressure.

But how does the presence of the fluid medium modify the particle weight as calculated above? Consider in Figure 3.3 an empty cube-shaped space with vertical and horizontal sides of length $d$ arranged at a depth $h$ below the free surface of a stagnant liquid of density $\varrho$. It is axiomatic that the hydrostatic pressure in a fluid

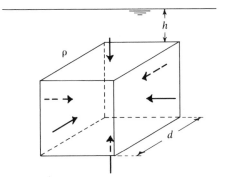

**Figure 3.3** Definition sketch for a simple proof of Archimedes' principle.

acts uniformly in all directions. Hence there is no net horizontal pressure force acting on the sides of the cube, since the pressures on opposite vertical faces are equal but opposed and the two pairs of faces comprise a symmetrical arrangement. Taking the downward direction as positive, the pressure force acting on the upper face of the cube is equal to $d^2(\varrho hg)\,\mathrm{N}$, that is, the product of the area of the face with the weight of a unit column of fluid above (atmospheric pressure being neglected). The force on the lower face of the cube acts upwards, and is of value $-d^2\varrho(h+d)g\,\mathrm{N}$, atmospheric pressure again being neglected. Subtracting forces, we obtain

resultant force (upthrust)
$$= d^2(\varrho hg) - d^2\varrho(h+d)g \qquad (3.1)$$

$$= -d^3\varrho g \quad \mathrm{N} \qquad (3.2)$$

The resultant force is equal in magnitude to the weight of fluid that can be contained in the cube and is called an upthrust because it always acts upwards (Archimedes' principle).

Applying these arguments to the unit reference column in Figure 3.2, the upthrust on the grains is $-\varrho Chg\,\mathrm{N\,m^{-2}}$, whence the effective body force driving the motion is the immersed weight of the grains spread over the base of the column, that is

$$\sigma Chg - \varrho Chg = (\sigma - \varrho)Chg \quad \mathrm{N\,m^{-2}}$$

This force acts downwards, since the solids density is the greater. Had the fluid been more dense than the solids, the effective body force would have acted upwards, as in the case of a balloon filled with hot air or hydrogen in the atmosphere.

### 3.2.3 Fluid forces

Figure 3.4 shows the experimental laminar flow of water around a solid cylinder, equivalent to a sphere sliced very thinly at its equator. The streamlines, interpreted in the line drawing, are made visible by streaks of colour liberated from potassium permanganate crystals. The pattern is symmetrical about that streamline which, continued through the cylinder, coincides with the diameter drawn parallel to the flow. On this axial streamline, the water gradually slows as it approaches the body and finally stagnates where it divides and attaches at the point A. Bernoulli's equation (Sec. 1.5.3) states that the stagnation-point pressure exerted by the water stream on the cylinder at A is equal to $\frac{1}{2}\varrho U^2\,\mathrm{N\,m^{-2}}$, where $U$ is the undisturbed velocity of the stream. The divided streamlines rejoin at the point S on the trailing surface,

41

**(a)**

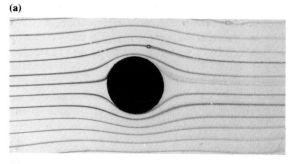

**(b)**

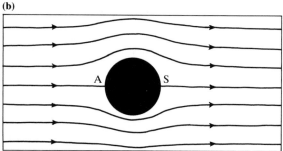

**Figure 3.4** Flow round a sphere (equatorial section) at low Reynolds number: (a) experimental; (b) reconstructed streamlines.

the water thereafter accelerating to the former value $U$ as it moves along the axial streamline leaving the body. It will be noticed that the distortion of the streamlines extends to at least one body diameter either side of the cylinder. Bernoulli's equation can be used again, to show that this distortion is equivalent to a resultant pressure force opposed to the stream flowing around the cylinder. Since a boundary layer will exist on the surface of the body, a frictional drag force must be added to the pressure force in compiling the total force opposing the aqueous flow.

Recalling the quadratic stress law (Eqn. 1.26), one formulation of the fluid force opposing the settling of particles in the reference volume of Figure 3.2 is

$$
\begin{array}{c}
\text{total} \\
\text{total} \qquad \text{number of} \quad \text{projection} \quad \text{stagnation-} \\
\text{fluid} = \begin{array}{c}\text{drag} \\ \text{coefficient}\end{array} \times \begin{array}{c}\text{grains in} \\ \text{reference} \\ \text{volume}\end{array} \times \begin{array}{c}\text{area of} \\ \text{one grain}\end{array} \times \begin{array}{c}\text{point} \\ \text{pressure}\end{array}
\end{array} \quad (3.3)
$$

$$
= C_D \times \frac{Ch}{(\pi/6)D^3} \times (\pi/4)D^2 \times \tfrac{1}{2}\varrho V_{rel}^2 \quad (3.4)
$$

$$
= \tfrac{3}{4}C_D \frac{\varrho Ch V_{rel}^2}{D} \quad \text{N m}^{-2} \quad (3.5)
$$

where $C_D$ is the drag coefficient, and where the volume and maximum cross-sectional area of a sphere are respectively $(\pi/6)D^3$ and $(\pi/4)D^2$. The non-dimensional drag coefficient describes the total drag (form plus skin friction), and must be either calculated theoretically or determined experimentally.

### 3.2.4 Final expression for settling

Since the particle array in Figure 3.2 is falling steadily, the immersed weight is by Newton's first law equal and opposite to the total fluid drag (Eqn. 3.5), whence

$$
(\sigma - \varrho)Chg = \tfrac{3}{4} C_D \frac{\varrho Ch V_{rel}^2}{D} \quad (3.6)
$$

and so

$$
V_{rel}^2 = \frac{4}{3} \frac{(\sigma - \varrho)gD}{C_D \varrho} \quad \text{m}^2\,\text{s}^{-2} \quad (3.7)
$$

with the reference-column height $h$ cancelling out and the particle concentration $C$ disappearing explicitly. Hence the square of the relative particle velocity increases linearly with the density difference and particle diameter, but varies inversely with the drag coefficient and the fluid density.

As our aim is to predict the terminal fall velocity $V$ measured relative to the ground, what is the relationship between $V_{rel}$ and $V$? Superimpose on the grains in Figure 3.2 an upward velocity equal to $V$, so bringing them to rest relative to the ground. Consider the unit cross-sectional area on the plane ABCD and imagine that now there are no grains above this plane. Because grains are lacking, the upward velocity of the fluid above the plane (called the superficial velocity) is also equal to $V$. Just below the plane, where grains exist at the concentration $C$, the fluid is streaming at the upward velocity $V_{rel}$ through pore spaces of concentration $(1 - C)$. Now continuity demands that the fluid discharge is a constant across the unit area on ABCD, whence $V_{rel}(1 - C) = V$. Substituting for $V_{rel}$ in equation 3.7, the terminal fall velocity of the particles in the array becomes

$$
V^2 = \frac{4}{3} \frac{(1 - C)^2(\sigma - \varrho)gD}{C_D \varrho} \quad \text{m}^2\,\text{s}^{-2} \quad (3.8)
$$

whence, for other things equal, the velocity decreases with increasing particle concentration or, alternatively, increases with $(1 - C)$, the fractional porosity.

It is interesting to use Equation 3.8 to compare the terminal fall velocity of the particles in an array with the terminal velocity $V_0$ of a solitary particle of the same diameter and density settling in an infinite extent of the same fluid. In the latter case, we set $C = 0$ and write the drag coefficient as $C_{D,0}$. Writing out the equation for each case, and dividing the first by the second, gives

$$\frac{V^2}{V_0^2} = \frac{C_{D,0}(1 - C)^2}{C_D} \tag{3.9}$$

Richardson and Zaki (1954) found experimentally that the terminal settling of arrays of uniform spherical particles could be described by

$$\frac{V}{V_0} = (1 - C)^n \tag{3.10}$$

where $n$ is a positive exponent related to the particle Reynolds number $Re = \varrho D V_0/\eta$. Maude and Whitmore (1958) concluded that $2.33 \leqslant n \leqslant 4.65$ and gave a theoretical curve relating $n$ and $Re$, with which Richardson and Zaki's experimental results agree quite well (Fig. 3.5). By squaring Equation 3.10, and substituting for the velocity ratio in Equation 3.9, we find that

$$C_D = \frac{C_{D,0}}{(1 - C)^{2n-2}} \tag{3.11}$$

Hence the drag coefficient of a particle in an array always exceeds that for the solitary particle, and is a steeply increasing function of the particle concentration. Therefore from Equation 3.8 the fall velocity of

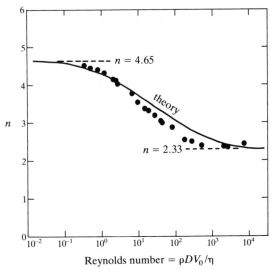

**Figure 3.5** The exponent in the equation for the terminal settling of arrays of uniform spherical particles as a function of particle concentration. Theory (Maude & Whitmore 1958) compared with experiment (Richardson & Zaki 1954).

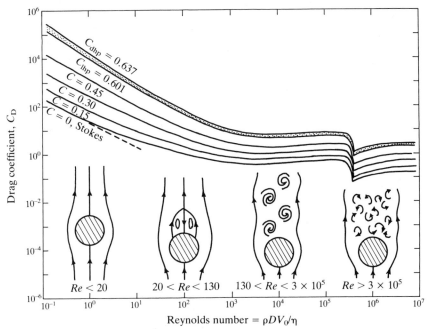

**Figure 3.6** Calculated drag coefficient of spherical particles as a function of Reynolds number and particle concentration. The inset diagrams show how the pattern of motion around a solitary sphere changes with Reynolds number.

grains in an array is a steeply decreasing function of their concentration. The same follows from the purely empirical result, Equation 3.10.

An important practical consequence is that the terminal fall velocity of sediment particles in a turbidity current, in the suspended load of a flooded river, and in an earthquake-induced quicksand must be less than the fall velocity of the same particles when alone in an infinite medium. Moreover, the difference will increase as the sediment becomes more concentrated.

Figure 3.6 shows how $C_D$ varies with $C$ and $Re = \varrho D V_0 / \eta$. The graph was calculated from Equation 3.11 using the experimental curve in Figure 3.5, and the theoretical and best available experimental data on $C_{D,0}$ as a function of $Re$ (Clift *et al.* 1978). Values for $V$ can be obtained either using this graph or from Equation 3.10.

## 3.3 Settling and fluidization

The term fall velocity used above refers to the motion of fluid and particles in a particular frame of reference, that in which the movement of the particles is compared to the ground. But if we change the frame of reference, either as in our second experiment or by imposing an appropriate superficial velocity as in the analysis which led to Equation 3.8, we find that the particles can be kept stationary while the fluid is kept moving. As before, the upward-acting fluid drag just balances the downward-acting immersed particle weight. A dense array of particles suspended in this way is described as fluidized, and the general process, much studied by chemical engineers and fuel technologists, is called fluidization (Leva 1959). The process can also occur naturally, where gases sweep violently upwards through the debris trapped in the vent of a volcano, and where ground water is forced upwards through sand beds, as in springs emerging beneath the sea or along the sides of river bars at low stage.

The form of Equation 3.8 appropriate to fluidization is obtained simply by substituting the superficial velocity $V_s$ for the terminal fall velocity $V$. By substituting for $C_D$ from Equation 3.11, the expression becomes

$$V_s^2 = \frac{4}{3} \frac{(1-C)^{2n}(\sigma - \varrho)gD}{C_{D,0}\varrho} \quad \text{m}^2\text{s}^{-2} \quad (3.12)$$

giving the superficial velocity in terms of the drag coefficient and exponent $n$, each of which is a function of Reynolds number.

Equation 3.12 is true up to and including a limiting concentration $C_{\lim}$ describing the mode of packing of uniform spheres when they are so close that they must touch and, in effect, form a static granular bed. The superficial velocity then corresponding to $C_{\lim}$ is that just able to fluidize a static granular bed of that particular packing mode. Let us call this critical value the fluidizing velocity $V_{s,cr}$. There is a wide choice for the limiting packing (Ch. 2). We may select either one of the geometrical arrangements, or one of Scott's (1960) random beds. The latter are particularly interesting because of their similarity to natural particle packings. Scott's experiments gave $0.601 \leqslant C_{\lim} \leqslant 0.637$ for uniform spheres, where the smaller value is for loose haphazard packing ($C_{lhp}$) and the larger for dense haphazard packing ($C_{dhp}$). Equations 3.12 and 3.11 then read

$$V_{s,cr}^2 = \frac{4}{3} \frac{(1-C_{\lim})^{2n}(\sigma - \varrho)gD}{C_{D,0}\varrho} \quad \text{m}^2\text{s}^{-2} \quad (3.13)$$

and

$$C_{D,\lim} = \frac{C_{D,0}}{(1-C_{\lim})^{2n-2}} \quad (3.14)$$

in which $C_{D,\lim}$ is the limiting value of the drag coefficient corresponding to $C_{\lim}$. The range of $C_{D,\lim}$ is plotted in Figure 3.6. These are as well the lower limiting curves applicable to the terminal fall velocity, since uniform spheres cannot be randomly packed more closely (without being crushed) than the range $C_{lhp} \leqslant C \leqslant C_{dhp}$. Hence from the derivations of Equations 3.8 and 3.13, the terminal fall velocity of spherical particles in an array of concentration $C$ is in fact equal and opposite to the superficial fluid velocity necessary to fluidize the same array. An important implication of Equations 3.12 and 3.13 is that a granular bed will suffer an increasing expansion, with a consequent decrease in $C$, as the superficial velocity is increased above the fluidizing velocity. The bed will expand infinitely when $V_s$ is increased to $V_0$, the situation modelled in our second experiment.

## 3.4 Flow in porous media

Chemical engineers measure fluidizing velocity by means of an, in principle, very simple experiment. A column of the loose granular material of interest is supported on a porous grid arranged across a vertical cylindrical tube fitted with side tappings leading to manometers (Fig. 3.7). The fluid is driven up through the column at a known rate, increasing in small steps. Using the manometers, the pressure drop across the

44

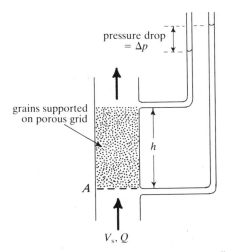

pressure drop
= $\Delta p$

grains supported
on porous grid

$h$

$A$

$V_s, Q$

**Figure 3.7**  Definition sketch for flow in porous media.

column of granular material is measured at each flow rate, until no further increase is detectable on increasing the fluid discharge. At this point, the grains are just fluidized, and the flow through the parts of the tube above and below the column is taking place at the fluidizing velocity. The upward-acting fluid drag within the granular column then exactly balances its downward-acting immersed weight. Notice in this experiment how the weight, distributed over the cross-sectional area of the tube, is measured explicitly in terms of pressure.

Hydrogeologists and petrophysicists also make experiments identical in principle to that shown in Figure 3.7, but with a view to measuring the permeability of a porous material, that is, its general ability to conduct a fluid (Dullien 1979). The instrument, called a permeameter, may use either a gas or a liquid as the test fluid. The material may be either a loose sand, in which case it is formed into a column firmly confined between porous grids (compare the chemical engineer's experiment), or a solid cylindrical plug cut from the rock. Whereas the chemical engineer is interested in the unique discharge for an unconfined loose material at which the one pressure drop–discharge law (increase with discharge) is replaced by the other (pressure drop independent of discharge), the petrophysicist and hydrogeologist require the general drop–discharge law for confined samples. Their experiments are so similar, however, that we must now ask what is the connection between, on the one hand, permeability and, on the other, settling and fluidization?

An answer can be given if we confine ourselves to slow, unidirectional, steady flow through a porous

granular material, which turns out to be the most important practical case. The mid-19th century civil engineer H. Darcy found empirically that this flow could be described by

$$Q = -\frac{kA}{\eta}\frac{\Delta p}{d} \quad \text{m}^3\,\text{s}^{-1} \qquad (3.15)$$

where $Q$ is the total fluid discharge through a confined sample of the material, $A$ the sample cross-sectional area normal to the flow, $\Delta p$ the pressure drop across the sample, and $d$ the sample length measured parallel to the flow. The negative sign appears to the right because the pressure declines in the flow direction, as shown in Figure 3.7. Equation 3.15, called Darcy's law, is a fundamental relationship in hydrogeology and petroleum reservoir engineering. It states that the discharge of a fluid at low Reynolds numbers through a porous material is linearly proportional to the pressure gradient $\Delta p/d$. The coefficient $k$ is called the specific permeability (or, simply, the permeability) and is an intrinsic property of a porous material. It is measured in square metres (m²), as you may easily check by examining the dimensions of the other quantities in Equation 3.15.

## 3.5  Controls on permeability

What controls the permeability of a loose granular material described by Darcy's law? Darcy's law for the fluidization experiment shown in Figure 3.7 is

$$V_s = -\frac{k}{\eta}\frac{\Delta p}{h} \quad \text{m}\,\text{s}^{-1} \qquad (3.16)$$

remembering that $Q = AV_s$, where $V_s$ as before is the superficial velocity, and the vertical height $h$ of the sample is substituted for the previous length $d$. Suppose that the discharge is just sufficient to fluidize the grains. We can then write $V_s = V_{s,cr}$ and, using the arguments of Section 3.2.2 above, put $\Delta p = (\sigma - \varrho)C_{lim}hg$. These substitutions into Equation 3.16 give

$$V_{s,cr} = \frac{k}{\eta}(\sigma - \varrho)C_{lim}g \quad \text{m}\,\text{s}^{-1} \qquad (3.17)$$

the negative sign being dropped as no longer necessary. Here is a new and most interesting equation for fluidization. It states that at low Reynolds numbers the fluidizing velocity is linearly proportional to the permeability of the granular material. Equation 3.17 differs considerably from the previous relationship

(Eqn. 3.13), but does share the term $(\sigma - \varrho)g$. But as both of Equations 3.13 and 3.17 can be rearranged as expressions for $(\sigma - \varrho)g$, this term can be eliminated between them. A further rearrangement gives

$$k = \frac{4}{3}\frac{\eta D(1 - C_{\lim})^{2n}}{\varrho C_{\lim} C_{D,0} V_{s,cr}} \quad m^2 \qquad (3.18)$$

as a statement for the permeability. Some simplification has resulted, but our equation still includes quantities describing the fluid as well as the particles and packing. However, because of the equivalence of settling and fluidization, $V_{s,cr}$ can be eliminated with the help of Equation 3.10, written in the form $V_{s,cr} = V_0(1 - C_{\lim})^n$. The $V_0$ thus introduced, together with the drag coefficient $C_{D,0}$, can themselves be eliminated by calling upon the relationship $C_{D,0} = 24/Re$, where $Re = \varrho DV_0/\eta$. The use of this relationship, applicable only at low Reynolds numbers, is justified below. On eliminating $V_{s,cr}$ and $C_{D,0}$ in this way, we finally obtain

$$k = \frac{1}{18}\frac{(1 - C_{\lim})^n D^2}{C_{\lim}} \quad m^2 \qquad (3.19)$$

showing that at low Reynolds numbers the permeability depends only on the particle size and degree of packing. There is no dependence on fluid properties, since at sufficiently small Reynolds numbers $n = 4.65$ is a constant (Fig. 3.5).

This result allows us to interpret the permeability as essentially the area of the pores controlling the flow. The permeability increases steeply as the particle size increases, as Krumbein and Monk (1942) found in experiments with synthetic sand mixtures (Fig. 3.8). Since $n$ is so large, it decreases very rapidly with increasing concentration, even within the narrow range bounded by loose and dense haphazard sphere packings. This extreme sensitivity is confirmed by the experimental work of von Engelhardt and Pitter (1951), who measured the permeability of the same sand at various concentrations. We can even use Equation 3.19 to predict the qualitative influence of size sorting on permeability. The most important effect of worsening the sorting of a sand of constant average grain size is to increase the particle concentration (Fig. 2.21). The permeability should therefore decline as the sorting becomes poorer, as Krumbein and Monk (1942) were able to demonstrate experimentally (Fig. 3.8).

## 3.6 Settling of a solitary spherical particle in a stagnant fluid

### 3.6.1 Terminal fall velocity

A solitary smooth spherical particle settling in a stagnant unbounded fluid (i.e. $C = 0$) represents a limiting case of Equation 3.8, which then reads

$$V_0^2 = \frac{4}{3}\frac{(\sigma - \varrho)gD}{C_{D,0}\varrho} \quad m^2\,s^{-2} \qquad (3.20)$$

where $C_{D,0}$ is the drag coefficient for the particle falling alone. Now when $Re = \varrho DV_0/\eta$ is very small, the viscous forces predominate over the inertial ones, which may therefore be neglected, and we can calculate from the equations of fluid motion that $C_{D,0} = 24/Re$. The graph of this coefficient appears in Figure 3.6 as the line labelled $C = 0$ sloping down to the right. Substitution into Equation 3.20 yields

$$V_0 = \frac{1}{18}\frac{(\sigma - \varrho)gD^2}{\eta} \quad m\,s^{-1} \qquad (3.21)$$

a most important generalization known as Stokes's law, valid for practical purposes for $Re \lesssim 1$.

According to Stokes's law, the terminal fall velocity of a smooth spherical particle settling alone in an unbounded fluid is proportional to the density difference and to the square of the diameter, but inversely propor-

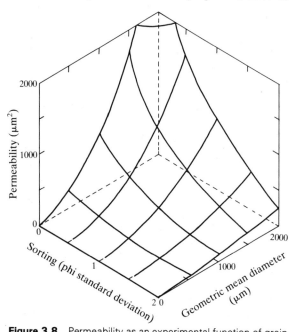

**Figure 3.8** Permeability as an experimental function of grain size and sorting in synthetic sands. Data of Krumbein and Monk (1942).

tional to the fluid viscosity. The striking practical implications of the law are revealed by considering the time it takes for quartz spheres ($\sigma = 2650$ kg m$^{-3}$) of various sizes to fall through the ocean (average depth $= 4$ km, $\varrho = 1025$ kg m$^{-3}$, $\eta = 9.05 \times 10^{-4}$ N s m$^{-2}$ at $10°$C), after having been introduced at the surface by a river or the wind (Table 3.1). The sand particle settles in a matter of a few days, but the silt takes of the order of a year, and the clay grains of the order of a century! If the ocean were in horizontal motion at a speed of $0.1$ m s$^{-1}$ – a value typical of its thermohaline circulations – the silt particle could travel several hundred kilometres before it reached the ocean bed, while the clay particle could circle the Earth (radius 6371 km) approximately 1.2 times in making the same downward journey! Terrigenous clay and silt entering the oceans may therefore become spread far and wide, and where found to be concentrated around the source, some special factor, such as flocculation or pelletization by the zooplankton, must be invoked (Scheidegger & Krissek 1982).

**Table 3.1** Fall and travel of spherical quartz-density particles in a representative ocean current.

| Particle | | Terminal fall velocity (m s$^{-1}$) | Time for vertical fall of 4 km (days) | Horizontal distance travelled in 0.1 m s$^{-1}$ current (km) |
|---|---|---|---|---|
| Name | Diameter ($\mu$m) | | | |
| clay | 1 | $9.79 \times 10^{-7}$ | 47 300 | 40 900 |
| silt | 10 | $9.79 \times 10^{-5}$ | 473 | 409 |
| very fine sand | 100 | $9.79 \times 10^{-3}$ | 4.73 | 4.09 |

The experimental data reviewed by Clift *et al.* (1978) on the drag coefficient for a solitary sphere falling in an unbounded stagnant fluid show that $C_{D,0}$ continues to decline with increasing $Re$ but that the form of the relationship changes beyond the Stokes region (Fig. 3.6). The coefficient declines increasingly gradually, to become sensibly constant at 0.45 over a range of Reynolds number between $10^3$ and $3 \times 10^5$. The drag coefficient falls steeply by almost an order of magnitude at about $Re = 3 \times 10^5$, thereafter increasing only slowly to another constant value of approximately 0.19 at $Re = 10^7$. Over the two ranges of the Reynolds number in which the drag coefficient is essentially constant, we can rewrite Equation 3.20 as

$$V_0 = k \left( \frac{(\sigma - \varrho)}{\varrho} gD \right)^{1/2} \quad \text{m s}^{-1} \quad (3.22)$$

where $k = (4/3 C_{D,0})^{1/2}$ is a new constant. The terminal fall velocity within these ranges of $Re$ is therefore independent of the fluid viscosity and proportional only to the square root of the particle diameter.

Unfortunately for sedimentologists, Stokes's law (Eqn 3.21) and the square-root law (Eqn 3.22) are useful only for limited parts of the ranges at which natural sedimentary particles occur. In aqueous media, Stokes's law is invalid for quartz-density solids larger than about very fine sand (approximately 100 $\mu$m), while the square-root law is applicable only where the diameter exceeds a few millimetres (i.e. gravel). Most sands therefore fall into that awkward range in Figure 3.6 where $C_{D,0}$ changes from being inversely proportional to $Re$ to being independent of $Re$. However, these practical difficulties can be overcome by using the graphical theoretical correlation shown in Figure 3.9 as the basis for calculating the terminal fall velocity of an unknown particle. The group $Ar = 4\varrho(\sigma - \varrho)gD^3/\eta^2$ plotted on the abscissa is a non-dimensional measure of the particle diameter, sometimes called the Archimedes or Best number. On the ordinate appears the group $3\varrho^2 V_0^3/4(\sigma - \varrho)g\eta$, a non-dimensional fall velocity. Given the properties of the fluid and solid, the steady fall velocity of a spherical particle can be obtained simply by entering the graph with the non-dimensional diameter and reading off the non-dimensional velocity.

What explains the important changes in the way the drag coefficient in Figure 3.6 varies with Reynolds number? The answer lies in the way the flow field around the sphere (see Fig. 3.6) responds to the increasing importance of the inertial forces signified by the growing Reynolds number (Taneda 1956, 1978). Up to $Re = 20$ the flow field is like that depicted in Figure 3.4. A wake containing a steady separation bubble is present for $20 < Re < 130$, the separated flow increasing in size with the Reynolds number. The wake becomes increasingly unsteady in the range $130 < Re < 400$, and large vortices begin to be shed from the sphere. A regular shedding is observed for $130 < Re < 3 \times 10^5$, but the wake cannot be said to be turbulent. The sharp fall in $C_{D,0}$ at $Re \simeq 3 \times 10^5$ marks the change in the boundary layer formed on the sphere from laminar to turbulent. Large turbulent eddies are shed partly randomly and partly spirally from the sphere at higher Reynolds numbers.

### 3.6.2 Settling from rest

Our theoretical discussion has so far been confined to particles in terminal fall, that is, moving without accelerating, but it is commonly of interest to predict the more general motion of a grain. We may, for example,

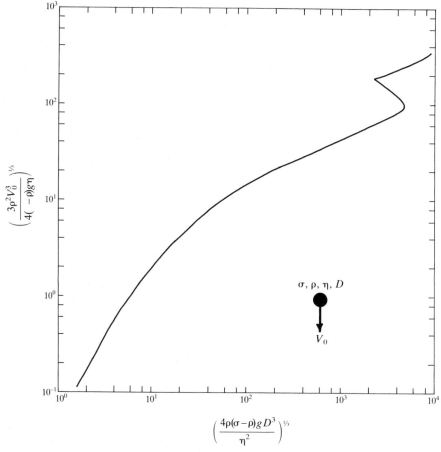

**Figure 3.9** Graph for the calculation of the terminal fall velocity of a solitary smooth spherical particle in a stagnant fluid.

want to know the path of a volcanic ash particle settling through a moving atmosphere, or of a sand grain swept into the recirculating separated flow to lee of a ripple or dune. These problems are in fact much more difficult to solve than is generally appreciated (Clift *et al.* 1978), but we can get some insight into their nature by examining the comparatively simple case of a solitary spherical particle falling from rest through an unbounded stagnant fluid.

To approach this problem, we apply Newton's first two laws of motion, stating that a body continues in a state of uniform motion unless acted on by a force, and affirming that the change of motion of a body is proportional to the impressed motive force. We may therefore say of the descending sphere at any instant that

$$\text{resultant force} = \text{body force} - \text{fluid drag force}$$
$$\text{(3.23)}$$

resultant force = mass of sphere ×
acceleration of sphere

$$\text{(3.24)}$$

and combining the two

mass of sphere × acceleration of sphere
= body force − fluid drag force     (3.25)

Using the results of Section 3.2, and dividing through by the mass of the sphere, the last statement becomes the equation of particle motion

$$\frac{dV}{dt} = \frac{(\sigma - \varrho)}{\sigma} g - \frac{3}{4} C_D \frac{\varrho V^2}{\sigma D} \quad \text{m s}^{-2} \quad \text{(3.26)}$$

in which $t$ is time, and $V$ and $C_D$ are to be understood as the instantaneous values of the fall velocity and drag coefficient, respectively.

48

The differential equation (Eqn 3.26) has no general solution because $C_D$ is such a complex function of the Reynolds number, which includes $V$, already appearing in the equation. When $V_0$ is sufficiently small as to put the particle in the Stokes range, however, substitution of $C_D = 24/Re$ into Equation 3.26 gives a first-order linear differential equation

$$\frac{dV}{dt} = \frac{(\sigma - \varrho)}{\sigma} g - \frac{18\eta V}{\sigma D^2} \quad \text{m s}^{-2} \qquad (3.27)$$

which can be solved using a simple integrating factor, to yield the exponential relationship

$$V = \frac{(\sigma - \varrho)gD^2}{18\eta} [1 - \exp(-18\eta t/\sigma D^2)] \quad \text{m s}^{-1} \qquad (3.28)$$

for the instantaneous fall velocity, where exp stands for the base of natural logarithms. A further integration with respect to $t$ gives the displacement–time equation for the particle. These equations show that a particle in the Stokes range effectively reaches its terminal fall velocity after a descent of a number of diameters just a few times the Reynolds number $Re = \varrho D V_0/\eta$. For a particle whose terminal fall velocity places it in the large Reynolds number range, we can put $C_D$ in Equation 3.26 equal to a constant and solve the equation numerically, ignoring the small period spent by the grain at low Reynolds numbers. The attainment of the terminal velocity is less rapid than for a small particle.

The only body force included in the simple analysis that gave Equation 3.26 is the force due to acceleration of the solid sphere. In fact, two more body forces contribute to the rectilinear acceleration of a particle in a fluid, as Hjelmfelt and Mockros (1967) remind us. From the streamline pattern shown in Figure 3.4, it is evident that a certain volume of the fluid must also be accelerated in order to make way for the accelerating particle. This volume, called the added mass, is equivalent to one-half of the volume of the solid sphere. The effect of added mass becomes increasingly important as the fluid approaches the particle in density. The other body force relates to the fact that the sphere accelerates at a changing rate as it approaches its terminal fall velocity. This contribution, known as the Basset history term, can only be represented by a complex integral, making the full equation of particle motion into a difficult integro-differential equation! The Basset force too becomes increasingly important as the density contrast declines. When the particle path is curved, as in most practical cases, a third body force should be in-cluded in the equations describing the motion. This force is the centrifugal force.

In sedimentological studies, it is safe to neglect the added mass and history effects only in the case of quartz-density solids in air. A significant underestimate of the time for the attainment of the terminal condition results if these effects are neglected in the case of solids in water, particularly so if the Basset term is ignored. The centrifugal force can safely be ignored if the radius of curvature of the particle trajectory is very large compared to the particle diameter.

## 3.7 Settling of a solitary non-spherical particle in a stagnant fluid

### 3.7.1 Regular geometrical solids

The fact that naturally occurring sediment particles can in many instances be approximated as regards shape by regular geometrical solids has prompted numerous studies of the terminal settling of discoidal, cylindrical, spheroidal and other regular bodies. Laboratory experiments are the chief means of study, but mathematical models are possible for some shapes at low and moderate Reynolds numbers.

There is much variety in the terminal settling behaviour of non-spherical particles. It is therefore necessary to distinguish carefully between the motion of the particle centre relative to the ground (particle trajectory) and the movement of the particle relative to its own axes (secondary motion). The trajectory may be either a straight line (not necessarily vertical), a plane curve, a zig-zag composed of short connected arcs, a spiral, or some combination of these. Particles move unsteadily over some kinds of trajectory. The secondary motion observed with certain trajectories is divisible between oscillations and rotations. In the simplest kind of oscillation the particle rocks about a single horizontal axis. Borrowing nautical vocabulary, the movement may be described as either rolling, pitching or yawing, according as the motion takes place about the long, intermediate or short axis. A more general oscillation, or gyration, is expressed when particle axes describe double cones. Rotation is of two kinds. A particle tumbles when it makes one complete revolution after another with respect to a horizontal axis. Spinning is rotation with respect to the vertical axis.

At least three non-dimensional parameters are necessary when correlating these complex modes of behaviour: a Reynolds number, a stability number related to particle shape and relative density, and a number describing the frequency of the secondary motion. Usually, the Reynolds number is based on the vertical component of the terminal fall velocity and the diameter of the circle having the same area as the particle projection normal to the general line of motion. A common form of the stability number relates the moment of inertia of the particle (with respect to an appropriate axis) to the moment of inertia of a fluid sphere. The moment of inertia of a body is a statement of how its mass is distributed with respect to a chosen plane or axis. In the case of the regular solids under discussion, the moment of inertia is the integral over the whole volume of the body of the product of the local mass with the square of the distance of that mass from a reference axis. The frequency of any secondary motion is described by the Strouhal number, which for a disc is defined as $Sr = fD/V_0$, where $f$ is the frequency, $D$ the diameter and $V_0$ the vertical component of the terminal fall velocity.

The terminal behaviour of settling discs is summarized in Figure 3.10 based largely on the work of Willmarth *et al.* (1964) and Stringham *et al.* (1969). The stability number used is $I = \sigma h/\varrho D$, where $\varrho$ is the fluid density, $h$ the thickness of the disc, and $\sigma$ the density of the disc. It is proportional to the ratio of the moment of inertia of the disc to the moment of inertia of a fluid sphere of the same physical diameter. At very small Reynolds numbers $Re = \varrho DV_0/\eta$, a disc falls vertically with its initial orientation, and at a steady terminal velocity. Steady but broadside-on terminal fall is

observed in the region $0.1 \lesssim Re \lesssim 100$. By painting the disc with a dye soluble in the fluid, it can be shown that a substantial separation bubble becomes trapped to its upper side. The motion becomes unsteady at Reynolds numbers larger than approximately 100. At first the disc oscillates about a diameter while continuing to fall broadside-on, the centre making horizontal displacements that are small compared to the size of the disc. The frequency and amplitude of the oscillation both decline with increasing stability number, but the amplitude increases with ascending Reynolds number. At larger *Re* the disc glides from side to side, to define a distinctly zig-zag trajectory composed of connected, concave-up arcs. The trapped separation bubble is released as a vortex into the fluid at the end of each glide. Discs tumble on a rectilinear but inclined trajectory, rotating about a diameter at a constant angular velocity, at sufficiently large Reynolds and stability numbers. In our first experiment, this kind of motion was shown by the aluminium circles as they fell through the air.

The laboratory experiments of Jayaweera and Mason (1965) show that the settling behaviour of cylinders resembles that of discs. At very small Reynolds numbers the initial attitude is maintained during fall, while at small to moderate *Re* the fall is steady and with the long axis maintained horizontal. A cylinder pitches about a horizontal axis normal to the long dimension at Reynolds numbers in excess of approximately 100. A gliding motion similar to that shown by discs is exhibited at rather larger Reynolds numbers. Other studies show that, at very large Reynolds numbers, long cylinders may oscillate or even spin about a vertical axis, while stubby ones may tumble during fall.

Stringham *et al.* (1969) made several observations on the settling behaviour of oblate spheroids (resembling discs) and prolate spheroids (resembling cylinders), shapes which closely resemble many naturally occurring sediment particles. Both kinds of body tend to fall broadside-on. At sufficiently large Reynolds numbers, oblate spheroids either tumble or gyrate about a vertical axis, while prolate spheroids either gyrate or pitch.

One application of these experimental results is to the fabric of the stones dropped from melting icebergs into muds on the sea bed below. The smaller stones, tending to fall broadside-on, should accumulate with their long and intermediate axes nearly parallel with the bedding. Stones that tumble as they sink, expected to be the larger ones, will accumulate within the mud in all orientations, affording an initially random fabric (consolidation may subsequently change this into a rather disordered girdle fabric).

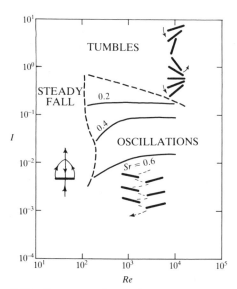

**Figure 3.10** Summary of experimental results on the behaviour of solitary discs settling in a stagnant fluid. Data of Willmarth *et al.*(1964) and Stringham *et al.*(1969).

50

### 3.7.2 Bivalve shells

The separated shells of bivalve molluscs are common sedimentary particles and can be found in a wide range of particularly marine deposits. How do these common particles behave in currents and when settling through water?

Rather than measure individually the fall velocity of the bivalve shells encountered in a sedimentological study, it is often preferable to refer to a simple correlation employing more readily ascertained shell properties. General principles will help us to choose an appropriate parameter. Referring to Sections 3.2–3.4 the force balance on a single shell falling steadily in a stagnant fluid can be written

$$Mg = \frac{(\sigma - \varrho)}{\sigma} = \tfrac{1}{2}A C_{D,0\varrho} V_0^2 \quad \text{N} \qquad (3.29)$$

in which $M$ is the dry mass of the shell and $A$ its projection area normal to the direction of fall. The left-hand side, representing the driving body force, is written so that the particle volume appears implicitly as $M/\sigma$. The right-hand term is evidently the fluid drag. Which quantities in the equation can be treated as constant? Bivalve shells on account of their size should yield Reynolds numbers so large that the drag coefficient is independent of the Reynolds number (see Fig. 3.6 for spheres). This being the case, Equation 3.29 becomes

$$V = k \left( \frac{M}{A} \frac{(\sigma - \varrho)}{\sigma} \right)^{1/2} \quad \text{m s}^{-1} \qquad (3.30)$$

in which

$$k = \left( \frac{g}{\tfrac{1}{2}C_{D,0\varrho}} \right)^{1/2} \quad \text{m}^2 \, \text{kg}^{-1/2} \, \text{s}^{-1} \qquad (3.31)$$

The drag coefficient, gravitational acceleration and fluid density have been lumped into the single dimensional constant $k$, to be determined empirically. Like Equation 3.22 for spheres in the large Reynolds number regime, Equation 3.30 states that the fall velocity of bivalve shells is proportional to the square root of their size. Here 'size' is measured as the effective mass per unit projection area of the shell, that is, the unit mass of the shell as if it had been weighed in the fluid. The explicit quantities in Equation 3.30 are each easy to measure, since it turns out that bivalve shells settle broadside-on. To measure the projection area of the shell, for example, lay the valve convex-up on a piece of paper, trace around the shell with a sharp pencil, and then cut out and weigh on a precision balance the paper representing

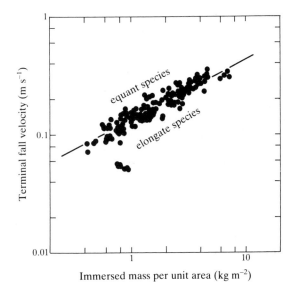

**Figure 3.11** Terminal fall velocity of single valves of common British bivalve sea shells. Data of Allen (1984a).

the shape of the valve. The projection area is this weight multiplied by the unit weight of the paper.

Similar reasoning to the above was used by Futterer (1978) as the basis for correlating laboratory measurements of the terminal fall of bivalve shells with their unit weight. Figure 3.11 shows the results of new experiments (Allen 1984a) using 176 bivalve shells representing 16 species to be found on British coasts (Table 3.2). For this set of shells $k = 0.14 \, \text{m}^2 \, \text{s}^{-1} \, \text{kg}^{-1/2}$. The scatter is due to the fact that the drag coefficient varies noticeably from one species to another. Elongate

**Table 3.2** Bivalve shells (single valves) studied experimentally.

| Species | No. of shells |
|---|---|
| *Nucula tenuis* Montagu | 7 |
| *Glycymeris glycymeris* (Linnaeus) | 3 |
| *Mytilus edulis* | 20 |
| *Chlamys (Chlamys) varia* (Linnaeus) | 13 |
| *Arctica islandica* (Linnaeus) | 4 |
| *Acanthocardia tuberculata* (Linnaeus) | 6 |
| *Cerastoderma edule* (Linnaeus) | 18 |
| *Venus striatula* (da Costa) | 11 |
| *Venerupis pullastra* (Montagu) | 16 |
| *Mactra corallina* (Linnaeus) | 17 |
| *Donax vittatus* (da Costa) | 14 |
| *Tellina tenuis* (da Costa) | 4 |
| *Macoma balthica* (Linnaeus) | 10 |
| *Scrobicularia plana* (da Costa) | 13 |
| *Pharus legumen* (Linnaeus) | 10 |
| *Mya arenaria* (Linnaeus) | 10 |

species (e.g. *Pharus legumen*) and flat species (e.g. *Scrobicularia plana*) tend to fall below the curve averaging the points, because their drag coefficient is significantly greater than for swollen valves (e.g. *Cerastoderma* spp.) and equant valves (e.g. *Glycymeris glycymeris*) which tend to plot above the line.

How do the shells of bivalve molluscs behave while settling? The first of our experiments revealed that cockle shells (*Cerastoderma edule*) fall broadside-on and concave-up. This attitude is in fact assumed by the shells of every one of the species listed in Table 3.2. Even such large and weighty forms as *Arctica islandica* and *Mya arenaria* show not the slightest tendency to tumble as they fall. Evidently a concavo–convex shape is highly stabilizing with regard to behaviour during fall. The fact that the shell invariably sinks concave-up as well as broadside-on means that its centre of mass must lie vertically below the point through which the upward fluid drag acts. When this is not the case the shell is acted on by a force couple tending to rotate it about a horizontal axis. As our first experiment will have shown, a shell released convex-up rapidly turns so that its concave side becomes permanently upwards.

Although all the shells listed in Table 3.2 fall concave-up, the steadiness of the motion varies with their degree of elongation. It should therefore be possible to correlate the behaviour of bivalve shells using a Reynolds number and the simple stability number $I = (1 - b/a)$, where $a$ and $b$ are respectively the long and intermediate

dimensions of the valve. The Reynolds number is calculated on the basis of the diameter $D_A$ of the circle equal in area to the projection of the shell. Figure 3.12 shows the results obtained using the 176 shells of the species listed in Table 3.2. Generally speaking, the shells sink on a steady vertical–spiral trajectory, provided that the stability number is less than 0.4. As with the smaller discs, a separation bubble is held captive to each shell (Fig. 3.13). However, more elongate shells, for example, *Donax vittatus*, *Mya arenaria* and *Pharus legumen*, pitch strongly while spiralling down. A vortex is partly released into the fluid at the end of each segment of the zig-zag downward path (Fig. 3.14). Viewed in the direction of motion, the right valves of certain species tend to spiral clockwise and the left valves anticlockwise, reflecting the fact that bivalve shells are

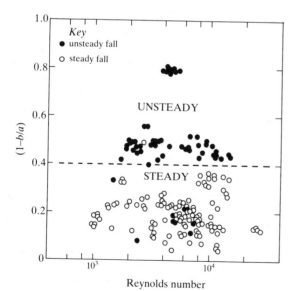

**Figure 3.12** Summary of experimental results on the behaviour of single valves of British bivalve sea shells settling in a stagnant fluid. Data of Allen (1984a).

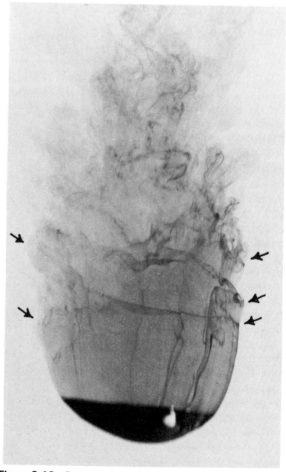

**Figure 3.13** Pattern of motion round a single valve of *Mactra corallina* falling in stagnant water at $Re \simeq 5500$. Arrows point to vortices developing in the mixing layer surrounding the separation bubble.

**Figure 3.14** Pattern of motion round a single valve of *Donax vittatus* falling in stagnent water at *Re* ≃ 3300.

only symmetrical bilaterally. For other species this relationship between the spiralling and handedness is reversed.

A bivalve shell settling through a current will therefore arrive concave-up on the bed, but without any preferred horizontal orientation. Only if the current is sufficiently strong will the shell become overturned into the more stable convex-up position. Allen (1984a,b) showed that a shell which settled through a cloud of suspended sand, as in a turbidity current, would trap grains as it fell to the bed. It would then be even more difficult for a current to turn the shell over from the concave-up position.

### 3.7.3 Quartz sand

Quartz grains are the non-spherical particles of greatest sedimentological importance. Their terminal fall velocity $V_0$ can be estimated using the theoretical correlation for spheres shown in Figure 3.9 but, because the grains are not perfectly spherical, most workers much prefer an empirical correlation which takes particle shape into account. Hallermeier (1981), reviewing many experimental results, suggests a simple scheme based on the Reynolds number $Re = \varrho D_{ms} V_0 / \eta$ and another form of the Archimedes number $Ar = \varrho(\sigma - \varrho) g D_{ms}^3 / \eta^2$, in

which $D_{ms}$ is the median sieve diameter of the grains. According to the value of $Ar$, the terminal fall velocity is calculated using one of the three empirical equations.

$$Re = \tfrac{1}{18} Ar \qquad Ar \leqslant 39 \qquad (3.32a)$$

$$Re = \tfrac{1}{6} Ar^{0.7} \qquad 39 < Ar \leqslant 10^4 \qquad (3.32b)$$

$$Re = 1.05 \, Ar^{0.5} \qquad 10^4 < Ar \leqslant 10^6 \qquad (3.32c)$$

Notice that Equation 3.32a reduces to Stokes's law, indicating that the departure of sand grains from the perfect spherical form has no practical significance at small Reynolds numbers. The predictions of Equations 3.32 become increasingly different with increasing Archimedes number from those of Figure 3.9. This is probably because of the earlier transition to turbulence promoted by the rough surfaces of natural particles.

## Readings

Allen, J. R. L. 1984a. Experiments on the settling, overturning and entrainment of bivalve shells and related models. *Sedimentology* **31**, 227–50.

Allen, J. R. L. 1984b. Experiments on the terminal fall of the valves of bivalve molluscs loaded with sand trapped from a dispersion. *Sed. Geol.* **39**, 197–209.

Clift, R., J. R. Grace and M. E. Weber 1978. *Bubbles, drop and particles.* New York: Academic Press.

Dullien, F. A. L. 1979. *Porous media. Fluid transport and pore structure.* New York: Academic Press.

Futterer, E. 1978. Untersuchungen über die Sink- und Transport-geschwingkeit biogener Hartteile. *Abh. N. Jb. Geol. Paläont.* **155**, 318–57.

Hallermeier, R. J. 1981. Terminal settling velocity of commonly occurring sand grains. *Sedimentology* **28**, 859–65.

Hjelmfelt, A. T. and K. L. F. Mockros 1967. Stokes flow behaviour of an accelerating sphere. *J. Engng. Mech. Div. Am. Soc. Civ. Engrs* **93**, 87–102.

Jayaweera, K. O. L. F. and B. J. Mason 1965. The behaviour of freely falling cylinders and cones in a viscous fluid. *J. Fluid Mech.* **22**, 709–20.

Krumbein, W. C. and G. D. Monk 1942. Permeability as a function of the size parameters of unconsolidated sand. *Petrolm Technol.* **5**, 1–11.

Leva, M. 1959. *Fluidisation.* New York: McGraw-Hill.

Maude, A. D. and R. L. Whitmore 1958. A generalized theory of sedimentation. *Br. J. Appl. Phys.* **9**, 477–82.

Richardson, J. F. and W. N. Zaki 1954. Sedimentation and fluidisation. *Trans. Inst. Chem. Engrs* **32**, 35–53.

Scheidegger, K. F. and L. A. Krissek 1982. Dispersal and deposition of eolian and fluvial sediments off Peru and Chile. *Bull. Geol. Soc. Am.* **93**, 150–62.

Scott, G. D. 1960. Packing of spheres. *Nature* **188**, 908–9.

Stringham, G. E., D. B. Simons and H. P. Guy 1969. *The behaviour of large particles falling in quiescent liquids*. Prof. Pap. US Geol. Surv., no. 562-C.

Taneda, S. 1956. Studies on wake vortices. (III) Experimental investigation of the wake behind a sphere at low Reynolds numbers. *Rep. Res. Inst. Appl. Mech. Kyushu Univ.* **4**, 99–105.

Taneda, S. 1978. Visual observations of the flow past a sphere at Reynolds numbers between $10^4$ and $10^6$. *J. Fluid Mech.* **85**, 187–92.

Von Engelhardt, W. and H. Pitter 1951. Über die Zusammenhänge zwischen Porosität, Permeabilität und Korngrösse bei Sanden und Sandsteinen. *Heidelb. Beitr. Min. Petrog.* **2**, 477–91.

Willmarth, W. W., N. E. Hawk and R. L. Harvey 1964. Steady and unsteady motions and wakes of freely falling discs. *Phys. Fluids* **7**, 197–208.

# 4 Sliding, rolling, leaping and making sand waves

Entrainment of sedimentary particles − rate of sediment transport − sediment load − sediment transport as work − modes of transport − transverse bedforms in water and beneath the wind − bedform movement and internal structures − cross-stratification − hydraulic controls on bedforms − wave theory of bedforms.

## 4.1 Some field observations

Have you ever stood by a flooded mountain stream? Pebbles and cobbles are probably being audibly carried over the bed. From the evidence of their frequent collisions amongst themselves and with debris stationary on the bed, these stones must lie in dense array close to the stream bottom, to form what is called the bedload. However, this load is most unlikely to be visible, on account of the turbidity of the water. The smaller and more uniformly dispersed particles which make the current turbid are evidently being transported in a different way than the stones keeping close to the bed. They constitute the suspended load of the stream (Ch. 7).

A related experience follows on watching and listening to storm waves as they comb a shingle beach, the shrill rattle of the transported shingle being especially striking. The particles can now be seen in motion, as there is generally little or no suspended sediment. The more spherical stones can be seen either to roll or to take short leaps (saltations) through or even out of the water. Look out for discoidal stones, the larger of which may slide over the stationary shingle. When driven by a particularly powerful backwash, however, discoidal particles will bound and cartwheel along, spinning as they saltate over flat trajectories that keep close to the bed. Finally, pick up some of the stones. Their smooth and rounded forms are just such as would be expected under the action of a myriad of violent collisions. Direct evidence of repeated impact is in some cases given by dense surface arrays of percussion marks (Fig. 4.1).

**Figure 4.1** Percussion marks on the surface of a flint pebble. Scale bar 0.01 m.

To study particle transport on sandy shores, look out for the channels draining off the tide. It is generally possible in these streams to find places where sand grains can be seen in motion, and small stones and shells may also be noticed rolling over the bed. One is now most struck by the fact that the flow is heaping the grains into flow-transverse ridges, which are slowly but inexorably advancing beneath the current. A great variety of transverse wave-like features called bedforms can in fact be fashioned from sand and even gravel by the action of streams of wind as well as water. These ridges and hollows are repeating features which generally occur in areally extensive trains. Most varieties are rather flat and strongly asymmetrical in cross section.

They range in wavelength from about a decimetre to tens and even hundreds of metres.

These observations represent experience of aspects and manifestations of sediment transport. Under what conditions can heavy sediment particles be set in motion? What is involved when particles are transported by a unidirectional wind or water stream? What governs the rate of particle transport? Why should particles become shaped into wave-like features?

## 4.2 Setting particles in motion

### 4.2.1 An experiment

How were the various particles impelled by the stream and over the beach set in motion?

Try the following experiment. Arrange a uniform layer of clean fine-grained quartz sand over the bottom of a crystallizing basin or pneumatic trough half-filled with clean water. Stir gently, using a smooth circular motion. A current will be present but, provided the stirring was sufficiently gentle, no grains will be moved. Now increase the rate of stirring until a value is found at which the grains are just in a state of continuous motion. Measure and note this critical stirring rate. Repeat with a very coarse sand or fine gravel, using the same container, stirrer and water depth. To move the coarse grains will require a higher stirring rate than was necessary for the fine sand. As the current increases with the stirring rate, the larger grains evidently need the faster current for their entrainment.

What does the experiment mean? In Section 3.2 we saw that a fluid moving relative to solid particles exerts a drag force on those particles. A horizontal drag must have been exerted on our sand beds by the currents circling above them. But why did this force fail to move the grains until in each case a critical speed was reached? Newton's first law of motion demands that there be a force opposing sand movement. As the sands can be freely poured from one container to another, the opposing force is not granular cohesion. However, the particles are more dense than the impelling fluid, and so had a certain downward-acting weight. This is the required force preventing entrainment. A critical state in our experimental system is reached only when the drag force has come into a specific quantitative relation with the immersed grain weight. This critical state is the threshold of particle movement.

### 4.2.2 Analysis of threshold conditions

Consider in Figure 4.2 the threshold of motion of a spherical particle of diameter $D_2$ resting on a flat bed of

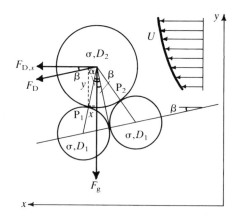

**Figure 4.2** Definition sketch for particle entrainment.

spherical particles of uniform diameter $D_1$ tilted at an angle $\beta$ from the horizontal $x$-direction, where $\beta$ is measured positively downwards in the flow direction. The grains are all cohesionless, of a uniform density $\sigma$, and the bed is subjected to a uniform steady fluid stream of density $\varrho$ flowing towards the left.

One of the two forces acting on the grain, to judge from the experiment, is its immersed weight

$$F_g = \frac{4}{3}\pi\left(\frac{D_2}{2}\right)^3(\sigma - \varrho)g \quad \text{N} \qquad (4.1)$$

where $g$ is the acceleration due to gravity. This force, acting vertically downwards parallel to the $y$-direction, holds the grain on the bed. The current exerts on the bed a tangential bed shear stress $\tau_{cr}$ N m$^{-2}$, the subscript showing that the value is the critical one for entrainment. As this drag acts over unit bed area, the total force on the grain parallel to the $x$-direction is

$$F_{D,x} = \pi\left(\frac{D_2}{2}\right)^2\tau_{cr}\cos\beta \quad \text{N} \qquad (4.2)$$

that is, the product of the grain projection area with the unit force.

But the grain in moving off pivots about the point $P_1$ in Figure 4.2. To balance the forces, as Newton's first law requires, we take moments about $P_1$, writing

$$yF_{D,x} = xF_g \qquad (4.3)$$

where $x$ and $y$ are the normal distances from $P_1$ to the forces. Assuming that the drag acts through the particle centre, the geometry dictates that

$$x = D_2 \sin(\alpha - \beta) \qquad (4.4)$$

$$y = D_2 \cos(\alpha - \beta) \qquad (4.5)$$

where $\alpha$ is the semi-angle subtended at the particle centre by the supporting points $P_1$ and $P_2$.

On substituting into Equation 4.3 from the other formulae and rearranging, the unit stress at the entrainment threshold becomes

$$\tau_{cr} = \frac{2D_2(\sigma - \varrho)g}{3 \cos \beta} \tan(\alpha - \beta) \quad \mathrm{N\,m}^{-2} \qquad (4.6)$$

where

$$\tan(\alpha - \beta) = \sin(\alpha - \beta)/\cos(\alpha - \beta) \quad \text{and}$$
$$\alpha = \tan^{-1} D_1(D_2 + 2D_1D_2)^{1/2}.$$

This equation states that the threshold stress increases linearly with the particle diameter, as suggested by the experiments described above, and with the excess density. Equation 4.6 can be rearranged in the non-dimensional form

$$\theta_{cr} = \frac{\tau_{cr}}{(\sigma - \varrho)gD_2} = \frac{2 \tan(\alpha - \beta)}{3 \cos \beta} \qquad (4.7)$$

where $\theta_{cr}$ is the critical value of the Shields–Bagnold non-dimensional boundary shear stress. Notice that, for the particular packing mode implied by Figure 4.2, the critical non-dimensional stress is a constant for each bed slope and diameter ratio $D_2/D_1$.

Equation 4.6 points to some interesting effects of slope on entrainment. First consider entrainment on a bed sloping down in the current direction. As $\beta$ approaches $\alpha$, $\tan(\alpha - \beta)$ becomes progressively smaller, and the critical stress declines. When $\beta = \alpha$, the stress is zero, and the particle tumbles down the bed under gravity alone. Next consider a bed sloping upwards with the current. The slope angle $\beta$ is now negative, so that $\tan(\alpha - \beta)$ increases as the slope steepens. The critical stress is therefore larger in upslope than downslope flow.

How does bed slope affect the entrainment of particles of contrasted size on the same bed? Suppose that grains of diameters $D_2$ and $D_1$ are being entrained together from the same bed of particles of diameter $D_1$. It follows from Equation 4.6 that $\tau_{cr,2} \leqslant \tau_{cr,1}$, where the subscripts refer to entrained particles of each size. Rewriting the equation for each size of particle, and comparing one expression with the other, simultaneous entrainment requires that

$$\frac{D_2}{D_1} \leqslant \frac{\tan(\alpha_1 - \beta)}{\tan(\alpha_2 - \beta)} \qquad (4.8)$$

where the subscripts again distinguish the entrained particles. When $D_2 > D_1$ we find from the geometry that $\alpha_2 < \alpha_1$, whence there is a downslope value of $\beta$ which just satisfies the relationship. The unusually large particles of size $D_2$ can then be entrained from and transmitted over the bed, the phenomenon being called overpassing. But $\alpha_2 > \alpha_1$ when $D_2 < D_1$ and there is then a sufficiently large upslope value for $\beta$ satisfying Equation 4.8. Sufficiently small grains can be entrained from the bed, but the larger ones are left behind. This could be the process responsible for armouring with coarse debris the upstream sides of gravel bars.

### 4.2.3 Empirical specification of threshold conditions

Our analysis is in fact much oversimplified and only qualitatively correct. The chief omission is an experimentally demonstrable upward-acting lift force (Chepil 1958, 1961), which almost equals the drag in amount, reducing the numerical value of the right-hand side of Equation 4.7 by almost an order of magnitude. Unfortunately, the lift force varies in a complicated way with flow and bed conditions and cannot be straightforwardly predicted. In practice, entrainment conditions are estimated using empirical correlations based on laboratory experiments.

Many of these empirical correlations are critically discussed by Miller et al. (1977), and Mantz (1977) has added further useful data. Figure 4.3 shows the entrainment threshold for solids in various liquids and applicable to particles in water, where the grain diameter is given in the non-dimensional form of the square root of the Archimedes number $Ar = \varrho(\sigma - \varrho)gD^3/\eta^2$ in which $\eta$ is the fluid viscosity. The graph also shows an experimental curve for mineral-density solids in air. For a number of reasons the experimental data scatter widely, allowing considerable latitude in the choice of threshold value. At sufficiently large grain sizes, however, the threshold stress seems to be linearly proportional to particle diameter (Eqn 4.7). At small sizes, it is a steeply increasing function of grain size. Figure 4.4 shows another widely used entrainment curve (Bagnold 1966). It applies only to quartz-density solids in water, and shows medium and coarse sands as more easily entrained than any other grade.

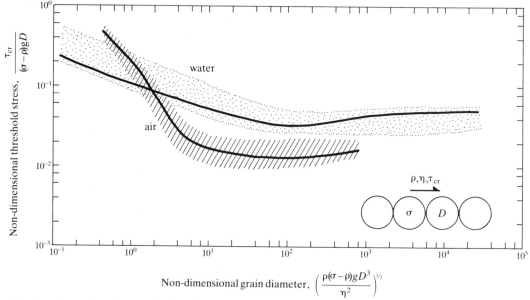

**Figure 4.3** Experimental entrainment threshold for solids in liquids including water and for mineral-density solids in air, together with the range of experimental scatter. Data of Miller *et al.* (1977) and Mantz (1977).

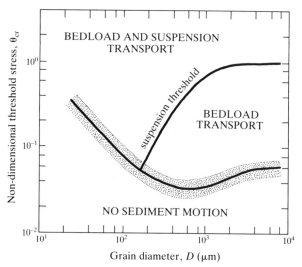

**Figure 4.4** Shields-type entrainment curve for quartz-density solids in water, showing the range of experimental scatter and the expected modes of sediment transport. Adapted from Bagnold (1966).

## 4.3 Defining the rate of sediment transport

### 4.3.1 *What should we measure?*

Consider the field observations outlined above. Sediment transport is evidently the carriage of quantities of

sediment particles past a fixed station. Essential to this concept are just two properties of the particles. One is the quantity or concentration of particles moving at each level in the flow. This could be measured by introducing into the current a sampling device suitable for drawing off a known volume of the particle–fluid mixture, without unduly disturbing the system. From this sample can be measured the total fractional volume concentration of the grains and their mean density. Assuming for convenience that the solids are of uniform density, the quantity of sediment in transport at a given level is simply the product of the fractional volume concentration and the solids density. Now the other relevant property is the streamwise motion of the grains. On looking into a clear stream carrying sand, we see that the particles at each level are moving with the current at a characteristic velocity.

These two properties – the quantity of particles and their rate of travel – together define the particle flux at any given level in the flow.

### 4.3.2 *Formal derivation of the sediment transport rate*

Consider the uniform steady transport of grains of solids density $\sigma$ by a uniform steady current of depth $h$. Let the particle fractional volume concentration in the flow be $C(y)$, where $y$ is measured upwards from the bed. Since the transport is uniform and steady, $C(y)$ varies neither with time nor with distance. Now consider

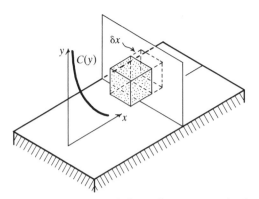

**Figure 4.5** Definition sketch for sediment transport rate.

relative to a fixed plane perpendicular to flow the movement of the cluster of grains contained in a cubical element of unit volume at an arbitrary height $y$ above the bed (Fig. 4.5). At some initial time the cluster has reached the flow perpendicular plane. After a small time $\delta t$ the grains have travelled onwards a small streamwise distance $\delta x$. Now as the mixture of fluid and grains is continuous, the element singled out is followed without a break by an identical mixture, making the total quantity of grains transmitted through the plane equal to $\sigma C \, \delta x$. The grain flux through unit area of the plane is therefore $(\sigma C) \, \delta x / \delta t$, which in the limit is $\sigma C U_s \, \mathrm{kg \, m^{-2} \, s^{-1}}$, where $U_s$ is the particle transport velocity relative to the ground at the given $y$.

So far we have derived the local flux but not yet calculated the transport rate. Recognizing that $U_s$ must also vary with height above the bed, the transport rate over the flow depth is clearly

$$J = \sigma \int_0^h (CU_s)(y) \, \mathrm{d}y \quad \mathrm{kg \, m^{-1} \, s^{-1}} \text{ (unit transport rate)}$$
$$(4.9)$$

that is, the dry mass of sediment passing above a lane of unit width in unit time. We need only integrate Equation 4.9 across the whole flow width to obtain the total sediment discharge ($\mathrm{kg \, s^{-1}}$ dry mass). Compare these three measures of sediment flow with the fluid discharges discussed in Section 1.2. Equation 4.9 is the practical definition of the transport rate.

## 4.4 Physical implications of sediment transport

### 4.4.1 Concept of sediment load
Whether by wind or water, sediment transport sees heavy grains being sustained in motion against gravity

by the action of a lighter fluid stream. We should therefore define the transport rate in terms of immersed particle weight rather than the dry mass, so introducing the notion of force. Equation 4.9 has the alternative form

$$J = U_s \int_0^h \sigma C(y) \, \mathrm{d}y \qquad (4.10)$$

where $U_s$ is from now on understood as the depth-averaged grain transport velocity, and the integral term is the total dry mass of particles contained in a fluid column above unit bed area. On multiplying the integral term by the conversion factor $(\sigma - \varrho)g/\sigma$, to introduce the notion of force and correct for buoyancy, we obtain the immersed weight of grains over unit bed area, with the units $\mathrm{N \, m^{-2}}$. This important quantity, the sediment load, has the quality and dimensions of a force per unit area (stress). Referring back to Equation 4.10, the revised transport rate is evidently

$$I = mU_s \qquad (4.11)$$

where $I$ is the immersed-weight unit transport rate and $m$ the load.

So far we have not distinguished as to mode of sediment transport, whereas it was found above that particles could either slide, roll, or leap or be more or less permanently suspended. Clearly, Equation 4.11 can also be written

$$I = I_1 + I_2 + \ldots + I_n = m_1 U_{s,1} + m_2 U_{s,2} + \ldots + m_n U_{s,n}$$
$$(4.12)$$

where the subscripts refer to particular transport modes, and $I$ is then properly called the unit total-load transport rate.

### 4.4.2 Necessity for supporting stresses
The sediment load is a true force and represents the downward pressure exerted on the bed by the immersed weight of the transported grains. It was Bagnold (1966) who pointed out that this downward pressure, in order to satisfy Newton's first law, must be balanced by an equal and opposite upward force if uniform steady sediment transport is to be maintained. Where does this balancing force come from? Ultimately, it must originate in the impelling stream, although it may comprise contrasting components, each corresponding to the fraction of the total load represented by a particular sediment transport mode.

### 4.4.3   Sediment transport as a rate of doing work

Bagnold (1966) pointed to another fundamental implication of sediment transport. As defined by Equations 4.11 and 4.12, the unit immersed-weight sediment transport rate is clearly the product of a force per unit area times velocity. The transport rate therefore has the dimensions and quality of a rate of doing work (Sec. 1.5.1). It is not yet an actual work rate, because the force involved, acting normal to the bed, is not in the same direction as the velocity of its action, parallel with the bed. The transport rate becomes an actual work rate, with the units $J\,s^{-1}m^{-2}$, when multiplied by a numerical conversion factor defined as the ratio of the tangential stress needed to maintain transport of the load to the normal stress due to the load's immersed weight.

### 4.4.4   Fluid stream as a transporting machine

The doing of work implies a time rate of expenditure of energy (power), and any mechanical system in which energy expenditure is traded for work done is a machine. Bagnold's (1966) third major contribution is his recognition that any sediment-transporting current is in principle a transporting machine. Now the performance of any machine is described by

rate of doing work = available power − unused power

or in an equivalent alternative form

rate of doing work = efficiency × available power

$$(4.13)$$

where the power $\omega$ appropriate to a sediment-transporting fluid flow is measured in $W\,m^{-2}$ (equivalent to $J\,s^{-1}m^{-2}$). The unit immersed-weight rate of sediment transport is therefore predictable knowing (1) the appropriate efficiency, and (2) the factor converting the measured transport rate into a work rate. The appropriate power may be defined as the product of a mean boundary shear stress times a velocity, and for each transport mode there is a proper choice of these quantities.

With these ideas in mind, the immersed-weight unit sediment transport rate finally becomes

$$I = \frac{e_1}{a_1}\omega_1 + \frac{e_2}{a_2}\omega_2 + \ldots + \frac{e_n}{a_n}\omega_n \qquad (4.14)$$

where $e$ is the efficiency, $a$ the conversion factor, and the subscripts again describe transport modes.

### 4.4.5   Practical transport formulae

Although Bagnold's model is theoretically most attractive, it cannot be used for prediction without appealing to empirical data. Most modern transport formulae use the general concept of stream power, but present a wide choice of parameters and coefficients by which to correlate measured transport rates in a predictive formula (Ackers & White 1973, Bridge 1981a, Hardisty 1983, Mantz 1983). All the parameters are in some measure justifiable, depending on the job to be done and the kind(s) of load and transport stages expected.

The simplest scheme correlates measured rates on $U_m^3$, where $U_m$ is the mean flow velocity, and the cube exponent makes it equivalent to a stream power. The correlation is inaccurate at low transport stages, however, as threshold conditions are ignored. An improvement is to use either $(U_m - U_{cr})^3$ or $(U_m^2 - U_{cr}^2)U_m$, where $U_{cr}$ is the velocity equivalent to the threshold stress and the bracketed terms may also be written as stresses. Perhaps more attractive theoretically is $(U_m^2 - U_{cr}^2)(U_m - U_{cr})$, taking full account of threshold conditions. The first bracket is equivalent to the excess stress, and the second to the excess velocity. Unfortunately, each field or laboratory data set when correlated on one of these parameters yields a unique coefficient. Indeed, published coefficients vary so much as to yield transport rates ranging over two orders of magnitude for nominally the same conditions.

### 4.4.6   Effect of changing the sediment transport rate

The accumulation of detrital sediments is impossible without sediment transport. In uniform steady transport, however, the unit transport rate changes neither with time nor with distance, and intuitively the bed beneath cannot vary vertically. Only when the transport rate changes in some manner does it seem possible for the bed to shift vertically in response to either erosion or deposition. Change may occur in space as well as time.

First examine spatial changes. Figure 4.6 shows as path lines the movement of sediment over an erodible surface of the same material, where $x$ is streamwise distance, $z$ lies normal to $x$, and $J\,kg\,m^{-1}s^{-1}$ is the local dry-mass unit transport rate, changing only with $x$. Consider what happens between two adjacent path lines. The total transport rate at an arbitrary width $w$ is simply $wJ\,kg\,s^{-1}$, that is, the product of the width times the unit rate. At a small distance $\delta x$ down stream, however, the width is $(w + \delta w)$ and the unit transport rate is increased to $(J + \delta J)\,kg\,m^{-1}s^{-1}$, where $\delta w$ and $\delta J$ are small increments. The new total transport rate is

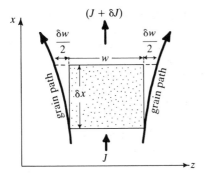

**Figure 4.6** Definition sketch for sediment transport rate in one variety of non-uniform flow.

the product of the new width times the new rate, that is, $(w + \delta w)(J + \delta J)$ kg s$^{-1}$. As the downstream rate exceeds the upstream value, and sediment cannot cross the path lines, the excess must have come from within the bed area $(\delta x\, w + \frac{1}{2}\delta x\, \delta w)$ m$^2$. Hence the rate of sediment loss from unit bed area is

$$R = - \frac{(w + \delta w)(J + \delta J) - wJ}{(\delta x\, w + \frac{1}{2}\delta x\, \delta w)} \quad \text{kg m}^{-2}\text{s}^{-1} \qquad (4.15)$$

which expanded and in the limit becomes

$$R = - \left( \frac{J}{w} \frac{\partial w}{\partial x} + \frac{\partial J}{\partial x} \right) \qquad (4.16)$$

where $R$ is the areal rate of change, and the minus sign signifies a loss. Whereas $J$ measures sediment movement parallel to the bed, $R$ is a rate of transfer, recording sediment motion normal to the surface. Erosion, that is, a movement from bed to flow, is denoted by $R < 0$, whereas $R > 0$ records deposition, a transfer from flow to bed. Most sedimentary sequences record the alternation of deposition and erosion, but for all deposition prevailed in the long term.

Before examining Equation 4.16 further, consider the effect of temporal changes in the transport rate at a fixed station. At time $t$ the unit transport rate is $J$ kg m$^{-1}$s$^{-1}$ but at a slightly later time $(t + \delta t)$ it is $(J + \delta J)$, where $\delta t$ and $\delta J$ are small increments. But in the time $\delta t$, the fluid column of unit basal area containing the transported sediment will have moved a downstream distance $\delta t\, U_s$, where $U_s$ is the depth-averaged sediment transport velocity. The extra sediment in the flow represented by the increment $\delta J$ has therefore come from a bed area of unit width but of length $\delta t\, U_s$, making the change per unit bed area

$$R = - \frac{(J + \delta J) - J}{\delta t\, U_s} \quad \text{kg m}^{-2}\text{s}^{-1} \qquad (4.17)$$

which in the limit is

$$R = - \frac{1}{U_s} \frac{\partial J}{\partial t} \qquad (4.18)$$

where the minus sign is again introduced to denote a loss.

Now put Equations 4.16 and 4.18 together. The total change of a bed in response to non-uniform and unsteady sediment transport is

$$R = - \left( \frac{J}{w} \frac{\partial w}{\partial x} + \frac{\partial J}{\partial x} + \frac{1}{U_s} \frac{\partial J}{\partial t} \right) \quad \text{kg m}^{-2}\text{s}^{-1} \quad (4.19)$$

Equation 4.19 is the sediment continuity equation, analogous to the continuity relation for a fluid stream (Sec. 1.3). It is a necessary accounting tool, permitting a 'balancing of the books' in the case of the sediment-transporting flow. Each term in Equation 4.19 describes a distinct and independent effect. The first shows how flow convergence and divergence in the horizontal plane affect the bed response. Expansion causes erosion but contraction ensures deposition. The effect is strong only in very narrow flows and in those changing rapidly in width. The second term reveals the effect of a transport rate changing with downstream distance, in response to a varying flow depth and/or velocity. For example, a downstream decrease in the rate is associated with deposition. The final term describes the effect of unsteady transport, as during a river flood. Erosion results from an increase with time, whereas deposition is ensured by a declining rate. The effect varies inversely with the sediment transport velocity.

The final effect of the terms in Equation 4.19 is their algebraic sum, and the individual terms can differ in sign. In a river, for example, the flow deepens down current in some places, whence the transport rate should decrease, but shoals in others, whence an increase in the rate should occur. A rise of the river will therefore affect these places in different ways, but with the shoaling reach eroding most.

Finally, by multiplying the right-hand side of Equation 4.19 by $1/\gamma$, where $\gamma$ is the dry bulk density of the bed sediment, the transfer rate is obtained in the form of a velocity. This formulation yields the speed with which a sedimentary surface is built up or lowered beneath a current.

## 4.5 Sediment transport modes

### 4.5.1 Defining the modes

It is now time to look more closely at the different sediment transport modes and the nature of the upward-acting normal forces that sustain the load. The two issues are linked because, in thinking about sediment transport dynamics, it is best to define each kind of transport mode in terms of the source of the normal force involved. To define transport mode in terms of either grain size or the location of the load within the flow, as many investigators have done, has a strong practical appeal but can obscure the underlying mechanics.

### 4.5.2 Bedload

Bagnold (1954) showed experimentally that when densely arrayed cohesionless grains are sheared together in a fluid, the time-averaged force generated by the repeated impact or close encounter between particles can be resolved into tangential and normal components, respectively $T$ and $P$, where $T/P = \tan \alpha$ is called the dynamical friction coefficient. Confirmation and amplification have recently come from Savage (1979), Savage and Jeffrey (1981) and Savage and McKeown (1983). Bagnold (1966) proposed that the normal stress $P$ also arose when coarse particles were impelled in dense array over a river bed or beneath the wind, and defined the bedload as that quantity of sediment whose weight is directly supported by this upward-acting inter-granular pressure related to grain-shearing over the static bed. Therefore $\tan \alpha$ is the numerical factor $a$ required in Equation 4.14 to convert the bedload transport rate into an actual work rate. Bagnold (1966) estimated the efficiency $e$ to be of the order of 0.14.

Is Bagnold's insight true? Try the following experiment. Obtain a large circular plastic washing-up bowl with a near-vertical side, and a quantity of cheap ping-pong balls. Swirl a few balls steadily round in the bowl. At a large enough speed, the balls travel continuously around the bowl. Centrifugal force presses them against the side, but there is evidently no strong force requiring them to move towards the centre. A different result is obtained after gluing a quantity of the balls cut in half over the inner side of the bowl. Balls now swirled around the bowl repeatedly bounce inwards off the ones glued to the side (Fig. 4.7). There now acts a strong inward-directed stress, corresponding to Bagnold's $P$. The other component of the intergranular force, the stress $T$, is exerted tangentially on the fixed balls.

The leaping motion of fluid-impelled particles, of which some impression can be gained from the above

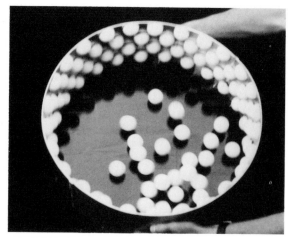

**Figure 4.7** Ping-pong balls swirling at speed in a rough-walled container.

experiment, is called saltation and is common in both water streams and beneath the wind. Mineral-density particles saltating in water take rather short and flat trajectories (Gordon *et al.* 1972, Abbott & Francis 1977), of the order of 10–20 grain diameters long and two diameters high (Fig. 4.8a). The colliding particles are somewhat cushioned by the effects of fluid viscosity and the moderate density difference. Mineral grains in air are about a thousand times denser than the impelling fluid, and the impacts between the moving particles and those stationary on the bed are violent, as anyone whose face or body has been stung by flying sand on a windy day can testify. Saltation trajectories in air (White & Schulz 1977) are steep and several to many orders of

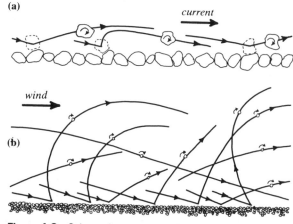

**Figure 4.8** Schematic representation of the paths of irregular particles saltating over a bed of similar grains (a) in water and (b) in the wind.

magnitude greater in length and height than the particle diameter (Fig. 4.8b). Particles spin while saltating in both air and water (Abbott & Francis 1977, White & Schulz 1977). The spin combines with the steep velocity gradient in the near-bed fluid to create a significant lift force helping to draw the saltating particle upwards.

Two other modes of grain motion can be important in aqueous bedload transport. An insight into these can be gained by turning the washing-up bowl used in the previous experiment on its side, so that it can be rolled along the edge of a table. A discoidal pebble placed in the very slowly turned bowl will slide unevenly over the balls, but without losing contact with the bed. Above a certain critical rolling speed, a ping-pong ball placed in the bowl begins to saltate. This occurs when the downward-acting particle weight is exceeded by the centrifugal force acting upwards on the ball as it travels over the convex-up surfaces of the fixed balls beneath. For particles of a given size and density, the sequence of transport modes with increasing driving force is therefore sliding, then rolling, and finally saltation.

### 4.5.3 Suspended load

Bagnold's (1966) dynamical bedload turns out to be (1) mainly the coarser particles in transport, (2) limited chiefly to a relatively thin zone immediately above the static bed, and (3) densely arrayed. He showed experimentally (Bagnold 1955) that bedload transport could occur in laminar as well as turbulent fluids, whence fluid turbulence is unnecessary for either grain sliding, rolling or saltation, although it may influence these motions. In a turbulent flow, however, like the wind or a flooded river, fine particles can be found fairly evenly spread at a low concentration throughout the whole fluid. This debris is so widely dispersed that grain collisions can hardly if ever occur. Hence its transport must depend on a stress other than the intergranular force sustaining the bedload. As such dispersions do not occur in laminar flows, the required stress must in some way be related to the fluid turbulence.

Bagnold (1966) defined as the suspended load that part of the total sediment load supported by this turbulence-related stress. Detailed discussion will be left until Chapter 7, but some points may be noted here. Bagnold postulated that in shear turbulence there arises a normal stress directed away from any solid flow boundary, on account of the outward-directed velocity fluctuations being larger than the inward-directed ones. Hence for uniform steady suspension transport in rivers or by the wind, the asymmetrical turbulence is in effect constantly pushing the grains up a notional incline of slope $V/U_s$, where $V$ is a characteristic terminal fall velocity for the particles and $U_s$ the depth-averaged grain transport velocity, equal in practice to the mean flow velocity. This ratio corresponds to Bagnold's $T/P = \tan \alpha$ applicable to bedload, and is the numerical factor $a$ required to convert the suspended-load transport rate to an actual work rate (Eqn 4.14). Using experimental results, Bagnold (1966) calculated that the suspended load should be a moderate fraction of the mean bed shear stress, and estimated the suspension efficiency $e$ as of order 0.016.

## 4.6 Appearance and internal structure of bedforms

### 4.6.1 What kinds are known?

Bedload transport in rivers and by the wind finds visible expression in bedforms. These are more or less regularly spaced transverse 'waves' of sand or gravel which partake in the motion of the debris and acquire internally a structural imprint of that movement. Bedforms exist in great variety. How can they be classified? A dynamical classification, based on bedform mechanics, is ideally required, but this goal is far from being attained. All that a kinematic classification can achieve is a division between bedforms travelling down current with the grains and those moving up current. The most useful practical classification remains one based on geometry. Experience gained from straight laboratory flumes, reasonably straight river channels at times of roughly steady flow, and the open desert suggest that unidirectional currents can generate at least five kinds of transverse bedforms, of which all but the third advance down current. These are, in water, (1) current ripples, (2) dunes (including barkhan-like forms) and (3) antidunes and, in the wind, (4) ballistic ripples and (5) transverse dunes (including barkhans) (Allen 1982). Individual bedforms can be described by measuring their wavelength (crest-to-crest distance parallel to the flow), height (vertical interval between trough and crest), span (horizontal distance along crest) and asymmetry (various measures). Trains of bedforms are best characterized using ensemble averages (e.g. group mean wavelength).

By no means every granular surface supporting a bedload is wavy in one of the above ways. A sixth and important bedform should be recognized, namely, a plane bed with sediment transport, aspects of which are discussed in Chapter 6.

### 4.6.2 Current ripples

These transverse ridges abound beneath river and tidal currents, but are restricted to quartz sands finer than a

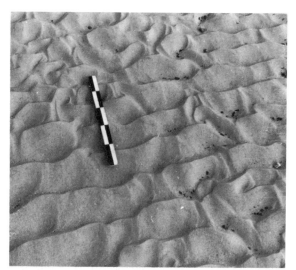

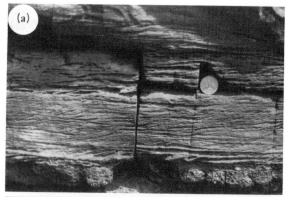

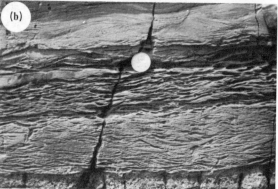

**Figure 4.9** Relatively two-dimensional current ripples in fine-grained sand, Norfolk coast, England. Current is towards top and scale is 0.5 m long.

**Figure 4.10** Strongly three-dimensional (linguoid) current ripples, Severn Estuary, England. Current is towards upper right and tape box is 0.05 m square.

**Figure 4.11** Cross-lamination due to the migration of current ripples (a) in a vertical streamwise profile and (b) in a vertical section normal to flow, Old Red Sandstone, South Wales. Current is right to left in (a) and into the face in (b). Coin is 0.028 m diameter.

diameter of about 600 $\mu$m. Current ripples in relatively slow deep flows have long, fairly regular crests of little curvature (Fig. 4.9). Shorter-crested forms, called linguoid, arise in relatively fast shallow currents (Fig. 4.10). Current ripples are strongly asymmetrical, with long convex-up upstream surfaces and short leeward sides inclined as steeply as 30–35°. Their wavelength varies between about 0.1 and 0.6 m and their height ranges up to approximately 0.04 m. Typically, the wavelength-to-height ratio is between 10 and 20, but can reach up to 40. The internal structure (cross-lamination) depends on ripple shape and the rate of net deposition on the bed. Streamwise and flow-normal vertical sections cut through the sediment beneath a rippled bed typically reveal inclined laminae arranged in erosively related sets (Fig. 4.11). These laminae represent the successive positions of the leeward sides of the ripples.

### 4.6.3 Ballistic ripples (small) and ridges (large)

These transverse mounds, arising where the wind blows freely over substantial expanses of sand and even fine gravel, occur chiefly in deserts and on the coast (Figs 4.12 & 13). Their wavelength increases with grain size and ranges from a few centimetres to several metres. Long regular crests typify those formed in the finest sands. As the sediment coarsens, the crests become shorter and more curved, tending to fade out rather than join up with neighbours. Ballistic ripples and ridges are rather flat, the wavelength-to-height ratio

**Figure 4.12** Ballistic ripples in fine sand, Norfolk coast, England. Wind is left to right and trowel is 0.28 m long.

**Figure 4.14** Dunes with shell–pebble lags on the stoss sides, Barmouth Estuary, Wales. Current is towards observer and dune wavelength is about 12 m.

**Figure 4.13** Large ballistic ripples in medium to coarse sand, Norfolk coast, England. Wind is towards upper right and trowel is 0.28 m long.

**Figure 4.15** Three-dimensional dunes shaped by a tidal current, Barmouth Estuary, Wales. Current is from bottom towards top and scale is 0.5 m long.

averaging about 20 and ranging as high as 70. The up-wind slopes are long and flat or slightly concave-up and the structures are moderately to strongly asymmetrical. The internal lamination is similar to that found within current ripples.

### 4.6.4 Dunes in water

These are perhaps the commonest bedforms of sand-bedded river and tidal channels and even occur in gravelly reaches. The structures are strongly asymmetrical in streamwise vertical profile and range from long-crested, in relatively deep and gentle currents, to short-crested, in the more vigorous and shallower flows (Figs 4.14 & 15). Dune wavelength exceeds 0.6 m and, in a deep river or tidal channel, can attain more than 100 m. The wavelength-to-height ratio is typically between 20 and 40 for structures of several metres wavelength, but is significantly larger for the biggest dunes. Isolated crescent-shaped dunes resembling desert barkhans arise where there is insufficient sand in transport to cover a hard surface completely (Fig. 4.16). Such dunes are being increasingly reported from rivers, shelf seas and

65

**Figure 4.16** Barkhan-shaped tidal dune travelling over eroded substrate of stiff mud, Severn Estuary, England. Current is from upper left and trowel is 0.28 m long.

**Figure 4.17** Cross-bedding in medium-grained sandstones, Coal Measures, South Wales. View of a vertical streamwise section with current from left to right. A geological hammer is shown for scale.

even from the deep ocean. Internally, a bed beneath dunes shows a pattern of inclined bedding in erosively related sets (cross-bedding), similar to that present beneath current ripples but on an appropriately larger scale (Fig. 4.17).

### 4.6.5 Transverse wind-formed dunes

Many deserts and most dry coasts yield examples. The structures are regularly spaced, strongly asymmetrical ridges aligned across the wind and with a wavelength of

**Figure 4.18** Transverse aeolian dunes on the Skeleton Coast, Namibia (South West Africa). Wavelength is several hundred metres, with wind from right to left. Photograph courtesy of Dr N. Lancaster, University of Cape Town.

10–100 m (Fig. 4.18). Crests are long and sinuous, the embayments generally coinciding with deep hollows in the troughs. Upwind slopes are gentle and convex-up, in contrast to the short steep leeward sides sloping as steeply as 30–35°. The wavelength-to-height ratio is about 20. An insufficient cover of sand over firm ground permits the formation of large crescent-shaped dunes called barkhans. The typical barkhan is about 10 m high, 100 m across the wind from one horn of the crescent to the other, and about 100 m long parallel to the wind. The downwind-pointing horns enclose a strongly curved leeward surface. Internally, transverse and barkhan dunes are cross-bedded in a similar manner to subaqueous dunes, but on an even larger scale.

### 4.6.6 Antidunes

These are common in gutters after rain, in steep streams draining sandy beaches, in some tidal currents and in flooded gravelly rivers. Antidunes are systems of essentially stationary in-phase bed and water-surface waves of sinusoidal streamwise profile (Fig. 4.19). The crests, lying transversely to flow, vary from long in some cases to short in others, when the bed consists of smoothly rounded oval hummocks and hollows beneath surface waves peaked up like rooster's tails. Many antidunes undergo with time a cyclic change, which involves a progressive steepening, culminating in the breaking and brief upcurrent migration of the surface wave. In response, laminae dipping gently up current accumulate on the upstream side of the bed undulation. As many antidunes also move very slowly down stream for short periods, their internal bedding is not uniquely oriented relative to current direction.

**Figure 4.19** Antidunes as (a) features on a free surface (flow is away from observer) and (b) bedforms in fine sand (flow was towards lower right and trowel is 0.28 m long).

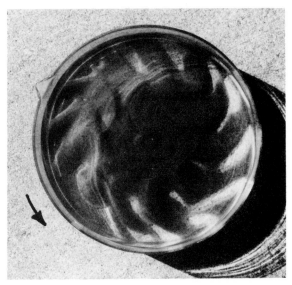

**Figure 4.20** Spiral asymmetrical bars produced on a sand bed by anticlockwise flow in a circular tank (diameter 0.22 m).

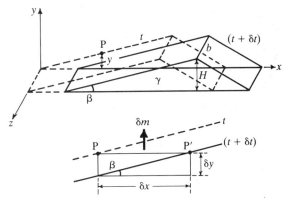

**Figure 4.21** Definition sketch for sediment transport in the course of bedform advance.

## 4.7 How do bedforms move?

### 4.7.1 Mechanics of movement

The bedforms described all take part in the transport of the sediment and therefore migrate beneath the current. An insight into this process can be obtained by stirring sand with water in a circular basin, when you will find that the grains become heaped into short bars (Fig. 4.20). These clearly resemble the bedforms described, and especially ripples and dunes. The bars have a long gentle upcurrent surface and a short steep leeward slope. You will also notice that they travel in the same sense as the current, as grains eroded from the upstream side are dumped to leeward. The movement involves no significant change in the size or shape of the bars. What are the implications of our observations?

Consider first the implications for sediment transport and transfer. Imagine as in Figure 4.21 a vertical streamwise slice of width $b$ through a bedform similar in shape to a ripple or dune and advancing with a steady current over a rigid surface. Suppose the bedform to be mature, so that it retains its size and shape. Because of erosion, a point P on the upcurrent side will advance down stream a small horizontal distance $\delta x$ in a small time increment $\delta t$. In the same time increment a dry mass of sand $\delta m$ will be lost through the surface lying between P and its new position P'. Introducing $\gamma$ as the dry bulk density, the volume of sand lost is clearly $\delta m/\gamma$. Now this volume, under our assumption of a constant bedform size and shape, is by the geometry of Figure 4.21 exactly equal to the volume of the

67

parallelepiped $b \, \delta x \, \delta y$, where $\delta y$ is the perpendicular change at P. Equating volumes

$$\frac{\delta m}{\gamma} = b \, \delta x \, \delta y \quad \text{m}^3 \qquad (4.20)$$

Again by the geometry $\delta y = \delta x \tan \beta$, where the slope $\beta$ of the surface is measured positive downwards in the current direction. Substituting into Equation 4.20, and recognizing that the changes occur in the same time increment $\delta t$, we find that

$$\frac{1}{b} \frac{\delta m}{\delta x \, \delta t} = \gamma \tan \beta \frac{\delta x}{\delta t} \quad \text{kg m}^{-2}\text{s}^{-1} \qquad (4.21)$$

becoming in the limit

$$\frac{1}{b} \frac{\mathrm{d}m}{\mathrm{d}x \, \mathrm{d}t} = \gamma \tan \beta \frac{\mathrm{d}x}{\mathrm{d}t} \qquad (4.22)$$

The left-hand term in Equation 4.22 has the same quality and dimensions as $R$ introduced in the equation of continuity (Eqn 4.19), and is the bedform transfer rate $R_{bm}$. Equation 4.22 states that the local rate (measured horizontally) on a ripple or dune due to its movement is proportional to the sediment bulk density, the surface slope and the quantity $\mathrm{d}x/\mathrm{d}t = U_{bm}$, the speed of advance of the bedform (measured horizontally). Our assumption that the bedform remains unchanged in size and shape means that, on the downstream side, where $\beta$ is positive, deposition takes place.

Now the back of the bedform loses grains in accordance with Equation 4.22 not only from P but from all stations up stream. The unit local dry-mass sediment transport rate at P is therefore the integral of Equation 4.22 from the toe of the bedform at $x = 0$ to P, that is,

$$J_P = - \gamma U_{bm} \int_0^x \tan \beta(x) \, \mathrm{d}x \quad \text{kg m}^{-1}\text{s}^{-1} \quad (4.23)$$

the negative sign making the transport rate positive. Hence at any point on a bedform travelling at constant speed, the local sediment transport rate depends on how the shape of the surface varies up stream. For a bedform of triangular profile, $\tan \beta$ is a constant over any one side, and Equation 4.23 has the simple solution

$$J_P = - \gamma U_{bm} x \tan \beta + C \qquad (4.24)$$

where the integration constant $C$ represents transported sediment not involved in the bedform. Now as P is at an

elevation $y$ above the base of the bedform, and the geometry dictates that, at P, $\tan \beta = -(y/x)$, Equation 4.24 can also be written

$$J_P = \gamma U_{bm} y + C \qquad (4.25)$$

This important result states that the transport rate at a point on a bedform is linearly proportional to the elevation of that point. Since $y$ cannot exceed the maximum bedform height $H$, the greatest local transport rate occurs at the crest and is

$$J_{crest} = \gamma U_{bm} H + C \qquad (4.26)$$

Over a train of bedforms, the local transport rate therefore varies in a spatially periodic way, rising to maxima at the crests.

Continuing with a simple triangular bedform, a local transport rate equal to the spatially averaged value clearly occurs when $y = H/2$, i.e.

$$J = \tfrac{1}{2} \gamma U_{bm} H + C \quad \text{kg m}^{-1}\text{s}^{-1} \qquad (4.27)$$

This equation, stating that the overall unit transport rate varies linearly with bedform speed and height, is the basis of a practical method for measuring bedload transport rates in the field. It can also be used to predict the movement of bedforms for known or predictable sediment transport conditions.

### 4.7.2 Movement expressed by inclined layering

Equations 4.22 and 4.23 usefully describe sediment transfer and transport on bedforms, but do not tell us how to interpret any internal features, such as cross-lamination and cross-bedding, that might have resulted from bedform movement.

Experimental studies (Allen 1965, 1968) show that grains carried over the crest of a bedform become dispersed over a substantial distance to leeward (Fig. 4.22), settling through a highly turbulent and sluggishly recirculating separated flow (Sec. 1.10). The intensity of particle deposition on the bed declines from the crest down slope into the trough, at a rate increasing directly with grain size but inversely with transport rate. The leeward slope therefore gradually steepens, appearing to rotate about its toe. But in Chapter 2 we saw that a surface underlain by cohesionless grains cannot be steepened beyond the angle of initial yield $\phi_i$, whence the leeward surface of a bedform receiving sand must eventually become unstable. An avalanche of grains, called a sand flow (Fig. 4.23), descends the leeward slope, reducing its steepness to the residual angle after

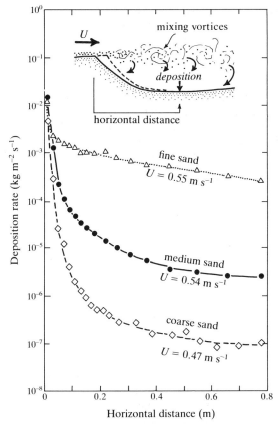

**Figure 4.22** Sediment deposition rate as a function of distance down stream from the crest of a laboratory dune 0.15 m high, in the case of three quartz sands. Data of Allen (1968).

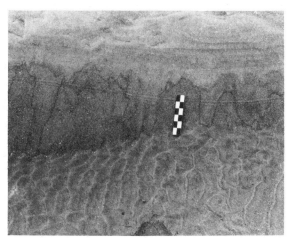

**Figure 4.23** Sand flows (avalanches) on the leeward face of a tide-shaped dune. Current is towards observer and scale is 0.15 m long.

shearing $\phi_r$. Steepening by settlement is then resumed, until instability is once more reached and another avalanche flows away. As these events occur at different times at different points along the crest, the bedform as the result of its forward march should consist internally of a stack of tongue-shaped layers at the residual angle after shearing, each layer recording one avalanche, as Hunter (1977) has described.

The layering is easily simulated. About half-fill a cylindrical glass jar with clean, dry ill-sorted sand, and firmly screw on the cap. Roll the jar over a table top through about one revolution. Now turn the jar very slowly, carefully observing the sand inside. As we saw in Section 2.9, its surface alternately steepens and collapses, only to steepen again. The steepening as the jar rotates simulates the effect of grain settlement in the lee of a bedform. In the jar, as on a bedform, the surface steepens until $\phi_i$ is reached, when an avalanche is triggered. The avalanche takes a short period to complete its downward flow, in the process gradually reducing the slope to the lower value represented by $\phi_r$. The process is repeated with continued rotation. As each avalanche flows away, the grains within it become sorted as to size. The larger ones migrate to the fast-moving top of the avalanche, and so are carried to the edges and toe of the avalanche. The resultant inclined layers should be clearly visible through the bottom of the jar. Figures 4.11 and 4.17 showed natural examples, picked out by weathering. On turning the jar faster, the avalanches will be found to become more frequent. The downward motion of the sand became continuous, and the grain sorting much less perfect, at a sufficiently large rotation speed.

What factors control avalanching on real bedforms? The jar experiment suggests that the avalanche interval varies with (1) the rate at which grain settling steepens the leeward slope, and (2) the opposing rate at which the flow of an avalanche lowers the slope. Consider a vertical streamwise slice of unit width through a bedform. If every avalanche travelled infinitely quickly, a sand flow would occur each time the current had driven over the crest a dry mass of sediment equal to the dry mass contained in a unit slice from a typical avalanche. The avalanche interval then equals $m_{av}/J_{crest}$ s, where $m_{av}$ kg m$^{-1}$ is the dry mass of grains in the unit avalanche slice and $J_{crest}$ the transport rate given by Equation 4.26. But an avalanche requires a characteristic time $p_{av}$ to descend and lower the leeward slope. The experiment showed that, at a certain speed of rotation, equivalent to a particular value for $J_{crest}$, the motion of the sand became continuous. For this condition $m_{av}/J_{crest} = p_{av}$ and in the general case

69

$$\text{avalanche period} = \frac{m_{av}}{J_{crest}} + p_{av} \quad \text{s} \qquad (4.28)$$

The avalanche interval therefore declines as the transport rate increases. The mass of the avalanche should remain unchanged, provided there is no change in the bedform and sediment.

### 4.7.3 Origin of cross-stratification sets

The preceding experiments and analyses help to explain the internal cross-stratification of bedforms, but fail to explain why the laminae should themselves be grouped between gentler erosional surfaces into larger packets or sets (e.g. Fig. 4.11). Evidently the cross-stratification structures formed by migrating bedforms comprise a hierarchy of bedding contacts. Whereas leeside avalanching creates the steep layering, expressing one hierarchical level, some other process, which we shall now examine, must account for the erosional surfaces at a higher level in the structural hierarchy.

The bedform we analysed in Figure 4.21 migrated over a rigid surface without change of either shape or size. The transfers on the two faces of the bedform were exactly balanced and there was neither net erosion beneath the bedform nor net deposition on top of it. What happens if we relax this severe restriction? After all, under field conditions, most bedforms exist beneath variable currents, so that either net deposition or net erosion prevails. Two cases are of interest: (1) net transfer between the bed and the suspended load, and (2) net transfer involving the bedload only. We can retain in the first the restriction that the bedforms remain constant in size and shape. A change of shape and/or size is inevitable in the second case, since grains are being transferred between the moving forms and the bed below.

The bedforms in the first case are advancing over a surface receiving deposits from a suspension above (Fig. 4.24a). They therefore climb upwards, but how steeply? In a small time $\delta t$ a point P moves to a new position P′, at a downstream distance $\delta x$ parallel to the $x$-direction but at $\delta y'$ perpendicularly upwards. But erosion in accordance with Equation 4.22 would have moved P parallel with the $x$-direction to the new position P″, and lowered the surface the small distance $\delta y''$ perpendicularly below P. The path of climb of P is defined by $\delta y'/\delta x$ which, since the component movements of P occur in the same time increment $\delta t$, can be written

$$\frac{\delta y'}{\delta x} = \frac{(\delta y'/\delta t)}{(\delta x/\delta t)} \qquad (4.29)$$

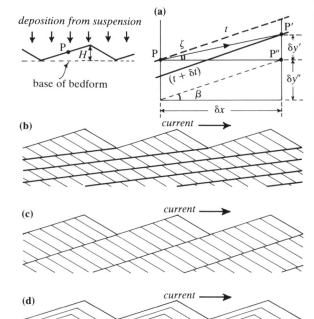

Figure 4.24 Origin of cross-stratification sets beneath uniform bedforms in the presence of a suspension load undergoing deposition:(a) definition sketch; (b) subcritical cross-stratification; (c) critical cross-stratification; (d) supercritical cross-stratification.

becoming in the limit

$$\frac{dy'}{dx} = \frac{dy'}{dt}\frac{dt}{dx} \qquad (4.30)$$

where $dy'/dx = -\tan \zeta$ is the required angle of climb (negative, as the slope is upwards in the current direction).

Now the rectangular area $\delta y' \, \delta x$ is proportional to the volume of sediment deposited from the suspension in the time $\delta t$, whence by continuity (Eqn 4.19) we can write $dy'/dt = R_s/\gamma$, where $R_s$ is the dry-mass deposition rate and $\gamma$ the sediment dry bulk density. Referring to the derivation of Equation 4.27, $dt/dx$ is the reciprocal of the $x$-directed bedform migration speed. Substituting into Equation 4.30, and ignoring sediment transport other than in the bedforms, we obtain

$$\tan \zeta = -\frac{HR_s}{2J} \qquad (4.31)$$

70

where $H$ is bedform height and $J$ the spatially averaged bedload transport rate. The bedform and its neighbours therefore climb at an angle increasing with the height and net transfer rate, but declining with the rate of bedload transport. Should net erosion prevail, making $R_s$ negative, $\tan \zeta$ becomes positive and the bedforms descend instead of climb.

We saw that $\delta y' \, \delta x$ is proportional to the deposition at P during the time $\delta t$, but it is also the case that the rectangular area $\delta y'' \, \delta x$ is proportional to the sediment lost from P due to $x$-directed bedform advance. The net change at P, that is, whether erosion or deposition occurred locally, consequently depends on the relative size of these two transfers. As above, $dy'/dx = -\tan \zeta$ and the first-mentioned area is proportional to $dy'/dt = R_s/\gamma$. Now the second area is proportional to $R_{bm}$, the local transfer rate due to $x$-directed bedform motion, whence by Equation 4.22 $dt/dx = \gamma \tan \beta / R_{bm}$. Making these three substitutions into Equation 4.30 gives

$$\frac{\tan \zeta}{\tan \beta} = -\frac{R_s}{R_{bm}} \qquad (4.32)$$

which states that the angle of bedform climb relative to the steepness of the upcurrent surface is proportional to the ratio of the transfers. Thus the upcurrent face is erosional when $\zeta < \beta$ and the packet of cross-strata due to one bedform is erosively related to the sets created by adjoining forms (Fig. 4.24b). When $\zeta > \beta$, however, laminae appear on both upcurrent and leeward slopes, and the bedform is preserved whole (Fig. 4.24d). Hence $\zeta = \beta$ represents a critical state (Fig. 4.24c), allowing us to describe cross-stratification as either subcritical (sets erosively related) (Figs. 4.11 & 4.17) or supercritical (whole bedforms preserved) (Fig. 4.25). The rock record abounds in examples of this range of patterns, which are readily simulated in the laboratory (Ashley *et al.* 1982).

In our second case net transfer only occurs between the bedforms and the deposit below their troughs, so that the forms inevitably change in size and shape as they advance. Their wavelength is unlikely to change, however, because this property is fixed by the neighbouring forms, so the most probable response is a change in height. Assuming constant wavelength, an analysis on the previous lines gives

$$\tan \zeta = -\frac{HR_b}{2J} \qquad (4.33)$$

at the bottom of the trough, where $R_b$ is the transfer rate between the bedforms and the deposit below. But at the

**Figure 4.25** Supercritical cross-lamination seen in vertical streamwise section, Uppsala Esker, Sweden. Current is from left to right and scale bar is 0.1 m.

trough $dy/dx = -dH/dx$, and as we can substitute from Equation 4.19 for $R_b$, Equation 4.33 becomes

$$\frac{dH}{dx} = \frac{H}{2J(x,t)} \left( \frac{\partial J}{\partial x} + \frac{1}{U_s} \frac{\partial J}{\partial t} \right) \qquad (4.34)$$

The solution of this differential equation depends on $J(x, t)$, but in the case of net deposition it implies that the sets decline down current in both thickness and steepness of climb.

## 4.8 Bedforms and flow conditions

Why are so many kinds of bedform possible? Nominally, the controls at the very least are the size and excess density of the sediment, and the fluid density, viscosity, velocity and depth. These quantities are nonetheless sufficiently numerous as to imply the possibility of more than one threshold condition and, therefore, more than one possible type of bedform. But how are bedforms initiated and how do they grow? What factors determine their equilibrium characteristics? Few answers can be offered but, in the case of bedforms in water, we can at least go some way towards describing the hydraulic conditions represented by the different kinds. Such descriptions are especially helpful in the interpretation of the rock record.

Seventy-five years of laboratory experimentation on quartz-density sediments transported through straight channels by uniform steady aqueous currents has resulted in a huge volume of data on the hydraulic relationships of bedforms. One way of summarizing these data appears in Figure 4.26 (613 experiments), in which bedform type is plotted in a graph of non-dimensional boundary shear stress against grain size. Another popular summary uses stream power instead of the stress (Simons *et al.* 1965), and a third plots bedforms on a triaxial graph of grain size, flow velocity and flow depth (Southard 1971, Costello & Southard 1981).

Each bedform type appears in a distinct part of Figure 4.26, called its existence field. Some of the bounding curves are clear-cut, with no more overlap than is explicable by experimental error. This is true of the threshold between current ripples on the one hand and dunes plus lower-stage plane beds on the other. In contrast, ripples and dunes overlap substantially with upper-stage plane beds. This is because the upper-stage plane beds developed locally near the crests of ripples and dunes are associated with significantly larger shear stresses than completely plane beds under the same hydraulic conditions (Bridge 1981b).

The data that gave Figure 4.26 are less complete than one would like. Grains exceeding 0.005 m in diameter are not represented, and there are few observations for silts. Those shown in the silt range are believed to be the

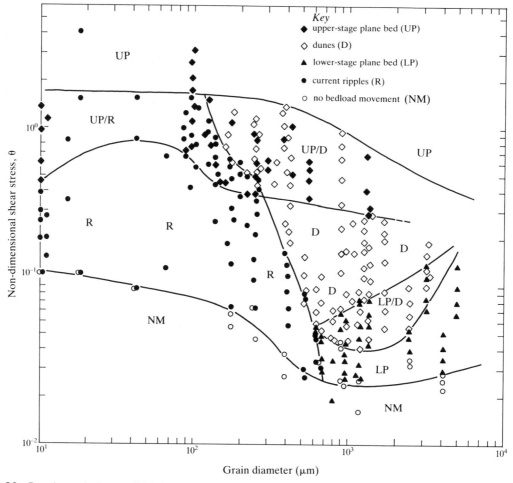

**Figure 4.26** Experimental existence fields for one-way aqueous bedforms in quartz sands under equilibrium conditions at 25°C, shown in the non-dimensional shear stress-grain size plane. The graph is based on 613 experiments, mostly listed in Allen (1982), but with only representative and critical points plotted.

72

more trustworthy available. Jopling and Forbes (1979) and Mantz (1983) extensively studied bedforms in this sediment range, but their ripples were restricted to quite exceptionally high $\theta$ values, perhaps because highly angular and possibly cohesive grains were used. The current-ripple boundary merits a closer look. Its steep decline with increasing shear stress implies that ripples may be restricted by some fixed ratio between the thickness of the viscous sublayer (Sec. 1.9) and grain size. Experimental data (Fig. 4.27) prove that ripples are replaced by either lower-stage plane beds (low stresses) or dunes (high stresses) when the particle diameter exceeds the sublayer thickness. Hence the character of ripples could depend only on the properties and behav-

iour of the viscous sublayer and the other inner regions of the flow. Significantly, the size of current ripples seems to be largely controlled by grain size. On a lower-stage plane bed, and over a substantial part of a dune, however, the grains protrude through the sublayer, putting these bedforms potentially under the control of flow depth and macroturbulence. Significantly, dune wavelength tends to be about five times the flow depth, and height of the order of one-sixth of the depth.

Trending horizontally across Figure 4.26 is a somewhat diffuse boundary between upper-stage plane beds above and dunes and current ripples below. This threshold sharpens considerably when examined by gradually reducing the stress on an upper-stage plane

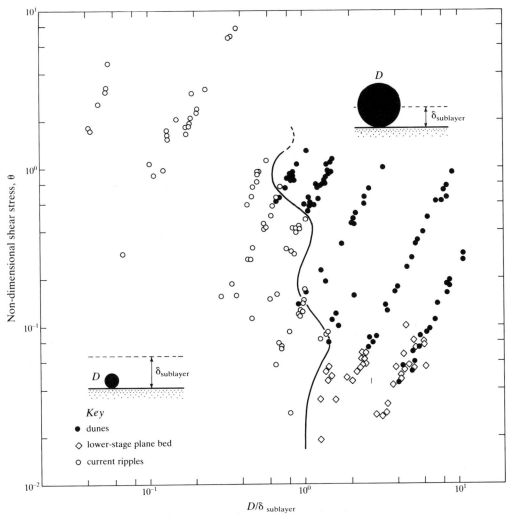

**Figure 4.27** Experimental occurrence of current ripples, dunes and lower-stage plane beds in quartz silts and sands, as a function of non-dimensional bed shear stress and relative grain size.

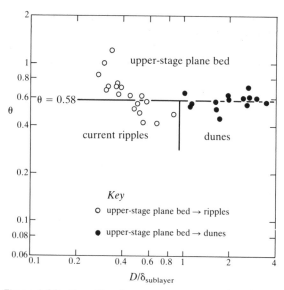

**Figure 4.28** Transition from an upper-stage plane bed to either current ripples or dunes, as determined experimentally by Hill *et al.* (1969).

bed until transition occurs. The experiments of Hill *et al.* (1969) suggest that the threshold takes the roughly constant stress value of $\theta = 0.58$ (Fig. 4.28), whether ripples or dunes result at transition again depending on relative grain size. This work confirms Bagnold's (1966) 'universal' criterion for the occurrence of an upper-stage plane bed:

$$\theta \geqslant C_b \tan \alpha \qquad (4.35)$$

where $C_b$ is the fractional grain concentration in the static bed and $\tan \alpha$ the dynamic grain friction coefficient. For Hill's sands, $C_b = 0.6$ and $\tan \alpha = 0.95$ are reasonable values, giving a minimum stress $\theta = 0.57$, in agreement with Figure 4.28. Physically, the transition coincides approximately with the mobilization of a single layer of bed grains, and therefore the full occlusion of the static bed by the bedload. Only at subcritical stresses do grains in the static bed directly experience the fluid force.

## 4.9 Making wavy beds

To ask how bedforms are initiated and grow is to enquire how a plane bed can be transformed into a transversely wavy one and vice versa. However, for either change to occur, certain inescapable geometrical conditions must be met by the sediment-transporting

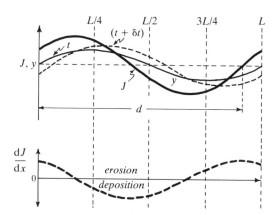

**Figure 4.29** Definition sketch for the migration of a wavy granular bed under bedload transport.

system, regardless of the particular mechanisms governing the transport intensity and mode.

Consider the steady transport of sediment over a two-dimensional regularly wavy bed of small amplitude (Fig. 4.29). The flow is therefore being alternately contracted and expanded, which implies that the flow properties, and hence the sediment transport rate, are varying down stream on the same scale as the bed undulations. If the transport rate is changing, then by continuity (Eqn 4.19) erosion and deposition must alternate along the flow, and on the same wavelength as the bed waviness. But what if the streamwise variations in transport rate and bed elevation differ in phase? In Figure 4.29 the transport-rate variation leads the bed wave by a lag distance $3L/4 < d < L$, where $d$ is the spatial lag and $L$ the common wavelength of variation of the two properties. The transport-rate maximum then occurs on the upcurrent face of the bedform, and nearer the crest than the trough. Deposition succeeds erosion where the transport rate begins to decline down stream, and vice versa. Therefore in a small time $\delta t$, a lamina becomes draped over the bed-wave crest, while a layer of equal volume is scoured from the trough. The undulation consequently grows in amplitude while advancing down current. Depending on the value of the spatial lag (Table 4.1), a bedform may either grow or dampen, or travel either up current or down current. Two cases in the table correspond to current ripples, ballistic ripples and dunes, and the remainder to antidunes. Notice the implied conditions for the maintenance of a constant amplitude, that is, for bedform equilibrium.

Thus wavy bedforms should arise only where a granular boundary carries prior defects, or has defects imprinted on it by the current, able to grow in amplitude because they shift the transport-rate maximum an

74

**Table 4.1** Effect of lag distance on behaviour of transverse sandy bedforms.

| Lag distance | Effect on bed waves | | Comments |
|---|---|---|---|
| | Amplitude | Translation | |
| $0 < d < L/4$ | decreasing | down stream | current ripples, dunes |
| $L/4 < d < L/2$ | decreasing | up stream | antidunes |
| $L/2 < d < 3L/4$ | increasing | up stream | antidunes |
| $3L/4 < d < L$ | increasing | down stream | current ripples, dunes |

appropriate streamwise distance. For example, current ripples can form on a plane bed either at tiny mounds due to the streak-bursting discussed in Chapter 6 (Williams & Kemp 1971) or at artificial features (Southard & Dingler 1971).

In harmony with the complexity of sediment-transporting systems, the lag distance $d$ in Table 4.1 turns out to be a resultant value, to which several mechanisms may simultaneously contribute (Allen 1983). A wavy bed so affects the fluid motion as to shift the maximum of the shear stress to a position up stream from the bed-wave crest. Since the sediment transport rate increases steeply with stress, this effect clearly favours bedform amplification (e.g. Fig. 4.29). Two further effects depend on local bed slope. As the threshold entrainment force is greater in upslope than downslope flow (Eqn 4.7), the transport-rate maximum must lie down stream from the crest, at which the slope is zero. Hence this effect promotes damping and a plane bed. The transport-rate maximum is also shifted down stream because, for the same hydraulic conditions, the downslope rate exceeds that up slope. In upslope flow, for example, work must be done to lift the sediment load up the slope, as well as to maintain its stage of transport. Again damping is favoured. Two further effects relate to transport mode. The transported grains attempt to follow the changes in flow as they are carried over the wavy bed, but do not do so immediately because they are dispersed in the fluid and cannot sink in the fluid faster than their terminal fall velocity. In aqueous transport, bedload grains lie within a few grain diameters of the bed, however, and so should respond over short streamwise distances to spatial changes of flow. A damping effect should nevertheless result. By contrast, suspended grains range through the whole flow, pointing to a much larger characteristic response distance. Whether a damping or amplification results

could depend on the relative size of the bedforms. The individual lag distances and strengths of these effects (there may be others) sum in each particular case to determine whether a wavy bed grows or decays in amplitude, as well as how fast the change occurs.

## 4.10 A wave theory of bedforms

### 4.10.1 Free-surface flow over a wavy bed

That some of the lag effects depend on bed steepness, and the striking regularity of such as ripples and dunes, combine to suggest that the hydraulic mechanisms leading to bedforms operate within, and are constrained by, the requirements of the theory of progressive water waves. Though seldom advocated from quite this basis, the idea is not new (e.g. Kennedy 1969, Hammond & Heathershaw 1981) and underpins an important body of bedform theory (e.g. Richards 1980). But bedforms move very slowly compared to water waves, so how can the two be related? Here we must anticipate some of the wave theory outlined in Section 13.6.

Figure 4.30a shows a train of uniform progressive sinusoidal waves advancing from right to left at a steady phase velocity $c$ across still water of depth $h_{cr}$ above a rigid plane horizontal bottom. We can halt the waves by superimposing on the fluid a velocity $U$ from left to right equal in magnitude to $c$. The internal streamlines representing this motion are also sinusoidal and in phase. Their amplitude decreases downwards, however,

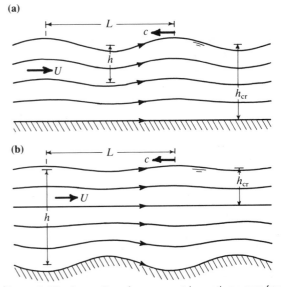

(a)

(b)

**Figure 4.30** Streamlines for a current beneath a wavy free surface during (a) supercritical and (b) subcritical flow.

and the streamline in the bottom is straight. Now any streamline can be replaced by a rigid surface of the same shape without disturbing the motion, permitting a smaller actual flow depth of value $h$ and making the original depth $h_{cr}$ the virtual depth to a virtual bottom. Milne-Thomson (1962) proved that the amplitude $a$ of the bed wave is then related to the amplitude $b$ of the surface wave by

$$\frac{b}{a} = \frac{1}{\cosh(2\pi h/L) - \dfrac{g}{(2\pi/L)U^2}\sinh(2\pi h/L)} \quad (4.36)$$

where $L$ is their common wavelength. The hyperbolic function $\cosh(2\pi h/L)$ increases from 1 to $\infty$ with increasing $2\pi h/L$, as shown in Figure 13.8. Now by wave theory, the phase velocity of waves on a fluid of real depth $h$ is

$$\frac{(2\pi/L)c^2}{g} = \tanh(2\pi h/L) \quad (4.37)$$

where $\tanh(2\pi h/L)$ is another hyperbolic function (Fig. 13.8), increasing from 0 to 1 with $2\pi h/L$. The virtual depth $h_{cr}$ is actually the least depth over which a water-surface profile of relative wavelength $2\pi h/L$ can propagate with phase velocity $U = c$, and is given by

$$h_{cr} = \frac{1}{(2\pi/L)}\tanh^{-1}[(2\pi h/L)Fr^2] \quad \text{m} \quad (4.38)$$

where $Fr = U/(gh)^{1/2}$ is the Froude number.

We now have all the data needed to explore Equation 4.36, by comparing the phase velocity $U$ of surface waves in water of virtual depth $h_{cr}$ to the phase velocity $c$, given by Equation 4.37, appropriate to a surface profile in a real depth $h$. When $U^2 > c^2$, it turns out that $h_{cr} > h$, $b/a > 0$ and $Fr^2 > 1$. In this case (fig. 4.30a) bed and surface profiles are in phase, and the surface wave has the greater amplitude (supercritical flow). Figure 4.30b shows the case where $U^2 < c^2$, so that $h_{cr} < h$, $b/a < 0$ and $Fr^2 < 1$. The virtual depth now lies within the limits of the real depth, and the bed wave has the larger amplitude (subcritical flow). Across the straight virtual bottom there occurs a phase change of one-half of a wavelength between the shape of the streamlines and between the bed and surface profiles.

If a wavy streamline can be replaced by a rigid surface of the same shape without affecting the flow, then it can also be substituted by a wavy erodible bed, provided that the bed is effectively stationary. This further

replacement is permissible because it is a matter of observation that bedforms travel at speeds much smaller than the driving currents. Wave theory has now presented us with two fundamental classes of bedform. Figure 4.30a represents antidunes, that is, in-phase bed and water-surface waves limited to supercritical and near-critical flows. Figure 4.30b corresponds to current ripples and dunes in water, formed in subcritical flows when the bed (including the separation bubbles) is out of phase with surface waves. The atmosphere has no well defined upper limit, and the flow filling a close conduit no free surface on which waves could develop, but the theory applies also to these cases (Kennedy 1969).

### 4.10.2 Bedforms and the wave-dispersion equation

Equation 4.37 describes waves spreading over a fluid surface, and is one form of the wave-dispersion equation. Introducing $h$ into the numerator and denominator on the left-hand side, and replacing $c$ by $U$, it can be rearranged into the general form

$$\frac{\tanh(2\pi h/L)}{2\pi h/L} = k\,Fr^n \quad (4.39)$$

where $k$ is a numerical coefficient and $n$ an exponent. To what extent do real bedforms obey it?

Since $\tanh(2\pi h/L)$ tends to 1 as $2\pi h/L$ tends to $\infty$, the approximation to Equation 4.39 for deep water is $(2\pi h/L)Fr^2 = 1$, putting $k = 1$ and $n = 2$, their normal values. Figure 4.31 plots data for laboratory antidunes in sands and fine gravels. The bedforms in all respects closely follow the deep-water approximation to the wave-dispersion equation. Such close agreement with the model suggests that for antidunes the surface wave strongly forces the bed wave.

Ripples and dunes in water also follow the generalized wave-dispersion equation, but not its deep-water approximation. Experimental data (Fig. 4.32) afford for ripples in silts and sands:

$$\frac{\tanh(2\pi h/L)}{2\pi h/L} = 2.69\,Fr^2 \quad (4.40)$$

but for dunes

$$\frac{\tanh(2\pi h/L)}{2\pi h/L} = 4.34\,Fr^2 \quad (4.41)$$

The asymptotic Froude number is therefore approximately 0.48 for ripples and 0.61 for dunes. There is

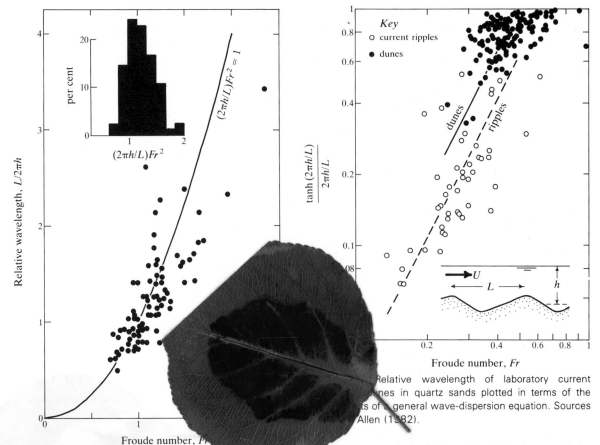

**Figure 4.31** Relative wavelength of laboratory as a function of Froude number. Sources of data in Allen (1982).

Relative wavelength of laboratory current dunes in quartz sands plotted in terms of the general wave-dispersion equation. Sources Allen (1982).

more scatter in these plots than in the case of antidunes, perhaps because ripples and dunes are prone to three-dimensional effects, but there is clearly some forcing of the bed wave by the surface wave.

### 4.10.3 Steepest waves

According to wave theory, the steepest waves occur for $2\pi h/L = 1$, that is, for $L = 2\pi h$. This could provide a natural limit on the upward growth of bedforms, which should also be most common at about $2\pi h/L = 1$.

Laboratory antidunes support the prediction, as the histogram in Figures 4.31 shows. The steepness of antidunes is not easily measured, however, so this case cannot be explored further. Figure 4.33 shows the measured steepness $H/L$ of experimental dunes in a range of sands, where $H$ is the dune height. The forms are commonest and also steepest at approximately $2\pi h/L = 1$, especially those in the finer grades. Some workers attribute the wavelength $L = 2\pi h$ of the steepest

dunes to a supposed but unexplained control by macroscale turbulence (Yalin 1972, Jackson 1976), but forcing by surface waves provides a simpler explanation. In fact, dune wavelength does not accord with the size of the large eddies of turbulence. These have a streamwise length of about 1.5 times the flow thickness (Corrsin & Kistler 1955, Ueda & Hinze 1975, Falco 1977, Thomas & Bull 1983) which, even if the eddies were *en echelon*, would yield a possible wavelength of only about $3h$.

The virtual bottom shown in Figure 4.30b should act as a lid on the lower part of the flow, constraining the upward growth of especially the larger features (dunes). The ratio $H/(h - h_{cr})$ compares dune height to the flow thickness between the virtual bottom and the mean bed level, and Figure 4.34 shows how this quantity varies with $1/(2\pi h/L)$. The ratio increases with relative wavelength, but for the larger wavelengths appears to level out at about 0.5, given dunes that are not too strongly three-dimensional. The virtual bottom, when sufficiently close to the real bed, does appear to limit dune height.

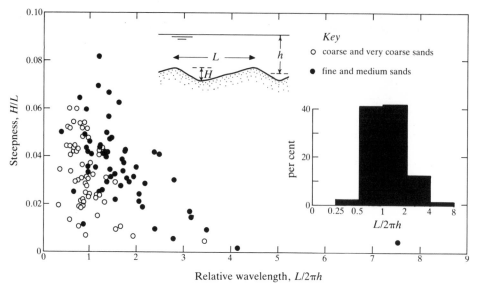

**Figure 4.33** Steepness of laboratory dunes in quartz sands as a function of relative wavelength. Sources of data in Allen (1982).

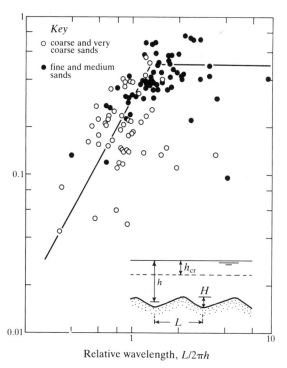

**Figure 4.34** Height compared to effective flow depth for laboratory dunes in quartz sands, as a function of relative wavelength. Sources of data in Allen (1982).

## Readings

Abbott, J. E. and J. R. D. Francis 1977. Saltation and suspension trajectories of solid grains in water streams. *Phil. Trans. R. Soc. Lond.* A **284**, 225–54.

Ackers, P. and W. R. White 1973. Sediment transport: a new approach and analysis. *J. Hydraul. Div. Am. Soc. Civ. Engrs* **99**, 2041–60.

Allen, J. R. L. 1965. Sedimentation in the lee of small underwater sand waves: an experimental study. *J. Geol.* **73**, 95–116.

Allen, J. R. L. 1968. *Current ripples*. Amsterdam: North-Holland.

Allen, J. R. L. 1982. *Sedimentary structures*, Vol. I. Amsterdam: Elsevier.

Allen, J. R. L. 1983. *River bedforms: progress and problems*. Spec. Publn Int. Assoc. Sedimentologists, no. 6, 19–33.

Ashley, G. M., J. B. Southard and J. C. Boothroyd 1982. Deposition of climbing-ripple beds: a flume simulation. *Sedimentology* **29**, 67–79.

Bagnold, R. A. 1954. Experiments on a gravity-free dispersion of large solid spheres in a Newtonian liquid under shear. *Proc. R. Soc. Lond.* A **225**, 49–63.

Bagnold, R. A. 1955. Some flume experiments on large grains but little denser than the transporting fluid, and their implications. *Proc. Instn Civ. Engrs* **4**, 174–205.

Bagnold, R. A. 1966. *An approach to the sediment transport problem from general physics*. Prof. Pap. US Geol. Surv., no. 422-I.

Bridge, J. S. 1981a. A discussion of Bagnold's (1956) bedload transport theory in relation to recent developments in bedload modelling. *Earth Surf. Processes Landforms* **6**, 187–90.

Bridge, J. S. 1981b. Bed shear stress over subaqueous dunes, and the transition to upper-stage plane beds. *Sedimentology* **28**, 33–6.

Chepil, W. S. 1958. The use of evenly spaced hemispheres to evaluate aerodynamic forces on soil surfaces. *Trans. Am. Geophys. Union* **39**, 397–404.

Chepil, W. S. 1961. The use of spheres to measure lift and drag on wind-eroded soil grains. *Proc. Soil Soc. Am.* **25**, 343–5.

Corrsin, S. and A. K. Kistler 1955. *Free-stream boundaries of turbulent flow.* Rep. Nat. Adv. Coun. Aeron., no. 1244.

Costello, W. R. and J. B. Southard 1981. Flume experiments on lower-flow-regime bed forms in coarse sand. *J. Sed. Petrol.* **5**, 849–65.

Falco, R. E. 1977. Coherent motions in the outer regions of turbulent boundary layers. *Phys. Fluids* **20** (10, II), S124–32.

Gordon, R., J. B. Carmichael and F. J. Isackson 1972. Saltation of plastic balls in a 'one-dimensional' flume. *Water Resources Res.* **8**, 444–59.

Graf, W. H. 1971. *Hydraulics of sediment transport.* New York: McGraw-Hill.

Hammond, F. D. C. and A. D. Heathershaw 1981. A wave theory for sandwaves in shelf seas. *Nature* **293**, 208–10.

Hardisty, J. 1983. An assessment and calibration of formulations for Bagnold's bedload equation. *J. Sed. Petrol.* **53**, 1007–10.

Hill, H. M., V. S. Srinivasan and T. E. Unny 1969. Instability of flat bed in alluvial channels. *J. Hydraul. Div. Am. Soc. Civ. Engrs* **95**, 1545–58.

Hunter, R. E. 1977. Basic types of stratification in small eolian dunes. *Sedimentology* **24**, 361–87.

Jackson, R. G. 1976. Sedimentological and fluid-dynamic implications of the turbulent bursting phenomenon in geophysical flows. *J. Fluid Mech.* **77**, 531–60.

Jopling, A. V. and D. L. Forbes 1979. Flume study of silt transportation and deposition *Geogr. Annlr* A **61**, 67–85.

Kennedy, J. F. 1969. The formation of sediment ripples, dunes and antidunes. *Annu. Rev. Fluid Mech.* **1**, 147–68.

Mantz, P. A. 1977. Incipient transport of fine grains and flakes by fluids – extended Shields diagram. *J. Hydraul. Div. Am. Soc. Civ. Engrs* **103**, 601–15.

Mantz, P. A. 1983. Semi-empirical correlations for fine and coarse cohesionless sediment transport. *Proc. Instn. Civ. Engrs* (2) **75**, 1–33.

Miller, M. C., I. N. McCave and P. D. Komar 1977. Threshold of sediment motion under unidirectional currents. *Sedimentology* **24**, 507–27.

Milne-Thomson, L. M. 1962. *Theoretical hydrodynamics*, 2nd edn. London: Macmillan.

Richards, K. J. 1980. The formation of ripples and dunes on an erodible bed. *J. Fluid Mech.* **99**, 597–618.

Savage, S. B. 1979. Gravity flow of cohesionless granular materials in chutes and channels. *J. Fluid Mech.* **92**, 53–96.

Savage, S. B. and D. J. Jeffrey 1981. The stress tensor in granular flow at high shear rates. *J. Fluid Mech.* **110**, 255–72.

Savage, S. B. and S. McKeown 1983. Shear stresses developed during rapid shear of concentrated suspensions of large spherical particles between concentric cylinders. *J. Fluid Mech.* **127**, 453–72.

Simons, D. B., E. V. Richardson and C. F. Nordin 1965. *Sedimentary structures generated by flow in alluvial channels.* Spec. Publn Soc. Econ. Palaeont. Mineral., no. 12, 34–52.

Southard, J. B. 1971. Representation of bed configurations in depth–velocity–size diagrams. *J. Sed. Petrol.* **41**, 903–15.

Southard, J. B. and J. R. Dingler 1971. Flume study of ripple propagation behind mounds on flat beds. *Sedimentology* **16**, 251–63.

Thomas, A. S. W. and M. K. Bull 1983. On the role of wall-pressure fluctuations in deterministic motions in the turbulent boundary layer. *J. Fluid Mech.* **128**, 283–322.

Ueda, H. and J. O. Hinze 1975. Fine-structure turbulence in the wall region of a turbulent boundary layer. *J. Fluid Mech.* **67**, 125–43.

White, B. R. and J. C. Schulz 1977. Magnus effect in saltation. *J. Fluid Mech.* **81**, 497–512.

Williams, P. B. and P. H. Kemp 1971. Initiation of ripples on flat sediment beds. *J. Hydraul. Div. Am. Soc. Civ. Engrs* **97**, 505–22.

Yalin, M. S. 1972. *Mechanics of sediment transport.* Oxford: Pergamon.

# 5 Winding down to the sea

River channel patterns – drag, mean velocity and power of a river – cross-sectional and long profiles – flow and sediment movement in channel bends – meandering – point-bar deposits.

## 5.1 Introduction

Rivers are channelized flows of water which drain from the continents the precipitation falling as rain or snow, together with the dissolved and entrained particulate products of rock weathering. Today rivers carry annually to the oceans and seas approximately 14 km$^3$ of weathered material, a substantial proportion of it in solution. Many rivers reach the sea or ocean only after traversing tidal mud-flats and estuaries, where further channels shaped by the mixed fresh and salt waters can be found. Indeed, by their ebbing and flooding, tidal waters can themselves create impressive channel systems in areas of saltmarsh and mud-flat. Confinement within laterally restricted channels seems to be an inevitable and universal feature of virtually all surface waters. River and tidal channels are unstable, however, and bend and shift over time, with the result that structured bodies of fluvial or tidal sediment arise to match the losses due to erosion. Why do river and tidal flows occupy channels? Why are those channels of a particular shape and size? What processes occur in channels, and how do they shape the sediment accumulations there?

Rivers are very variable in plan. That shown in Figure 5.1 occupies a single tortuous channel, enclosing within each bend or meander a large sediment accumulation called a point bar. Notice the vegetation patterns on these bars, which suggest successive earlier positions of the channel, and therefore its uneven lateral migration. The river depicted in Figure 5.2 also occupies a single channel but its course is less sinuous and, because the flow is constrained between resistant valley walls, the

bends are markedly asymmetrical. Their sense of asymmetry, and the accretionary scrolls on the point bars, all proclaim the general direction of river flow. Many rivers occupying single channels are even less winding in plan, so much so that some workers describe them as straight. A quite different river appears in Figure 5.3. Here the flowing water repeatedly divides and rejoins around innumerable lozenge-shaped sediment accumulations called braid bars. Hence the channel consists of many short connected segments, most of which are slightly curved in plan.

A simple classification of river-channel planimetric patterns can be erected using the degree of channel winding, or sinuosity (ratio of channel length to valley length), and the degree of braiding, that is, the number of channels or threads of current visible in a cross profile. But whether channels are single-thread or multi-thread, their curvature in plan is a persistent theme demanding explanation.

River channels also vary greatly in cross-sectional shape (Fig. 5.4). The cross section in a bend tends to be triangular, particularly where the bend is 'tight' and the radius of curvature similar to the channel width. The steeper bank lies on the outside of the bend, and is eroded in the course of the lateral and downstream wandering of the stream. A gentle slope typifies the inner side of the bend, where sediment accumulates in harmony with removal from the outer bank. At a crossing, where one bend turns into another, the profile tends to be rectangular and relatively shallow, perhaps including an island or shoals of accumulated sediment. The relative dimensions of channel cross sections also vary considerably. The width of highly sinuous single-channel

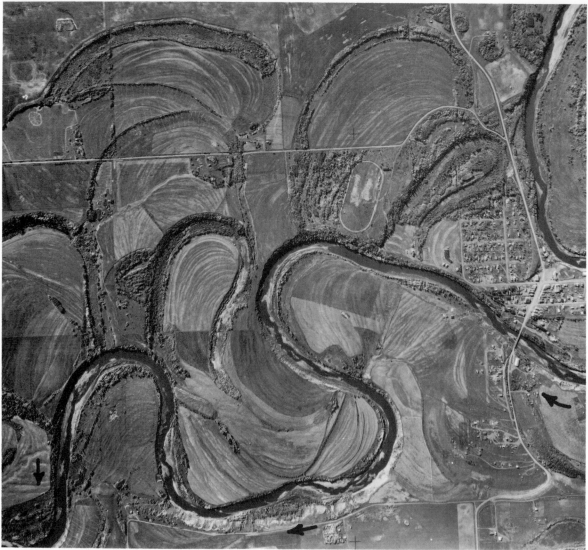

**Figure 5.1** The tortuous free meanders and cut-offs of the Notikewin River, Alberta, Canada. Area shown measures 2.3 × 2.6 km. Copyright Alberta Energy and Natural Resources.

**Figure 5.2** Constrained meanders of the Beaver River, Alberta, Canada. Area shown measures 2.4 × 5.00 km. Copyright Alberta Energy and Natural Resources.

**Figure 5.3** The braided Howse River, Alberta, Canada. Area shown measures 3.0 × 5.6 km. Copyright Alberta Energy and Natural Sources.

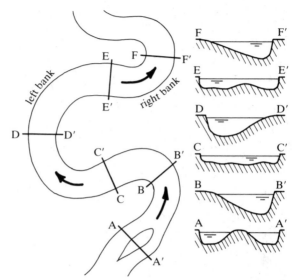

**Figure 5.4** Schematic changes in river cross section through a series of meanders.

few tropical and subtropical ones experience two low-water and two flood periods annually. Added to these annual or semi-annual variations are smaller discharge fluctuations which represent individual storms or rainy intervals. During sunny months, streams fed from glaciers and snow fields are subject as well to regular diurnal variations of discharge. The flow of rivers is therefore unsteady, and on more than one timescale. Tidal currents are even more unsteady, for their chief discharge fluctuation is either semidiurnal or diurnal, depending on geographical setting.

## 5.2 Drag force and mean velocity of a river

How can we model mathematically the velocity and drag of a river, in order to gain an insight into channel processes? The task at first looks hopeless, for we have just seen that real rivers vary in discharge with time, and that their channels are neither straight nor fixed in plan position. The only practical approach is to reduce a channelized flow to its most elementary form, and to attempt to build towards the more complex reality from there. In essence, a river or channelized tide is a flow of unchanging discharge and uniform depth contained in a straight, rigid channel of rectangular cross section and uniform slope and roughness. What force now drives the motion? What force opposes the flow and so defines a steady state? What in consequence is the flow velocity?

rivers is of the order of ten times the average depth. That of highly braided streams can be hundreds of times the depth.

We know much less about tidal than river channels. Tidal channels (Fig. 5.5) seem almost without exception to be of moderate to high sinuosity and unbraided, unless one proposes to count the large and periodically emergent sand shoals found in many estuaries. Their cross sections vary in a similar manner to river channels.

What patterns of discharge typify river and tidal channels? The vast majority of rivers exhibit a single flood period and a single low-water period each year. A

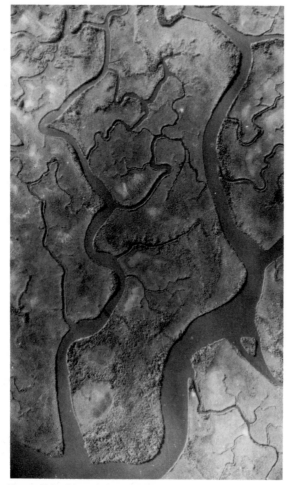

**Figure 5.5** Meandering tidal channels in mangrove swamps, Niger Delta, Nigeria. Area shown measures approximately 3.2×5.0 km.

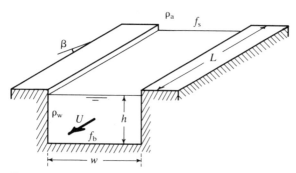

**Figure 5.6** Definition sketch for uniform steady flow in an open channel.

Experience shows that rivers flow from high to low ground, whence the driving force must be the down-slope component of the weight of the channelized water. Referring to the representative part of the flow shown in Figure 5.6, we can write

$$\text{driving force} = (whL)(\varrho_w - \varrho_a)g \sin \beta \quad \text{N} \quad (5.1)$$

where $w$ is the channel width, $h$ the flow depth measured normal to the water surface and bed, $L$ a representative flow length, $\varrho_w$ and $\varrho_a$ the density of the water and atmospheric air respectively, $g$ the acceleration due to gravity, and $\beta$ the angle of slope of the bed and water surface (equal since the flow is defined as uniform and steady). The first bracketed term is the flow volume contained in the reference element and the second its effective density.

What of the resisting force? As the channel is rigid, and the water in it viscous, a part of the total resisting force must be a frictional drag operating over the wetted solid boundary. But the river flows beneath the viscous atmosphere, assumed stagnant. Referring to Figure 5.6, the total resisting force is therefore

$$\text{resisting force} = (wL)\tau_s + L(2h + w)\tau_b \quad \text{N} \quad (5.2)$$

simplifying to

$$\text{resisting force} = L[w\tau_s + (2h + w)\tau_b] \quad (5.3)$$

in which $\tau_s$ and $\tau_b$ are the boundary shear stresses $(\text{N m}^{-2})$ averaged over the water surface and the wetted bed and banks respectively. The term $(wL)$ in Equation 5.2 is the area of the water surface over which the unit shear stress $\tau_s$ acts, while $L(2h + w)$ is the wetted area of the bed and banks acted on by the unit shear stress $\tau_b$.

Since the flow is assumed uniform and steady, Newton's first law requires us to balance the driving and resisting forces. Putting Equations 5.1 and 5.3 equal gives

$$w\tau_s + (2h + w)\tau_b = (wh)(\varrho_w - \varrho_a)g \sin \beta \quad (5.4)$$

from which only the length $L$ has so far been eliminated. Two unknown shear stresses remain together with some flow dimensions. Can Equation 5.4 be simplified further? Inspection immediately suggests one small simplification. Little error will result if we eliminate $\varrho_a$, for water is some 750 times more dense than air at the Earth's surface. What then of the relative magnitudes of $\tau_s$ and $\tau_b$? Here we can refer to Section 1.6 (Eqn 1.27) for the quadratic stress law $\tau = (f/8)\varrho U^2$, where $\tau$ is the unit shear stress, $f$ the Darcy–Weisbach friction co-efficient, $\varrho$ the fluid density, and $U$ the relative velocity. Because the overall mean velocity of the water in the channel is the same relative to the stagnant atmosphere as to the stationary bed and banks, the quadratic law gives the relative stress as

$$\frac{\tau_s}{\tau_b} = \frac{f_s \varrho_a}{f_b \varrho_w}$$

where $f_s$ and $f_b$ represent friction at the water surface and over the solid boundary respectively. But we already know that $\varrho_a/\varrho_w$ is very small, so unless $f_s/f_b$ is very large, little error will be introduced into Equation 5.4 by ignoring $\tau_s$. Now $f_b$ is of order 0.1 for a river whereas $f_s$ is of order 0.01 when $U$ is of order $1 \text{ m s}^{-1}$, that is, at a relative velocity representative of a river flowing beneath stagnant air. Hence the neglect of $\tau_s$ is fully justified, whence Equation 5.4 simplifies to

$$\tau = \frac{wh}{(2h + w)} \varrho g \sin \beta \quad \text{N m}^{-2} \tag{5.5}$$

all subscripts being dropped. The quantity $wh/(2h + w)$, called the hydraulic radius, is the ratio of the cross-sectional area of the channel to the cross-sectional length of the wetted perimeter. It can also be written $h \cdot w/(2h + w)$, effectively equal to $h$ when the width is very large compared to the depth. A further simplification is permissible, for as $\beta$ is very small for real rivers, $\sin \beta$ is approximately equal to the conventional slope $S = \tan \beta$. Equation 5.4 finally reduces to the simple approximation

$$\tau = h\varrho gS \tag{5.6}$$

describing the shear stress exerted on its bed and banks by a uniform steady channelized flow of large width : depth ratio. The larger shear stress should therefore be exerted by the deeper of two rivers of the same slope and by the steeper of two rivers of the same depth.

Two expressions for the average boundary shear stress of a channelized flow have now been introduced, namely the quadratic stress law

$$\tau = \frac{f}{8} \varrho U^2 \quad \text{N m}^{-2} \tag{5.7}$$

and Equation 5.6

$$\tau = h\varrho gS \quad \text{N m}^{-2}$$

Eliminating the stress between them gives

$$U = \left(\frac{8S}{f}\right)^{1/2} (gh)^{1/2} \quad \text{m s}^{-1} \tag{5.8}$$

for the overall mean velocity of flow in the channel. A comparison of Equations 5.6 and 5.8 shows that the mean velocity increases more slowly with depth and slope than the boundary shear stress. Finally, on re-arranging Equation 5.8 to read

$$\frac{U}{(gh)^{1/2}} = \left(\frac{8S}{f}\right)^{1/2} \tag{5.9}$$

the left-hand side, now dimensionless, becomes the Froude number of Section 1.5. This number increases with slope but decreases with friction.

## 5.3 Energy and power of channelized currents

In moving from high to low ground, the water in a river exchanges potential energy for kinetic energy, a part of which is dissipated in viscous friction at the bed and banks and in transporting sediment. Thus a river has the capacity to do work.

What is the power of a river? Figure 5.7 depicts, in a uniform steady flow contained in a straight channel, an element composed of a column of water of normal depth $h$ and unit basal area, moving down stream at the overall mean flow velocity $U$. At time $t$ the element lies at position 1, where its centre of gravity is at a vertical height $y_1$ above an arbitrary horizontal datum. The potential energy of the element relative to the datum is therefore $E_1 = (h\varrho g)y_1 \text{ J m}^{-2}$. At the later time $(t + \delta t)$, where $\delta t$ is a small increment, the element reaches position 2 down stream, where its potential energy is reduced to $E_2 = (h\varrho g)y_2$. The energy loss is therefore

$$\delta E = (h\varrho g) \, \delta y \tag{5.10}$$

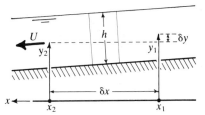

**Figure 5.7** Definition sketch for the power of an open-channel flow.

where $\delta E = (E_1 - E_2)$ and $\delta y = (y_1 - y_2)$. Now the loss occurs in time $\delta t = \delta x / U \cos \beta$ s, where $\delta x$ is the small horizontal distance between the successive positions of the element and $\beta$ the bed slope. Therefore the time rate of energy change, that is, the power $\omega$ of the flow, is

$$\frac{\delta E}{\delta t} = \omega = (h \varrho g) \left( \frac{\delta y}{\delta x} \right) U \cos \beta \quad \text{W m}^{-2} \quad (5.11)$$

$$= (h \varrho g) U \cos \beta \tan \beta \quad (5.12)$$

since $(\delta y / \delta x) = \tan \beta$ in the limit. Equation 5.12 further reduces to

$$\omega = (h \varrho g S) U = \tau U \quad \text{W m}^{-2} \quad (5.13)$$

as $\tan \beta = S$ and $\beta$ for a river is very small, making $\cos \beta$ indistinguishable from unity.

The unit power of a river is therefore the product of the boundary shear stress and the rate of its action. It is proportional to $U^3$, since by Equation 5.7 the stress can be expressed in terms of $U^2$. Hence small discharge fluctuations cause relatively large changes of power. The onset of a flood should increase the power especially steeply, and it is during floods that most sedimentologically useful work can be expected. River power is very variable but lies broadly in the range $1-100 \text{ W m}^{-2}$.

## 5.4 Why flow in a channel?

Why do river and many tidal currents occupy channels? This question is widely ignored, yet it involves perhaps the most fundamental of the interactions between water and the deformable rocks, soils and sediments at the Earth's surface. The channel-occupying habit of rivers is strikingly evident in the river drainage network, that arrangement of confluent channels, increasing in size but decreasing in number down stream, occupying the river drainage basin. The structure of river drainage networks is interpreted as purely random by many geomorphologists, but there are no sound physical reasons why this should be so. There is to other observers considerable order evident in drainage basins, and particularly between and within basins in the same physiographic region. The group of rivers – the Nueces to the Sabine – draining the Gulf of Mexico coastal plain in Texas is an excellent example. Their tributaries decrease in areal density as they join seawards. The trunk streams, of a similar size and discharge, have a remarkably uniform transverse spacing of about $50-100$ km.

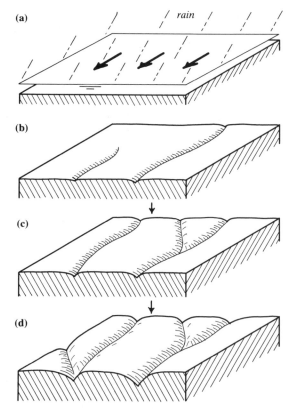

**Figure 5.8** Hypothetical sequence of morphologies illustrating the development of a channel network through instability of sheet flow over an erodible substrate.

In an attempt to answer the question, imagine that an extensive, uniformly smooth surface of mud sloping in one direction is suddenly exposed beneath the atmosphere, where rain falls on it at an areally uniform rate. As sketched in Figure 5.8, the water drains off in the form of a downslope-thickening sheet, for the discharge at any station is the sum of the rainfall at all the points up stream. Now imagine that the local slope of the surface is just that value causing the shear stress (Eqn 5.6) slightly to exceed the stress $\tau_{cr}$ at which the mud can be eroded. The surface is evidently in a critical state, but we should see uniform erosion only if it were mathematically smooth. A real surface will inevitably possess random irregularities, some of which will be wider and deeper than others. The deeper the irregularity, the more according to Equation 5.6 will the boundary shear stress be increased above $\tau_{cr}$, and the more rapidly will the irregularity be lowered through erosion. The local lowering of the bed will further increase the local flow depth, which by Equation 5.8 will further increase the local flow velocity. But as the water streaming off the surface is viscous, any local increase in velocity will spread diminishingly for some distance either side through the flow, and to a total extent that should scale on the flow depth. The larger depressions, deepening the most rapidly, will therefore begin to swamp the shallower ones, with the result that the larger runnels will begin to emerge at a characteristic depth-related transverse spacing. As the sequence of Figure 5.8 suggests, we are perhaps here beginning on the mud surface to develop downslope-connected channels whose spacing increases down slope because the flow depth also increases in this direction. The reality is likely to be much more complex, but it seems undeniable that a process similar to that described has an early role.

Speaking broadly, it appears that drainage networks record an instability. It is contended that (1) an erodible surface acted on by rain is unstable to small disturbances of shape (and, we might add, erodibility), and (2) there is a viscosity-dependent and depth-controlled mechanism at work selecting disturbances of a preferred spanwise wavelength.

## 5.5  Width : depth ratio of river channels

If channelization records the instability of earth materials to small disturbances when acted on by overland flow, what gives a river or channelized tidal current its characteristic width : depth ratio?

Consider a uniform steady flow of discharge $Q$ in a straight rectangular channel of slope $S$. Let $f_{bd}$ and $f_{bk}$ describe friction at the bed and banks respectively. Instead of being rigid, the boundary is now formed of an erodible material whose critical entrainment threshold is specified by the velocity $U_{cr}$. For a given discharge, slope, boundary friction and erosion threshold, there is only one combination of channel width $w$ and flow depth $h$ that will give us a stable channel, that is, one neither widening nor deepening. This is the equilibrium cross-sectional form towards which real channels will tend.

Let us now calculate the depth of the equilibrium channel. Ignoring air resistance, we first rewrite Equations 5.1 and 5.4 for the bed and banks treated separately, and then use the appropriate forms of the quadratic law (Eqn 5.7) to eliminate the stresses. The resulting expression may be put as

$$U_{cr}^2 = \frac{8whgS}{(2f_{bk}h + f_{bd}w)} \quad \text{m}^2\text{s}^{-2} \qquad (5.14)$$

equivalent to the squared form of Equation 5.8. Now

we can eliminate $w$ from Equation 5.14 by using the definition $Q = whU_{cr}$, that is, the discharge in the equilibrium channel is the product of the flow cross-sectional area with the overall mean flow velocity at the threshold of erosion (Sec. 1.2). Making the substitution, and then multiplying through by the depth $h$, we obtain

$$2f_{bk}U_{cr}^2h^2 - \frac{8QgS}{U_{cr}}h + f_{bd}U_{cr}Q = 0 \qquad (5.15)$$

This is a quadratic in $h$ of the general form $Ax^2 + Bx + C = 0$, in which

$$A = 2f_{bk}U_{cr}^2 \qquad B = -\frac{8QgS}{U_{cr}} \qquad C = f_{bd}U_{cr}Q \qquad (5.16)$$

and with the general solution

$$h_{cr} = \frac{-B \pm (B^2 - 4AC)^{1/2}}{2A} \quad \text{m} \qquad (5.17)$$

where $h_{cr}$ is the depth of the equilibrium channel.

The physical interpretation of Equations 5.15 to 5.17 is as follows. When $(B^2 - 4AC) < 0$, the roots are imaginary. A channel therefore cannot exist, and the flow must be presumed to be sheet-like. The condition that a channel can exist is therefore $B^2 \geqslant 4AC$. Substituting from Equation 5.16, one form of this condition is

$$S \geqslant \frac{U_{cr}^{5/2}}{g}\left(\frac{f_{bk}f_{bd}}{8Q}\right)^{1/2} \qquad (5.18)$$

showing that the limiting slope for the occurrence of a channel increases steeply with $U_{cr}$ but only gradually with boundary friction, and decreases slowly with increasing discharge. When $(B^2 - 4AC) = 0$ the roots of the equation are real but equal, and $h_{cr} = -B/2A$ by Equation 5.17, that is,

$$h_{cr} = \frac{2QgS}{U_{cr}^3 f_{bk}} \quad \text{m} \qquad (5.19)$$

on substituting from Equation 5.16. This is a special value of $h_{cr}$ because, on applying Equation 5.18, it is found to be the greatest possible depth for an equilibrium channel of specified discharge contained in a specified material. All the other possible channels, occurring at larger slopes, are shallower than this limiting channel.

Now repeat for $w$ the analysis which led to Equation 5.15, by substituting for $h$ from the definition of discharge and then multiplying through by $w$ itself. Corresponding to Equation 5.19, we find

$$w_{cr} = \frac{4QgS}{U_{cr}^3 f_{bd}} \quad m \quad (5.20)$$

and discover that all other possible channels, occurring at larger slopes, exceed this limiting width. Hence by Equations 5.19 and 5.20, the width : depth ratio of the limiting equilibrium channel is $w_{cr}/h_{cr} = 2(f_{bk}/f_{bd})$. The limiting channel is twice as wide as deep if the bed and banks are equally frictional.

## 5.6  Long profiles of rivers

The long profiles of rivers range between straight and variously concave to the sky. The equations developed in the preceding section help us to explore the long profile, on the supposition that a river tends to an equilibrium cross-sectional profile at least at some points along its length.

Consider the long profile of a river whose channel is of uniform bed and bank resistance and of limiting equilibrium form, as set by Equation 5.18 in the form of an equality. Because the whole river is now involved, rather than just one cross section, it is sensible to let both $U_{cr}$ and the steady discharge $Q$ vary down stream, say $U_{cr} = ax^m$ and $Q = bx^n$, where $a$ and $b$ are co-efficients, $m$ and $n$ exponents, and $x$ the horizontal distance measured along the channel from the source. Equation 5.18 then reads

$$S = \frac{a^{5/2} b^{-1/2}}{g} \left( \frac{f_{bk} f_{bd}}{8} \right)^{1/2} x^{5m/2} x^{-n/2} \quad (5.21)$$

Noting that $S = -dy/dx$, where $y$ (positive upwards) is the vertical elevation of the channel relative to the source, the long profile becomes

$$y = -\frac{a^{5/2} b^{-1/2}}{g} \left( \frac{f_{bk} f_{bd}}{8} \right)^{1/2} \int_0^x x^{(5m/2 - n/2)} dx \quad m \quad (5.22)$$

where all the constants are to the left of the integral sign.

Clearly, the profile shape given by Equation 5.22 is fixed by the exponents $m$ and $n$. Inspecting Equations 5.21 and 5.22, the profile is

(a)  concave-up when $(5m/2 - n/2) < 0$;
(b)  straight when $(5m/2 - n/2) = 0$;
(c)  convex-up when $(5m/2 - n/2) > 0$.

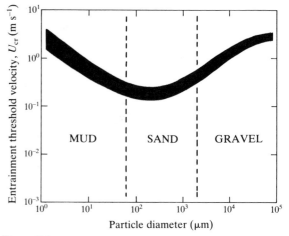

**Figure 5.9**  Hjulström-type entrainment curve for quartz-density solids in water. After Hjulström (1935).

What variations of $U_{cr}$ and $Q$ might occur along natural streams? Provided that the drainage basin is relatively long, and the long-term precipitation is areally uniform, $Q$ will increase approximately linearly with downstream distance, making $n \simeq 1$. The erosion threshold probably varies in a more complex way. To judge from Hjulström's (1935) summary (Fig. 5.9), it could be largest in the headwaters and far down stream, where coarse gravel and cohesive muds respectively may form the banks, but smallest in the sandy intermediate reaches. As a first shot, then, put $m = 0$. But note that quite a small change in $m$ has a profound effect on the governing exponent $(5m/2 - n/2)$, making the long profile of a river much more sensitive to downstream changes of bed and bank strength than discharge.

How well does our model square with observation? Figure 5.10 is a plot of the local slope against local mean annual discharge for a sample of 341 rivers and 18 laboratory streams (Begin 1981). The points define a line declining to the right with an exponent of about $-0.4$, close to the exponent of $-0.5$ predicted by Equation 5.21 with $m = 0$ and $n = 1$. A similar scatter in slope results from varying $U_{cr}$ in Equation 5.21 between $0.2 \, m \, s^{-1}$ representative of sand, through $0.6 \, m \, s^{-1}$ representative of mud, to $1.5 \, m \, s^{-1}$ representative of gravel (Fig. 5.9). As several workers noted (Leopold & Wolman 1957, Henderson 1961, Osterkamp 1978), most of the streams represented in the upper half of the graph have low-sinuosity single-thread or braided channels, whereas the majority falling in the lower half have single-thread channels of high sinuosity. Referring to Equation 5.21, these differences could partly reflect the influence of bed and bank strength. Equation 5.21 when

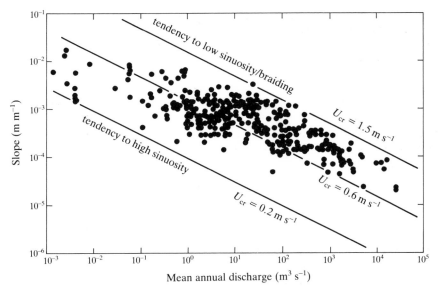

**Figure 5.10** Empirical variation of local channel slope with local mean annual discharge in rivers based on data of Begin (1981). The curves plotted are for Equation 5.21 at three different values for $U_{cr}$, and for $f_{bk} = f_{bd} = 0.15$.

integrated (Eqn 5.22) predicts with $m = 0$ and $n = 1$ that the long profile of rivers is concave-up. This is the general form of river long profiles (Hack 1957, Lee & Henson 1978, Wheeler 1979). In one of Hack's streams – Gillis Falls – the gravelly bed material increases down stream in coarseness and hence erosion resistance, giving a flatter long profile than the others, in accordance with Equation 5.22 when $m > 0$.

## 5.7 An experimental interlude

Aside from the preliminary sketch of real rivers, our exploration of channel character and processes has so far been almost entirely with the aid of simple models. Some experiments will help to prepare the way for the remainder of the exploration, which looks at processes in curved flows.

Flow curvature is one of the most striking features of real rivers, the current in plan either winding from side to side, as in single-channel streams (Fig. 5.1), or dividing and rejoining around lozenge-shaped bars and islands (Fig. 5.3). Does flow curvature arise gradually and inevitably in an initially straight channel, so that retention of a truly straight course is physically impossible? What happens to the flow and sediment in channel bends that causes those bends to migrate in the manner implied by the vegetation patterns visible in Figures 5.1 and 5.2?

It is easy to test the stability of flow in a straight channel with erodible boundaries. Take a shallow tray and fill it firmly and to a little higher than the top with clean dry sand. Smooth the sand surface and cut a shallow channel in it (Fig. 5.11a) by drawing crosswise along the edges of the tray a shaped wooden template. The channel should measure a centimetre or so wide and a few millimetres deep. Slightly raise one end of the channel on adjustable wooden blocks. Gradually admit water from the tap to the upper end of the channel. Water will appear in the channel once the sand as a whole is saturated. Now raise the discharge in the channel sufficiently to erode and transport the sand, by varying the tap and/or the tilt of the tray. As sand movement proceeds, a sequence of channel patterns similar to that shown in Figure 5.11b will appear. The straight channel is evidently unstable, the flow developing a winding course. These bends will be found to increase in amplitude, at least over an initial period, and to march down stream. Translation occurs by erosion of the outer bank, principally over its downstream part, and a harmonious deposition on the inner slope. Notice that the flow tends to be deepest on the outside of each bend, as reported from real rivers. There is nothing unrepresentative about these observations, despite their comparatively small scale. Ackers and Charlton (1970) found similar patterns to develop in their rather larger experiments. Artificially straightened rivers eventually develop meanders (Noble & Palmquist 1968, Lewin 1976).

Aside from the evident spatially distinct erosion and

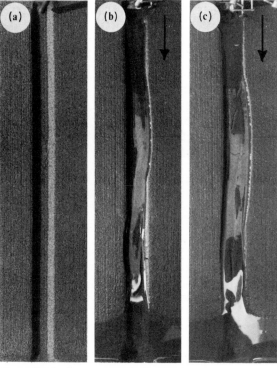

**Figure 5.11** Three stages in the development of a straight channel 0.03 m wide containing a water stream.

deposition, it is difficult to tell from sand-tray experiments what is happening to the flow in each bend. Another apparatus is required. Take a glass pneumatic trough or flat-bottomed crystallizing basin about half-full of water and stand it on white paper. Gently stir the water in one direction with a smooth circular motion and quickly remove the stirrer. The flow in the trough now closely resembles that in a continuous or 'infinite' channel bend. Speaking broadly, the current will be seen to increase in speed outwards from the axis of the trough. On looking through the side, it will be noticed that the water surface is no longer flat but mildly concave-up (the effect can be quite exaggerated in a stirred beaker of water or cup of tea) (Fig. 5.12a). To reveal the detailed pattern of flow, sprinkle a few crystals of potassium permanganate into the water near the side of the trough. Streaks of colour will spiral inwards over the bottom and towards the axis (Fig. 5.12b). If the flow is not too vigorous, the streaks will be seen to unite to form a vertical column at the axis of the trough, from the top of which the colour (now rather faint) will spread outwards just below the water surface. Evidently, water particles do not follow a simple circular path round the trough, but rather a horizontal helical-spiral or corkscrew. Flow at the base of the spiral is directed inwards over the bed, while that at the top is directed outwards just below the free surface. Such a motion is called a secondary flow (Ch. 11), and the velocity and stress characteristic of it each have tangential and radial components. Patterns of secondary flow are recorded from many laboratory

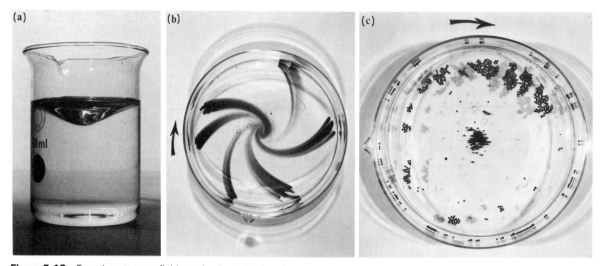

**Figure 5.12** Experiments on a fluid rotating in a circular channel. (a) Form of free surface. (b) Corkscrew motion revealed by inward spiral over bottom. (c) Effect of corkscrew motion on sediment particles more dense than fluid (axial cluster) and less dense than fluid (outer circle of floating grains). Clockwise flow in all cases. Dish is 0.22 m wide.

**(a)**                                           **(b)**

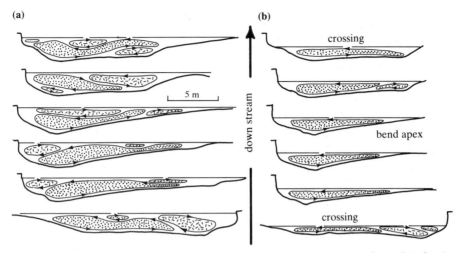

**Figure 5.13** Variation in pattern of secondary flow through a single meander bend in the River South Esk, Scotland, between (a) a near-bankfull stage and (b) a low stage. Based on velocity profiles by Bridge and Jarvis (1982).

bends and numerous rivers (Bridge & Jarvis 1982) (Fig. 5.13).

How would bedload particles respond to secondary flow? You have only to repeat the trough experiment with a little sediment stirred into the water to find that the heavy grains, under the influence of the inward radial component of the shear stress, drift towards the axis of the container (Fig. 5.12c). The deepest part of the remaining channel is thereby shifted outwards, while the cross-sectional form tends to the triangular shape observed from river bends (Fig. 5.4).

## 5.8 Flow in channel bends

### 5.8.1 Radial forces

What forces act in a river bend? There are clearly tangential forces related to the body force and to the opposing frictional resistance, but the fact that the average path of the water particles is curved horizontally suggests that radial forces are also present.

One is the centrifugal force. Let the flow in a circular trough (Fig. 5.14a) at a radial distance $r$ from its axis be characterized by a steady vertical profile of tangential velocity $U(y)$. Consider in Figure 5.14b the motion of a fluid element moving with a velocity $U$ on a circular path of radius $r$ and elevation $y$. Its velocity at A is $U$ directed along the tangent AS. The element at a small time later is at the point B, where its velocity is $U$ directed along the new tangent BT. The velocity at B is equivalent to a velocity $U \sin \theta$ along BD, where BD is parallel to AO, together with a velocity $U \cos \theta$ along

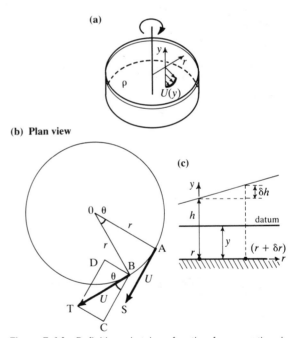

**Figure 5.14** Definition sketches for the forces acting in channel bends.

BC, where BC is parallel to the tangent AS. Because the angle $\theta$ is small, these are equivalent to $U\theta$ and $U$. At AO the component velocity was zero, so that the change in velocity in this direction is $U\theta$ and this has occurred in the small time taken to get from A to B at velocity $U$, i.e. AB/$U$ or $r\theta/U$. The element is therefore being accelerated along AO at the rate $U\theta/(r\theta/U) = U^2/r$. If the element is of unit volume, then the centrifugal force acting on it is $\varrho U^2/r$ N m$^{-3}$, where $\varrho$ is the fluid density. The depth-averaged centrifugal force per unit volume at a radial distance $r$ is therefore

$$\text{average centrifugal force} = \frac{\varrho}{rh}\int_0^h [U(y)]^2 \, dy \qquad (5.23)$$

$$= \frac{k\varrho}{r} U_m^2 \quad \text{N m}^{-3} \qquad (5.24)$$

where $h$ is the flow depth, $U_m$ the vertically averaged velocity, and $k$ a numerical coefficient depending on the form of $U(y)$.

The surface of the water stirred in the trough proved to be concave-up, that is, tilted radially inwards. The other radial force is therefore an inward-directed pressure gradient related to the tilt. Consider the flow in a radial section containing an arbitrary horizontal datum of elevation $y$ (Fig. 5.14c). At a radial distance $r$ the hydrostatic pressure in excess of atmospheric at the datum is $\varrho(h - y)g$ N m$^{-2}$, where $h$ is the flow depth. At the slightly greater outward distance $(r + \delta r)$ it is $\varrho(h + \delta h - y)g$, whence in the limit

$$\text{pressure gradient} = \varrho g \frac{dh}{dr} \quad \text{N m}^{-3} \qquad (5.25)$$

where $dh/dr$ is the radial slope of the water surface. Notice that the pressure gradient is independent of depth, whereas the *local* centrifugal force given above depends on the local, depth-related tangential velocity.

### 5.8.2 Shape of water surface

The motion in a channel bend being steady, Newton's first law requires that the pressure force (Eqn 5.25) and the average centrifugal force (Eqn 5.24) shall be equal and opposite, whence

$$\varrho g \frac{dh}{dr} = \frac{k\varrho}{r} U_m^2 \quad \text{N m}^{-3} \qquad (5.26)$$

$$\frac{dh}{dr} = \frac{kU_m^2}{gr} \qquad (5.27)$$

By integrating the last equation, we can calculate the shape of the water surface, provided we know the radial variation of the depth-averaged tangential velocity $U_m(r)$.

Consider a stirred circular trough of rectangular cross section. Here the depth is sensibly constant, and from

one of our experimental observations we may suggest that $U_m(r) = ar$, where $a$ is a coefficient. Substituting into Equation 5.27 and integrating yields

$$h = \frac{ka^2}{2g} r^2 + C \quad \text{m} \qquad (5.28)$$

where $C$ is a coefficient of integration, to be determined from the conditions on the problem. As the term in $r$ is positive, the water surface has the form of a concave-up parabola, similar to the observed shape (Fig. 5.12a). Integrating between $r = 0$ at the trough axis and $r = r_0$ at the outer margin gives us the superelevation of the water surface $\Delta h = (ka^2/2g)r_0^2$, that is, the difference in water-surface height across the trough. For $r_0 = 0.1$ m and $a = 5$ s$^{-1}$, corresponding to brisk stirring, $\Delta h$ is approximately 0.025 m for $k$ of order 1. This amount of superelevation is consistent with observation.

An erodible river bend is approximately triangular in cross section, with a greatest depth near the outer bank. The depth now increases approximately linearly with radial distance, in which case, by Equation 5.8, we may put $U_m(r) = br^{1/2}$, where $b$ is a new coefficient. Integrating after this substitution into Equation 5.27 gives

$$h = \frac{kb^2}{g} r + C \qquad (5.29)$$

where $C$ is a new unknown integration constant. The new water surface slopes linearly upwards towards the outer bank. Integrating between inner and outer banks, at radial distances $r_i$ and $r_0$ respectively, affords the superelevation $\Delta h = (kb^2/g)(r_0 - r_i)$. For a river of $r_i = 2000$ m, $r_0 = 3000$ m, and $U_m = 2$ m s$^{-1}$ at the outer bank, $\Delta h \approx 0.14$ m for $k$ of order 1. Superelevation in large rivers is therefore very small relative to their size (compare the trough).

Our conclusions regarding water-surface shape all rest on the implicit assumption that the motion is unaccompanied by a secondary flow. It is rather difficult to include the secondary flow, but the above results are not changed significantly.

### 5.8.3 Secondary flow

How can we explain the secondary currents present in curved channels? Because the water is viscous, and consequently there is bed friction, the tangential flow velocity necessarily increases vertically upwards at each radial distance. Hence the local centrifugal force $\varrho U^2/r$ also varies vertically. But the pressure force (Eqn 5.25) is independent of depth and equals the local centrifugal force at only one depth, that corresponding to $U = U_m$. At shallower depths, the centrifugal force exceeds the pressure force, whence the resultant force drives the fluid outwards. But deeper down the pressure force is the larger, causing the fluid to drift inwards. The secondary current is therefore a viscous effect, dependent on the slowing of the near-bed fluid.

We shall now calculate the radial shear stress characteristic of the secondary flow, using an idea suggested by Zimmerman (1977). An observer submerged at the axis of the flow and looking down stream sees the current turning about his line of sight. Above him the fluid is moving towards the outer bank, while below him it is drifting inwards towards the point bar. The spatially varying centrifugal force therefore promotes a torque, or twisting force, on the water. By Newton's first law, an equal and opposite torque must exist if the radial motion is steady. The opposing torque can only come from the radial bed friction. Remember in the following analysis that a torque is the product of a force times the distance from its centre of action.

Consider in Figure 5.15 a radial wedge-shaped segment from a channel of rectangular cross section and circular plan in which there is a uniform steady flow $U(y) = U_{max}(y/h)^n$, where $U_{max}$ is the maximum tangential velocity (at the water surface), $y$ distance above the bed, $h$ flow depth, and $n$ a small positive exponent. The local centrifugal force per unit volume is $\varrho U^2/r$ as before, whence the driving torque per unit bed area becomes

$$\text{unit driving force} = \frac{\varrho}{r} \int_0^h U^2 (y - \tfrac{1}{2}h) \, dy \quad (5.30)$$

where $(y - \tfrac{1}{2}h)$ is distance from the midplane of the flow. Now in the annular segment of Figure 5.15, the total driving force is the product of the unit torque (Eqn 5.30) with the basal area $r \, d\theta \, dr$, where $\theta$ is the central angle. For the whole of the given slice

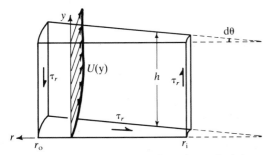

**Figure 5.15** Definition sketch for the calculation of secondary flow in a channel bend.

$$\text{total driving force} = \frac{\varrho}{r} \int_0^h \int_{r_i}^{r_o} r U^2 (y - \tfrac{1}{2}h) \, dy \, dr \, d\theta$$

$$(5.31)$$

in which $r_i$ and $r_o$ are the radii of curvature of the inner and outer banks respectively. This force must be balanced by friction over the wetted boundary.

The friction force giving the resisting torque acts only on the wetted boundary of the channel segment. The contribution from the bed is

$$\frac{\tau_r h}{2} \int_{r_i}^{r_o} r \, dr \, d\theta \quad (5.32)$$

remembering that $\tau_r$ is the mean radial stress acting over unit area. The inner bank, of area $h r_i \, d\theta$, contributes

$$\tau_r h (r_c - r_i) r_i \, d\theta \quad (5.33)$$

where $r_c$ is the radius at the channel centreline. Similarly, the contribution

$$\tau_r h (r_o - r_c) r_o \, d\theta \quad (5.34)$$

comes from the outer bank, of area $h r_o \, d\theta$. The total resisting torque is the sum of Equations 5.32, 5.33 and 5.34.

Equating the torques, and substituting $U_m$ for $U$ using the given velocity profile, leads after rearrangement and simplification to

$$\tau_r = \frac{\varrho}{3} \left( \frac{n+1}{2} - \frac{(n+1)^2}{4n+2} \right) \frac{h U_m^2}{r_c} \quad \text{N m}^{-2} \quad (5.35)$$

for the mean inward-acting radial component of the boundary shear stress. For each tangential velocity, the stress increases linearly with flow depth but is inversely proportional to the radius of centreline curvature. There is no transverse force acting in a straight channel, for which $r_c$ is infinitely large. The exponent $n$ increases gradually with the bed friction coefficient but for a typical river is comparable with 1/7. Equation 5.35 has some experimental support.

One way of measuring the strength of the secondary flow in a channel like that of Figure 5.15 is to calculate the ratio $\tau_r/\tau_x = \tan \alpha$, where $\tau_x$ is the tangential shear stress and $\alpha$ the angular deviation in the horizontal plane of the resultant shear stress $\tau$ from the tangential direction. This ratio turns out generally to be a small fraction. Alternatively, given the Darcy–Weisbach friction coefficient and knowing $\tau_r$, we can use the quadratic stress law (Eqn 1.27) to determine the mean radial component of the flow velocity $U_{m,r}$. The ratio $U_{m,r}/U_m$ is also small.

## 5.9 Sediment particles in channel bends

### 5.9.1 Condition for cross-sectional stability of a channel bend

We have now shown that the boundary shear stress exerted in a river bend can be resolved into two components due to the presence of secondary flow. The tangential component $\tau_x$ is directed down stream parallel to the channel centreline, whereas the radial component $\tau_r$ is directed inwards over the enclosed point bar. How will the bedload respond to these forces? Under what conditions can the channel transmit bedload without changing in shape, that is, remain stable in cross-sectional form? What is that stable cross-sectional form, and how might bed particles of different sizes be distributed on it? It is important to remember that the following analysis assumes a fully developed secondary flow in the channel.

Consider in Figure 5.16 a spherical bedload particle of diameter $D$ and density $\sigma$ travelling around a channel of circular plan at a radial distance $r$, where the flow is of depth $h$ measured positively downwards. The channel bed slopes outwards at the angle $\beta$ and the local shear stress acting on the particle can be resolved on the bed between the local tangential and radial components $\tau_x$ and $\tau_r$ respectively. The flow is steady and the channel is of negligible downstream slope as compared to radial slope. Because the particle is in the slow-moving bedload, its velocity is neglected compared to that of the fluid. Only two radial forces then act on the particle, the upslope-directed fluid drag and the downslope-directed immersed weight, as follows

$$\text{drag force} = \pi \left(\frac{D}{2}\right)^2 \tau_r \quad \text{N} \qquad (5.36)$$

and

$$\text{immersed weight} = \frac{4\pi}{3} \left(\frac{D}{2}\right)^3 (\sigma - \varrho)g \sin \beta \quad \text{N} \qquad (5.37)$$

Notice that the immersed weight varies with the bed slope. When the drag exceeds the immersed weight, the particle drifts up slope into shallower water under the action of the resultant upslope force. But because the depth decreases up slope, the flow velocity and resultant shear stress also decline (Eqns 5.6 & 5.8) and the particle is ultimately deposited. The transverse bed slope is

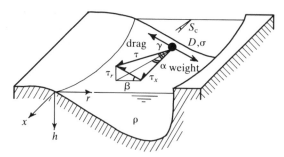

**Figure 5.16** Definition sketch for the equilibrium of a sediment bedload particle travelling round a channel bend.

thereby steepened and the whole cross section is unstable and changes shape. The opposite effect occurs when the immersed weight is the greater. The particle then drifts down slope as it rounds the bend, implying erosion of the higher slopes and upbuilding lower down. Hence the bed can be stable only when

$$\text{drag force on particle} = \text{immersed weight of particle} \qquad (5.38)$$

When this condition prevails everywhere, particles of a range of sizes follow concentric paths parallel with the channel banks, with no tendency to drift sideways and either deposit or erode.

### 5.9.2 Properties of a stable cross section

By the preceding argument, a stable cross section is obtainable when Equation 5.38 holds, i.e. equating Equations 5.36 and 5.37:

$$\pi \left(\frac{D}{2}\right)^2 \tau_r = \frac{4\pi}{3} \left(\frac{D}{2}\right)^3 (\sigma - \varrho)g \sin \beta \quad \text{N} \qquad (5.39)$$

which becomes after rearrangement

$$\sin \beta = \frac{3\tau_r}{2(\sigma - \varrho)gD} \qquad (5.40)$$

This equation states that the bed increases in transverse slope with increasing radial shear stress but decreasing particle size and excess density.

Equation 5.40 is further useful only if we can relate $\tau_r$ to the whole flow and also introduce simplifications. Rozovskii (1961) found theoretically for a fully developed secondary flow over a *horizontal* bed that $\tau_r/\tau_x = \tan \alpha$ varied with flow depth and radius as

$$\tan \alpha = 11 \frac{h}{r} \qquad (5.41)$$

where $\alpha$ is shown in Figure 5.16 as the horizontal deviation of the resultant shear stress from the mean flow direction. The geometry in Figure 5.16 shows that $\tau_r = \tau_x \tan \gamma$ and $\tan \gamma = \tan \alpha / \cos \beta$. From Equation 5.6

$$\tau_x = h \varrho g S \qquad (5.42)$$

where $S$ is the local water-surface slope in the mean flow direction. From the plan geometry of a winding channel

$$S = S_c \frac{r_c}{r} \qquad (5.43)$$

where $r$ is radial distance and $S_c$ and $r_c$ are the water-surface slope and channel radius respectively at the channel centreline. Introducing these supplementary relationships into Equation 5.40 yields

$$\sin \beta = \frac{1}{\cos \beta} \frac{33 \, \varrho r_c S_c}{2(\sigma - \varrho) D} \frac{h^2}{r^2} \qquad (5.44)$$

But as long as $\beta$ is a small angle, $\sin \beta \simeq \tan \beta = \mathrm{d}h/\mathrm{d}r$ and $\cos \beta \simeq 1$, whence Equation 5.44 simplifies to

$$\frac{\mathrm{d}h}{\mathrm{d}r} = \frac{33 \, \varrho r_c S_c}{2(\sigma - \varrho) D} \frac{h^2}{r^2} \qquad (5.45)$$

The local transverse slope of the channel bed therefore increases as the square of the flow depth but decreases with increasing particle size and radius squared.

Equation 5.45 is a differential equation with separable variables which may be rearranged as

$$\int \frac{\mathrm{d}h}{33 \varrho r_c S_c h^2} = \int \frac{\mathrm{d}r}{2D(\sigma - \varrho) r^2} \qquad (5.46)$$

and solved to give

$$h = \frac{2D(\sigma - \varrho) r}{33 \varrho r_c S_c [1 - 2D(\sigma - \varrho) Cr]} \quad \text{m} \qquad (5.47)$$

where $C$ is an unknown constant of integration (expected to be very small but positive), to be calculated using the channel discharge. The last equation can also be written as

$$D = \frac{33 \varrho r_c S_c h}{2(\sigma - \varrho)(r + 33 \varrho r_c S_c Crh)} \quad \text{m} \qquad (5.48)$$

Physically, Equation 5.47 means that the flow depth increases with radial distance outwards across the channel, provided the bedload particles are uniform. Inspection of Equation 5.47 suggests that the cross-sectional profile described is convex-up. This is easily proved by differentiating Equation 5.47 twice with respect to $r$, yielding

$$\frac{\mathrm{d}^2 h}{\mathrm{d}r^2} = \frac{2C[66 \varrho (\sigma - \varrho) r_c S_c D]^2}{\{33 \varrho r_c S_c [1 - 2D(\sigma - \varrho) Cr]\}^3} \qquad (5.49)$$

As $\mathrm{d}^2 h/\mathrm{d}r^2$ is seen to be positive, a convex-up profile obtains.

What use are these equations? Equation 5.47 will yield the cross-sectional shape of a channel bend, provided that the discharge, calibre of bedload, curvature and water-surface slope are prescribed. Equation 5.48 will predict the radial variation of $D$ in a channel bend of known geometry, discharge and slope. For example, when the transverse slope is constant, so that $h$ increases linearly with $r$, the permissible grain size also increases outwards, and therefore from shallower to deeper water. Equation 5.45 can be used for back-calculation. For example, from an exposure of a fossil channel, we could measure local values for $D$, $\mathrm{d}h/\mathrm{d}r$ and $h$ on each bedding surface representing the channel cross section. A good estimate could then be made of the product $(r_c S_c)$. The elimination of $r_c$, however, requires either further information from the rocks or assumptions about the full channel geometry.

But our analysis has one rather serious limitation. Real rivers transmit bed materials of a wide range of size, whereas our analysis assumes a uniformly coarse bedload. When more than one grain size is available, what will be the form of the stable cross section, and how will the different sizes be distributed across it? The question is too difficult to answer fully here, but a step forward is possible when we remember that the analysis is for a *dynamically* stable cross section. The particle distribution must therefore not only satisfy the balance of radial forces (Eqn 5.39) but also the sediment entrainment criterion.

Referring to Section 4.2.2, the entrainment threshold for bed particles at a radial position $r$ and depth $h$ in the channel of Figure 5.16 can be written

$$\theta_{cr} = \frac{\tau_{x,cr}}{(\sigma - \varrho) g D} \qquad (5.50)$$

where by Equation 5.6

$$\tau_{x,cr} = h_{cr} \varrho g S \quad \mathrm{N \, m^{-2}} \qquad (5.51)$$

and by Equation 5.43

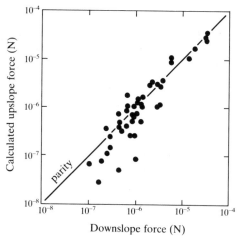

**Figure 5.17** Transport of bedload particles round a single meander bend in the River South Esk, Scotland. Calculated upslope drag force versus the measured downslope particle weight. Readjusted data of Bridge and Jarvis (1982).

$$S = S_c \frac{r_c}{r} \qquad (5.52)$$

in which $\tau_{x,\mathrm{cr}}$ is the tangential boundary shear stress at the threshold, $h_{\mathrm{cr}}$ the corresponding depth, and $\theta_{\mathrm{cr}}$ can be read from Figure 4.3. Substituting and rearranging, we obtain

$$h_{\mathrm{cr}} = \frac{\theta_{\mathrm{cr}}(\sigma - \varrho)Dr}{\varrho r_c S_c} \quad \mathrm{m} \qquad (5.53)$$

Since $r$ varies little in comparison with $r_c$, and $\theta_{\mathrm{cr}}$ is similar for gravel and sand, $h_{\mathrm{cr}}$ increases mainly with $D$. As the flow deepens outwards, the calibre of the equilibrium bed grains must also increase outwards. Hence a stable profile will show fine grains in shallower depths and at higher stratigraphic positions than coarse ones.

### 5.9.3 Empirical support for the equations
How well do the preceding equations square up to reality? The severest possible test, striking at the root of the entire analysis, is to compare using field data the drag and weight forces (Eqns 5.36 & 5.37) acting on grains rounding a bend. Figure 5.17 illustrates this test as applied to the River South Esk using Rozovskii's (1961) equation (Eqn 5.41) and Bridge and Jarvis's (1982) data. Considering the difficulties of field measurement, the agreement between the upslope drag and downslope weight forces is good and supports the analysis.

## 5.10 Migration of channel bends

### 5.10.1 Plan stability of channels
One conclusion demanded by our experiments (Fig. 5.11) is that a straight channel containing an eroding flow is unstable and gradually changes into one with more or less regular bends. Does our analysis of secondary flow and channel cross-sectional stability help to explain this observation, together with the universal fact that real rivers wind more or less conspicuously?

Consider a straight open channel of uniform cross section with banks of an erodible material of uniform entrainment threshold. Let the flow in the channel be steady and at this threshold, and the bed consist of a deep layer of grains capable of being transported as bedload. Such a channel is metastable. If we locally perturb the position of one of the banks, moving a portion of it outwards slightly to make a concavity, we immediately introduce a slight local curvature into the channel plan. Fluid drifting past the perturbed bank is at once acted on by a centrifugal force, which only on the average can be balanced by the simultaneously created inward pressure gradient. A secondary flow arises, as described by Equations 5.31 to 5.35. Because the channel bed is level, the secondary flow begins to transport grains inwards towards the centre of curvature of the perturbed channel. The bed steepens up transversely, until Equation 5.39 is satisfied. But the steepening deepens the channel below the perturbed bank, so that flow conditions there exceed the bank-material entrainment threshold. The perturbed bank therefore proceeds to erode, which accentuates the introduced curvature, and steepens the bed still further, inducing yet more losses, and so on. The opposite or inner bank meanwhile shoals and builds outwards in harmony.

Hence an unobstructed erodible channel turns out to be inherently and irreversibly unstable. Although in our model the banks consisted of a uniform material, and the channel shape was artificially disturbed, in real rivers many natural sources of perturbation are inevitably present. Probably the most important are variations in erosion resistance along a river bank, as conferred by small spatial changes in the succession and lithology of the river's earlier deposits.

Now it is one thing to demonstrate the inherent instability of a straight erodible channel, but quite another to show how the channel comes to meander as regularly as in Figures 5.1 and 5.2. Evidently there operates in an unstable channel a mechanism of wavelength selection. This causes bank perturbations of one particular wavelength to develop preferentially, that is, to be amplified more rapidly than and so suppress

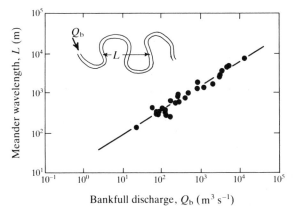

Figure 5.18 Average meander wavelength as a function of bankfull discharge for a sample of North American rivers. Data of Carlston (1965).

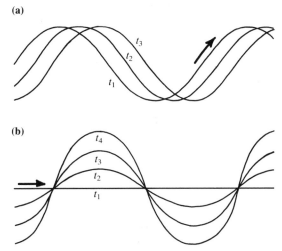

Figure 5.19 Successive positions of a meandering river channel changing (a) in phase only and (b) in amplitude only.

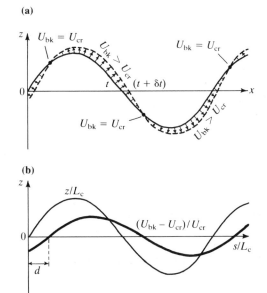

Figure 5.20 Definition sketches for the development of channel meanders.

other wavelengths. Many mathematical models describe this mechanism (Callander 1978), but Carlston (1965) shows empirically that

$$L = 24.5 Q_b^{0.62} \quad \text{m} \qquad (5.54)$$

where $L$ is the bend wavelength and $Q_b$ the bankfull river discharge (Fig. 5.18). Again empirically, bend wavelength turns out to be about 12 times channel width, regardless of discharge.

### 5.10.2 Horizontal movement of channel bends

Our experiment (Fig. 5.11) shows that the progressive distortion of an initially straight erodible channel involves the appearance of regular bends which move both sideways and down stream. To judge from the point-bar accretion patterns picked out by vegetation, bends of the river in Figure 5.1 spread less rapidly down stream than sideways, whereas those in Figure 5.2 appear to have only shifted down stream. What mechanism controls the pattern of horizontal movement? Whatever it is, the motion can be neatly resolved into two components if we liken a series of bends to a train of regular waves (Fig. 5.19). Bends that shift down stream without change of amplitude change only in phase. Bends that spread only sideways grow in amplitude alone. Most river bends combine marked growth in amplitude with some change in phase.

The control on the resultant motion of a channel bend can be seen by considering how the uniformly and regularly winding river of Figure 5.20a responds to bank erosion. Here the channel centreline at time $t$ is plotted in terms of the distance $x$, measured down stream along the bisector of the series of bends, and the distance $z$,

measured normal to $x$ and positive to the left of an observer looking down valley. Speaking broadly, we should expect each concave bank to erode, and for the erosion to be directed outwards, opposite to the sense of bank concavity. Hence there are repeating stations along the channel where the erosion direction alters in sign, that is, changes from left to right, or from right to left, of the flow direction. No erosion occurs at the stations themselves, and the channel at each is in equilibrium (Eqn 5.14). At all other stations, one of the banks is eroding, in response to the effects of the secondary flow. At these stations $U_{bk} > U_{cr}$, where $U_{bk}$ is the mean flow velocity adjacent to the eroding bank, and $U_{cr}$ as before is the bank-material erosion threshold. The difference $(U_{bk} - U_{cr})$ should rise to a series of maxima at stations related to, but not necessarily identical with, the places where the bends attain their maximum curvature. Of course, $U_{bk} = U_{cr}$ at the equilibrium stations. Now the rate of bank erosion should increase with $(U_{bk} - U_{cr})$ and will therefore range from zero at the equilibrium stations to maxima where $(U_{bk} - U_{cr})$ rises to maxima. After a small time increment $\delta t$, the channel in Figure 5.20a will have moved to the new position indicated. The relative erosion pattern was so chosen as to make the bends move in phase (down stream) and simultaneously increase in amplitude.

Evidently other movement patterns are possible by further changing the relationship between the channel and erosion schemes. We can best explore these by altering the co-ordinate system used to describe the channel and erosion patterns. Instead of plotting the channel in real space, we can show it in transformed space by plotting $z$ against $s$, the downstream distance along the channel centreline. Furthermore, we can normalize, that is, make non-dimensional, both $z$ and $s$ by comparing them to the bend wavelength $L_c$, also measured along the centreline. The quantity $(U_{bk} - U_{cr})$ can be normalized by $U_{cr}$. Figure 5.20b shows the previous channel and erosion patterns in transformed space. They are periodic functions with the same wavelength but different phase, as measured by the non-dimensional lag distance $d$. Table 5.1 summarizes the effect of varying this distance. Most channel bends, for example, those in

**Table 5.1** Effect of lag distance on behaviour of channel bends.

| Non-dimensional lag distance | Effect on bend | | Comments |
| | Amplitude | Phase | |
| --- | --- | --- | --- |
| $0 < d < 1/4$ | increases | moves down stream | most channel bends |
| $1/4 < d < 1/2$ | decreases | moves down stream | bends with concave-bank benches |
| $1/2 < d < 3/4$ | decreases | moves up stream | no known bends |
| $3/4 < d < 1$ | increases | moves up stream | no known bends |

Figure 5.1, represent a lag distance of less than one-quarter of a bend wavelength. There are no known channels which represent the limiting case of zero lag, when the bends change in amplitude only. Bends in Figure 5.2 represent another limiting case, when the pattern of bank erosion lags that of the channel by exactly one-quarter of a wavelength, resulting in phase changes only. Acute bends in some rivers, such as the Murrumbidgee (Page & Nanson 1982), possess what is called a concave-bank bench. These are the only known bends possibly compatible with a lag distance of more than one-quarter of a bend wavelength.

Why is the lag distance non-zero? Consider the behaviour of the water as it flows from one bend to the next of opposite curvature. The secondary flow must change its sense of rotation as the current moves from one bend to the other. But because a substantial fluid mass is involved, the response of the secondary flow to the change in channel curvature cannot be immediate. The inertia of the fluid, and the frictional resistance of the bed and banks, prevent immediate adjustment, and so introduce a lag between the change in channel orientation and the adjustment of the secondary flow in response to the new orientation.

Rozovskii (1961) calculated the lag distance for a flow passing from a straight channel into a circular bend. Using the present set of variables, his lag distance is

$$d = \frac{k}{2\pi} \left(\frac{8}{f}\right)^{1/2} \frac{h_m}{r_{c,m}} \tag{5.55}$$

where $k = 2.3$ is a numerical coefficient, $f$ the overall Darcy–Weisbach friction coefficient, $h_m$ the bend-averaged channel depth, and $r_{c,m}$ the bend-averaged radius of centreline curvature. Plausible values for a river of intermediate size are $h_m = 5$ m, $r_{c,m} = 200$ m and $f = 0.1$, yielding $d = 0.082$ at the stage of bend development implied. The bend should therefore move down stream while increasing in amplitude, as seems true of most freely meandering rivers.

### 5.10.3 How do bends evolve?

Equation 5.55 suggests how bends might evolve from an initially straight channel. In the earliest stages, $r_{c,m}$ is very large, whence the bend should grow mainly in amplitude. Growth should be slow, as Equation 5.47 implies that $U_{bk}$ will only slightly exceed $U_{cr}$. But as the bend increases in curvature, and the channel slope grows less, $h_m$ should increase by Equation 5.17 while the radius $r_{c,m}$ declines. Movement in phase becomes increasingly important relative to increase in amplitude, and growth is now more rapid. Hints of this trend towards increasing relative downstream movement occur in described bend accretion patterns (Brice 1974, Hickin 1974). One notable feature of point bars in an advanced stage of growth is their lack of bilateral symmetry (Carson & Lapointe 1983), a feature particularly clear in Figure 5.2. The spatial pattern of bend migration is therefore a non-linear process, as Equation 5.55 itself implies.

## 5.11  A model for river point-bar deposits

We now have enough information and ideas to suggest a simplified model for the deposits formed on the accretionary point bar of a river bend (Fig. 5.21).

The deposit is dominated by the relatively coarse-grained sediments associated with the channel and bedload, and takes a stratigraphic thickness comparable with the bankfull depth of the river in its deeper parts. The progressive outward and downstream migration of the bend, combined with deposition in harmony on the point bar itself, means that the bar deposits rest on a laterally extensive erosion surface recording scour at the successive positions of the channel deeps. In response to the progressive steepening of the point-bar surface, and simultaneous decrease in channel slope due to increasing sinuosity, this surface may dip outwards slightly in cases where the bend moved more rapidly in amplitude than phase. The bar deposits should be graded upwards from coarse to fine, in accordance with the implications of Equation 5.53. Amongst the coarser elements may be lumps of cohesive sediment (intraformational debris) resulting from the erosion of steep concave banks. A

hierarchy of bedding types is present. A 'master' bedding, otherwise called epsilon cross-bedding or lateral accretion bedding, records successive accretions on the convex channel bank. As demanded by the secondary flow (Eqn 5.40), this bedding dips towards the former concave bank, and ranges through the full thickness of the deposit. It is expressed by two kinds of surface: (1) non-erosional contacts across which grain size and/or small-scale sedimentary structures change more or less abruptly, and (2) discordant and commonly somewhat steeper scours. Referring to the sediment transport and continuity relationships (Eqns 4.9 & 4.19), the first kind of surface defines depositional episodes associated with flood decay, whereas the second results from an increase of discharge and may even represent the seasonal flood. Figure 5.21 suggests what sedimentary structures are present. Because of the outward increase of flow depth and velocity across the bend, the structures recording large stream power, for example, dunes and cross-bedding, are restricted to the channel deeps and the lower part of the point-bar sequence. Current ripples and cross-lamination are restricted to the shallows and the upper part, where the stream power and grain size are less.

Speaking broadly, the inferred pattern of sediment features is confirmed by studies of modern rivers. Figure 5.22 shows the upper part of a set of master bedding surfaces exposed by the growth of a modern meander into an older point bar. Studies by Bluck (1971), Jackson (1975, 1976), Dietrich *et al.* (1979), Nanson (1980) and Bridge and Jarvis (1976, 1982) suggest that in much of a point bar the vertical sequence of grain size and sedimentary structures is similar to the

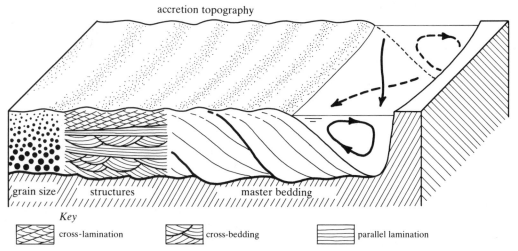

**Figure 5.21**  Model for the internal geometry, sedimentary structures and grain size of a river point-bar deposit.

**Figure 5.22** Cut bank on the River Twyi, South Wales, showing lateral accretion structure in the upper part of the point bar enclosed within a cut-off meander. Cut bank is about 2 m tall.

model. However, in the upstream sections of many bends, there is either no upward fining or a slight upward coarsening over a substantial interval.

Patterns similar to those in the model abound in the rock record. Figure 5.23 records a section through a Devonian point bar. The base is a laterally extensive erosion surface, and each of the above kinds of master bedding is represented. Grain size declines upwards and

the sedimentary structures recording the largest stream power lie near the base of the sequence. In favourable places, weathering has removed the softer rocks from around fossil river point bars, rendering them fully visible, even to details of the surface accretion topography (Puigdefabrigas 1973, Nami 1976, Padgett & Ehrlich 1976).

## Readings

Ackers, P. and F. G. Charlton 1970. The geometry of small meandering streams. *Proc. Instn Civ. Engrs* **S12**, 289–317.

Begin, Z. B. 1981. The relationship between flow-shear stress and stream pattern. *J. Hydrol.* **52**, 307–19.

Bluck, B. J. 1971. Sedimentation in the meandering River Endrick. *Scott. J. Geol.* **7**, 93–138.

Brice, J. C. 1974. Evolution of meander loops. *Bull. Geol. Soc. Am.* **85**, 581–6.

Bridge, J. and J. Jarvis 1976. Flow and sedimentary processes in the meandering River South Esk, Glen Clova, Scotland. *Earth Surf. Processes* **1**, 303–36.

Bridge, J. S. and J. Jarvis 1982. The dynamics of a river bend: a study in flow and sedimentary processes. *Sedimentology* **29**, 499–541.

**Figure 5.23** Vertical section through fluvial upward-fining sequence (base arrowed) to show master bedding recording lateral accretion from right to left, Catskill Formation, New York State.

Callander, R. A. 1978. River meandering. *Annu. Rev. Fluid Mech.* **10**, 129–58.

Carlston, C. W. 1965. The relation of free meander geometry to stream discharge and its geomorphic implications. *Am. J. Sci.* **263**, 864–85.

Carson, M. A. and M. F. Lapointe 1983. The inherent asymmetry of river meander planform. *J. Geol.* **91**, 41–55.

Dietrich, W. E., J. D. Smith and T. Dunne 1979. Flow and sediment transport in a sand bedded meander. *J. Geol.* **87**, 305–15.

Hack, J. T. 1957. *Studies of longitudinal stream profiles in Virginia and Maryland*. Prof. Pap. US Geol. Surv., no. 294-B.

Henderson, F. M. 1961. Stability of alluvial channels. *J. Hydraul. Div. Am. Soc. Civ. Engrs* **87**, 109–38.

Hickin, E. J. 1974. The development of meanders in natural river-channels. *Am. J. Sci.* **274**, 414–42.

Hjulström, F. 1935. Studies of the morphological activities of rivers as illustrated by the River Fyris. *Bull. Geol. Inst. Univ. Uppsala* **25**, 221–527.

Jackson, R. G. 1975. Velocity–bedform–texture patterns of meander bends in the lower Wabash River of Illinois and Indiana. *Bull. Geol. Soc. Am.* **86**, 1511–22.

Jackson, R. G. 1976. Depositional model of point bars in the lower Wabash River. *J. Sed. Petrol.* **46**, 579–94.

Lee, L. J. and B. L. Henson 1978. The longitudinal river, valley and regional profiles of the Arkansas River. *Z. Geomorph.* **22**, 182–91.

Leopold, L. B. and M. G. Wolman 1957. *River channel patterns; braided, meandering and straight*. Prof. Pap. US Geol. Surv., no. 282-B.

Leopold, L. B., M. G. Wolman and J. P. Miller 1964. *Fluvial processes in geomorphology*. San Francisco: W. H. Freeman.

Lewin, J. 1976. Initiation of bed forms and meanders in coarse-grained sediment. *Bull. Geol. Soc. Am.* **87**, 281–5.

Nami, M. 1976. An exhumed Jurassic meander belt from Yorkshire, England. *Geol. Mag.* **113**, 47–52.

Nanson, G. C. 1980. Point bar and floodplain formation of the meandering Beatton River, northeastern British Columbia, Canada. *Sedimentology* **28**, 3–29.

Noble, C. A. and R. C. Palmquist 1968. Meander growth in artificially straightened streams. *Proc. Iowa Acad. Sci.* **75**, 234–42.

Osterkamp, W. R. 1978. Gradient, discharge, and particle-size relations of alluvial channels in Kansas, with observations on braiding. *Am. J. Sci.* **278**, 1253–68.

Padgett, G. V. and R. Ehrlich 1976. Paleohydrologic analysis of a late Carboniferous fluvial system, southern Morocco. *Bull. Geol. Soc. Am.* **87**, 1101–4.

Page, K and G. Nanson 1982. Concave-bank benches and associated floodplain formation. *Earth Surf. Processes Landforms* **7**, 529–43.

Puigdefabrigas, C. 1973. Miocene point-bar deposits in the Ebro Basin, northern Spain. *Sedimentology* **20**, 133–44.

Richards, K. S. 1982. *Rivers*. London: Methuen.

Rozovskii, I. L. 1961. *The flow of water in bends of open channels*. Jerusalem: Israel Program for Scientific Translations.

Wheeler, D. A. 1979. The overall shape of the longitudinal profiles of streams. In *Geographical approaches to fluvial processes*, A. F. Pitty (ed.), 241–60. Norwich: Geobooks.

Zimmerman, C. 1977. Roughness effects on the flow direction near curved stream beds. *J. Hydraul. Res.* **15** (1), 73–85.

# 6 Order in chaos

Measurement and description of turbulence – ideal eddies – boundary-layer streaks – streak bursting – large eddies – sedimentary structures related to streaks and large eddies.

## 6.1 Introduction

In Section 1.9 we briefly introduced turbulence and described an experiment illustrating turbulent flow. Turbulence might thereby have been painted as a thoroughly disorderly time-dependent fluid motion, marked by velocity fluctuations of a wide range of period and of all wavelengths, between a viscosity-related minimum and a maximum controlled by flow dimensions. Is turbulence just a random stochastic phenomenon, susceptible only to elaborate statistical description? The experiment does not test the whole truth. Recent research on the origins and pathways of turbulent energy reveals that turbulence is orderly and deterministic, as well as stochastic. Turbulent flows comprise definable flow configurations – eddies or vortices are less fancy terms – which are coherent, that is, they retain their character over a substantial downstream transport distance (Laufer 1975, Cantwell 1981). Moreover, these coherent structures, occurring on two main scales, depend for their character on such flow properties as boundary shear stress (describing wall conditions) and the flow thickness and mean velocity (typifying the outer flow).

It is straightforward to demonstrate experimentally for oneself the smaller of these coherent structures. A glass-sided flume is needed, together with adhesive drafting tape and either potassium permanganate crystals or crystalline methylene blue and industrial alcohol. The potassium salt should be finely crushed and made into a paint with a little distilled water (keep in a small well sealed jar to prevent oxidation). The methylene blue should be similarly made into a paint

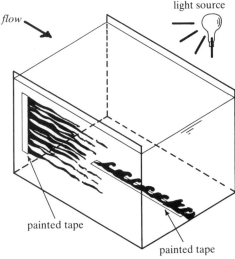

**Figure 6.1** Apparatus for the visualization of coherent structures in the turbulent boundary layer of a water channel.

with a little of the alcohol. These water-soluble paints will be used to visualize the flow structures.

For the first experiment (Fig. 6.1) attach a vertical strip of the tape to the inside wall of the flume, thickly painting the exposed side of the tape with the selected marker. Allow to dry. Arrange a diffuse light source on the far side of the flume. Now introduce a gentle water current and watch the thin sheet of dye released into the fluid immediately against the wall. Surprisingly, the dye forms wavering subparallel streamwise streaks with a characteristic centimetre-scale transverse spacing (Fig. 6.2). Bands of clear fluid separate the streaks of concentrated dye, which occasionally divide and rejoin.

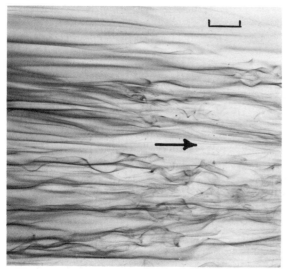

**Figure 6.2** Near-wall structure in the turbulent boundary layer, as seen in the plane of the wall. Scale bar is 0.01 m long.

velocity components, for what began as a sheet gradually changed into streamwise streaks of concentrated colour divided by zones of unmarked faster fluid which evidently had moved inwards to replace the sideways-travelling dye. From the fact that the dye lay in many streaks, the direction of action of these transverse velocity components at the bed must be alternately to left and right of the mean flow. Thus the near-bed flow comprises streamwise streaks of alternately low-speed (dyed) and high-speed (clear) fluid. The second experiment tells us that transverse rotational (vortical) motions affect the outer part of the flow, and suggests that a mechanism operates to cause slow-moving wall fluid to be either pulled outwards or thrust upwards from the bed. Could the intermittent washing away of streaks observed in the first experiment represent the start of this process?

Periodically, part of a streak is abruptly swept away from the wall, as a parcel of chaotic fluid. By slowing the current the streak spacing can be increased. An increase in flow speed has the reverse effect, decreasing their spacing and increasing the frequency with which parts of streaks are washed away.

In the second experiment, a streamwise piece of tape about 0.75 m long is fastened along the centreline of the flume bed, and the marker paint again applied and allowed to dry (Fig. 6.1). On admitting a current to the flume, the fluid becomes marked in what is effectively a vertical streamwise plane. Figure 6.3 illustrates some of the features brought to light. The bulk of the dyed fluid hugs the flume bed for a substantial distance down stream from the tape. From time to time and place to place, parcels of dyed fluid become rapidly drawn or pushed up from the bed into the outer flow. These parcels are shaped like shepherd's crooks, a voluminous top, curling over and round in the flow direction, being supported on spindly legs inclined up stream at about 45°. Even when the current is too fast for features of this sort to be distinguishable individually, one can generally still make out a 45° banding or streakiness within the coloured zone.

What do these experiments teach? The first allows us to judge the behaviour of marked fluid in the immediate vicinity of the wall, that is, in and near the viscous sublayer of Section 1.9. This fluid is certainly not turbulent, but neither is it straightforwardly laminar. We are bound to infer that the fluid is affected by transverse

## 6.2  Assessing turbulent flows – how to see and what to measure?

### 6.2.1  What to measure?

We are largely concerned here with turbulent boundary layers and channelized turbulent flows. Using the right-hand rule for the co-ordinate system (Fig. 6.4), the local velocity in such uniform and steady flows is described by

$$
\begin{array}{ll}
u = U + u' & \text{m s}^{-1} \\
v = V + v' & \text{m s}^{-1} \\
w = W + w' & \text{m s}^{-1} \\
V = 0,\ W = 0 & \text{m s}^{-1} \\
p = P + p' & \text{N m}^{-2}
\end{array}
\tag{6.1}
$$

where $u$, $v$ and $w$ are the instantaneous velocities measured in the positive $x$-, $y$- and $z$-directions, $U$, $V$ and $W$ the corresponding time-averaged components, and $u'$, $v'$ and $w'$ the fluctuating velocities representing turbulence. In the final equation, $p$ is the total fluid pressure, $P$ the mean total pressure and $p'$ the fluctuating value.

The overall scale of a turbulent flow is either its radius $r$ or thickness $h$, if occurring in respectively a circular conduit or open channel, or the boundary-layer thickness $y = \delta$, the perpendicular distance from the wall at which the local mean velocity $U$ reaches some specified fraction (say 0.99) of the external stream $U_o$ (Fig. 6.4). The velocity approximately corresponding to $U_o$ is $U_a$ at the axis of a conduit and $U_s$ at the surface

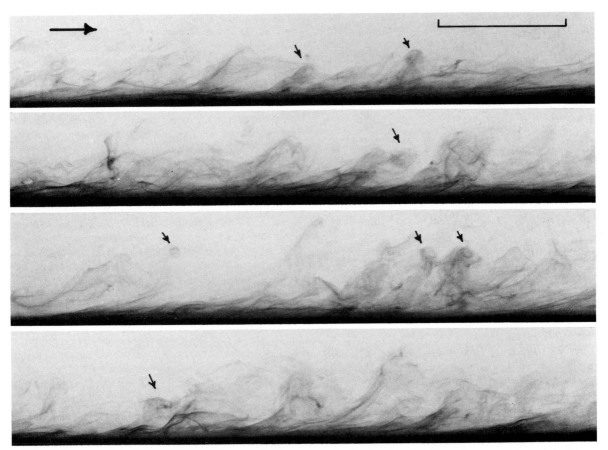

**Figure 6.3** Four different frames illustrating in a streamwise plane the near-wall structure in the turbulent boundary layer developed in a water channel. The arrowed structures (hairpin vortices) appear to have been visualized almost completely. Scale bar is 0.05 m long.

of an open channel. The turbulent velocity profile $U(y)$ is divisible into inner and outer regions. The innermost part of the inner region is the viscous sublayer, present when the flow boundary is hydraulically smooth or transitional. Here the non-dimensional velocity $U/U_\tau$ increases linearly as the non-dimensional distance from the wall $Y = \varrho y U_\tau/\eta$, where $U_\tau = (\tau/\varrho)^{1/2}$ is the shear velocity based on the boundary shear stress $\tau$, and $\varrho$ and $\eta$ are respectively the fluid density and viscosity. Its thickness is conventionally taken as $\delta_{sub} = 11.5\eta/\varrho U_\tau$. In the remaining part of the inner region, $U/U_\tau$ varies as the logarithm of $Y$, to limiting values of $Y$ (order of a few hundred to a few thousand) increasing with the Reynolds number $Re = \varrho \delta U/\eta$. The outer region, involving some 85% of the flow thickness, shows a more complex form of velocity variation conveniently scaled on $y/\delta$. But how was this and other information obtained?

### 6.2.2 Flow visualization

Flow visualization (Merzkirsch 1974) – the rendering visible of flow patterns by the introduction of some contaminant – is one of the oldest and most powerful means of studying turbulence. It underpins the experiments just described, as well as others in this book.

At the qualitative level, the most commonly used techniques involve either the dissolution into the flow of some coloured dye or the introduction of fine particles that will remain suspended. Smoke composed of fine oil droplets is widely used in air flows. The flow configurations present can be photographed in selected planes with the help of special lamps providing sheets of light. Flow patterns in water are commonly visualized using fine aluminium powder or crushed mica, both of which reflect light well. Soap powder has been successfully used for studying turbulence in thin oils. The general method becomes quantitative when the moving particles

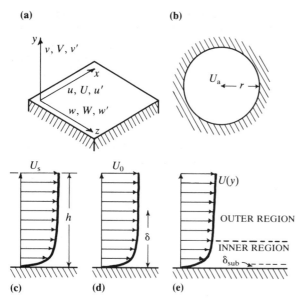

**(a)** **(b)** **(c)** **(d)** **(e)**

**Figure 6.4** Turbulent flows: (a) general co-ordinate system and notation; (b) in a pipe; (c) a free-surface flow; (d) boundary layer against a wall; (e) inner and outer zones in the boundary layer and the viscous sublayer.

are filmed in the field of view of a microscope. Analysis of the results is tedious, however, and only small areas and fluid volumes can be examined at any one time.

The hydrogen-bubble technique, limited to water, is perhaps the most powerful of the quantitative methods of visualization, providing at the same time a helpful qualitative picture of the motion. A conducting wire stretched across the flow acts as the cathode in an electrolytic system, the gas bubbles being released in discrete quantities in response to a regularly pulsed electric current. Normally, short lengths of the wire are coated with an insulator, so that a series of rectangular sheets of bubbles is released at each pulse. The wire can be given any desired orientation − normal to the bed as well as parallel to it − and the white bubble clouds are easily filmed over large areas of the flow (see Fig. 6.8). Traced from frame to frame, the clouds yield a detailed picture of both the flow pattern and its velocity structure. Related to the hydrogen-bubble technique are electrolytic methods that produce a coloured dye or titration indicator at the cathode, or an opaque colloidal solid at the anode.

### 6.2.3 Anemometry

The most important laboratory instrument for the study of turbulence is the hot-wire anemometer (Bradshaw 1971). This comprises a length (or lengths) of thin tungsten or platinum wire suitably oriented and supported on a strong probe sufficiently small as not to disturb unduly the fluid motion being measured. The technique depends on the fact that the electrical resistance of a wire conducting an electric current varies with the temperature of the wire. Such a wire loses heat when placed in a fluid stream, its temperature being correspondingly altered according to its orientation and the local flow velocity. Once the wire has been calibrated, the fluctuations in the strength of the current it passes can be measured and converted directly into velocity estimates. A single wire can only be used to measure one directional component of velocity at a time, but by using two wires arranged like an X, two velocity vectors can be estimated simultaneously, e.g. $u'_{forward}$, $u'_{backward}$, $v'_{up}$ and $v'_{down}$. This is possible because sophisticated electronic devices, combined with computer techniques, are now available for the analysis of fluctuating electrical signals. Velocity components can be added, subtracted, squared and multiplied; empirical functions that involve them can be either differentiated or integrated.

It is worth noting here that we can process either the whole of the continuous sequence of electrical signals coming from an anemometer or just some part of it. Selecting the latter, those parts of the signal to be processed may be chosen either randomly or conditionally. Conditional sampling means that we choose to process a predetermined length of the continuous signal only when a particular kind of signal, serving as a trigger, is recorded by the probe. In this way large numbers of examples of recurrent specific signal sequences, say, a steadily declining velocity component, can be assembled and treated to form ensemble averages. Conditional sampling is a valuable way of detecting and describing flow configurations in turbulent currents.

A cold-wire anemometer serves to measure the fluctuating temperature of a fluid stream into which heat is being introduced as a contaminant. If there is an organized structure, the electrical signal fluctuates with time in an orderly way as the variously warm parcels of fluid convected past the fine wire alter its resistance. Signal treatment is similar to that from a hot-wire probe.

Some investigators study turbulent flows by measuring pressure fluctuations, either at the flow boundary (wall pressure) or some station within the flow. All modern methods involve the use of a microphone, normally a piezoelectric transducer, sufficiently small in size as to respond to the shortest-lived eddies. The pressure signals can with some difficulty be interpreted in terms of velocity.

### 6.2.4 Measurements at a single point

A hot-wire probe inserted into a turbulent flow at some station will tell us about the strength and symmetry of the turbulence, and about how much flow energy is involved. Distinguishing between positive and negative fluctuating components, we can obtain true frequency distributions for $u'$, $v'$ and $w'$, and so establish the degree of isotropy of the turbulence. The quantities $\overline{u'^2}$, $\overline{v'^2}$ and $\overline{w'^2}$, formed from products, are the mean-square components of the turbulence intensity, where the bar denotes the average over time. Each of these intensities when multiplied by the fluid density becomes a normal stress − one of the so-called Reynolds stresses characteristic of a turbulent flow − in the co-ordinate system of Figure 6.4. The sum of these three components of the Reynolds stress is proportional to the turbulent energy per unit volume of the fluid. A particularly important Reynolds stress is the cross product $-\varrho u'v'$, that is, the stress necessary to shear a fluid element at the rate $\partial U/\partial y$, where $U$ is the local time-average streamwise velocity.

### 6.2.5 Spatial correlations

Simple measurements at a single point in space tell us much about the velocity structure of turbulence but little about flow patterns. Simultaneous measurement at two points, however, offers us the possibility of detecting pattern by exploring the recurrence in space of a particular velocity structure.

Suppose that we measure the fluctuating component $u'$ at a station $x_0$ along the flow and the same quantity at a place $(x_0 + x)$ further down stream. The quantity $\overline{u'(x_0).u'(x_0 + x)}$, called the covariance or, somewhat inaccurately, the correlation, measures the similarity of the values of $u'$ at the two stations. By varying the probe separation, we can find the value of $x$ for which the covariance is a maximum, that is, when the stream-wise velocity at the two stations is the same at the same time. This $x$-value measures the streamwise scale of the eddies present. By repeating the process parallel to the $y$- and $z$-axes, the flow configurations can be characterized three dimensionally.

### 6.2.6 Time correlations

Reliable values for the streamwise covariance can be difficult to obtain because of interference from the upstream probe on the downstream one. It is more convenient then to measure the autocovariance, that is, the product of the streamwise fluctuating velocities measured at the same point but two different times. Formally, the autocovariance is $\overline{u'(t_0).u'(t_0 + t)}$, where $t_0$ is the first time at which the velocity is measured and

$(t_0 + t)$ the later time. Notice that the time interval $t$ is equivalent to the streamwise separation $x = -Ut$, where $U$ is the local time-average velocity, making the techniques of spatial and time correlation interchangeable. Autocovariances cannot of course be measured other than parallel to the flow.

### 6.2.7 Spectral analysis

The probe signal is a continuous record in time of the fluctuating properties of the turbulent stream. These time series can be analysed by Fourier methods to establish the distribution of frequencies on which the time signal is varying. This distribution is the frequency spectrum, a description of the distribution of time scales representative of the turbulence. There is a mathematical relationship called a transform between the frequency spectrum and the autocovariance. As distance and time are convertible through $x = -Ut$, there should also be a transform between the covariance and some other kind of spectrum. It turns out that the covariance is equivalent to the distribution of spatial scales representative of the turbulence, that is, to the spectrum of eddy wavelengths.

## 6.3 Character of an ideal eddy

The above claim that turbulent flows consist of orderly, coherent eddies is based on observation but, because turbulence is such a complicated phenomenon, and the means of studying it are so restrictive (to points and planes), there remains considerable uncertainty about the real shape and dynamics of these configurations. The problem resembles that faced by the six blind wise men trying to describe an elephant; it is impossible at present to see an eddy whole, and what you can say of it depends on how and where you encounter the eddy. Can an ideal turbulent eddy be conceived from general principles? Such a concept might be helpful when interpreting the rather restrictive visualizations and probe measurements available.

In a turbulent boundary-layer flow there is a profile of velocity increasing away from the solid boundary, such that momentum is continuously being transferred from the outer flow towards the boundary. Suppose as part of this transfer mechanism that low-momentum fluid at the wall is periodically gathered into blobs, which are then thrust or pulled upwards into the faster outer current.

Figure 6.5a shows how, to an observer riding with it, the blob might appear in streamwise view at successive instants. Because the upper parts of the blob project

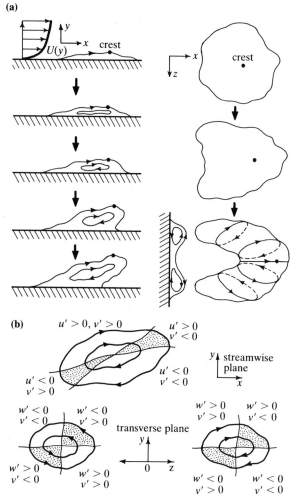

**Figure 6.5** Model of an ideal turbulent eddy in the wall region of a turbulent boundary layer, to show (a) the internal flow and changing shape of the eddy with time in the streamwise and wall planes, and (b) its instantaneous velocity structure in different flow planes.

stream, up and over the stagnating fluid to leeward (Fig. 6.5a). The flanks of the blob are also carried forwards and around the outside of this sluggish mass, but less rapidly, so that they become stretched out to form limbs. The blob late in its evolution could resemble a long-sided horseshoe; the internal circulation is in the same sense as the rotation implied by the gradient of mean velocity $\partial U/\partial y$. Presumably the limbs remain attached to the boundary, but low-momentum fluid may be drawn between them upwards into the blob.

What velocity and wall-pressure signals would be measured in the path of our eddy? The fluid in the leading portion of the vortex has a downward motion, so the total pressure here will exceed the long-term average. The fluid in its rear, however, has an upward component, causing a pressure drop when this part of the eddy passes by. The transit of the eddy should therefore be marked by a distinctive pressure signal: a rise followed by a fall. The period of this signal will increase with the size of the vortex, but decrease with its convection velocity. To an observer, the velocity structure varies with sampling position within the vortex (Fig. 6.5b). First consider the velocity components in the streamwise plane, that is, $u'$ and $v'$. There are four possible instantaneous combinations of $u'$ and $v'$ that might be observed, two with components having the same sign (product positive), and two where the signs are opposite (negative product). The time sequence in which these combinations are detected by a stationary probe clearly depends on the relative position of the measuring station. At a probe swept by the upper part of the vortex the sequence $(u' > 0, v' < 0)$ and then $(u' > 0, v' > 0)$ might be observed, whereas $(u' < 0, v' < 0)$ followed by $(u' < 0, v' > 0)$ is to be expected for a range of low positions. More combinations are possible in the flow-normal plane, depending not only on the relative position of the probe but also on whether, to an observer looking down stream, the sample of $v'$ and $w'$ comes from the clockwise-rotating or anticlockwise-circulating vortex limb (Fig. 6.5b).

Our ideal eddy is with little doubt oversimplified. This model nonetheless suggests how the shape, circulation and dynamics of real deterministic eddies might be reconstructed from the limited clues collectable at one or a few points within the flow (e.g. Blackwelder & Eckelmann 1979).

## 6.4 Streaks in the viscous sublayer

### 6.4.1 Fluid motion

The innermost region of the turbulent boundary layer on a smooth or transitional wall − the viscous sublayer

furthest into the flow, they travel faster than portions nearer the bed, and so gradually overtake this slower fluid. The blob consequently becomes wedge-shaped, eventually developing a tall overhanging downstream front. The process is helped by the fact that the blob, as it grows in height, increasingly shelters the already sluggish fluid in its lee, which consequently slows even more. As the blob is being sheared, the observer would see that a circulation, clockwise in the diagram, was developed internally.

But the blob has a third dimension. In plan, and again with the observer riding with the blob, it is the central portion that is uppermost and so moves furthest down

– used to be called the laminar sublayer, until it was found that vortical structures were present making the flow not laminar in a simple sense. Many studies have now been made of the streaky structure in the wall region of turbulent boundary layers. It is beyond reasonable doubt that low-speed and high-speed streaks similar to those in Figure 6.2 are a coherent structure to be found in all turbulent flows past smooth and transitional surfaces.

The streaks have been studied in many ways. Some workers analysed wall-pressure variations, and others the covariance and other aspects of the fluctuating velocity. The most telling work is based on flow visualization. Kline *et al.* (1967), in a seminal study, used dye injected as threads and sheets into the flow, and also the hydrogen-bubble method. Grass (1971) partly relied on using sand as a tracer. His grains became concentrated into streamwise bands (Fig. 6.6) corresponding to the low-speed streaks into which the

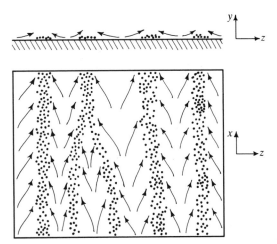

**Figure 6.7** Suggested fluid forces and motion associated with boundary-layer streaks as visualized by Grass (1971).

dye became concentrated (Fig. 6.2). What can be inferred from the fact that the grains are clustered and not randomly dispersed over the bed? As the only horizontal forces acting on them arise from the fluid, their concentration into streamwise bands can only be due to transverse components of the fluid force. But the bands vary little in spanwise position, whence over any substantial area these transverse forces must at any instant be roughly in balance (Fig. 6.7). Striking confirmation is provided by hydrogen-bubble studies, in which lines of bubbles are released from a straight transverse wire arranged close to the bed (Nakagawa & Nezu 1981, Smith & Metzler 1983). The lines and bubbles shown in Figure 6.8 have locally advanced different downstream distances according as they were released into low-speed or high-speed streaks. Simultaneously, the bubbles shed into high-speed streaks became thinned out and transported sideways into the low-speed ones, where they became concentrated in a similar manner to Grass's sand.

The streaks are known to be deterministic in character, taking a size and shape dependent on flow properties. Their transverse spacing can be measured from flow visualizations, and turns out to be a constant when scaled on the variables describing conditions at the flow boundary, namely,

$$\bar{Z}_s = \frac{\varrho \bar{\lambda}_{z,s} U_\tau}{\eta} \simeq 100 \qquad (6.2)$$

in which $\bar{Z}_s$ is the non-dimensional average streak spacing, $\bar{\lambda}_{z,s}$ the measured average transverse spacing, and $U_\tau$, $\varrho$ and $\eta$ are respectively the shear velocity, fluid

**Figure 6.6** Streaky structure in the near-wall region of the turbulent boundary layer in a water channel, as revealed by sand grains carried over the bed. Scale bar is 0.01 m. From Grass (1971) *J. Fluid Mech,* **50**, 233–55, with permission of Cambridge University Press. Photograph courtesy of Dr A. J. Grass, University College London.

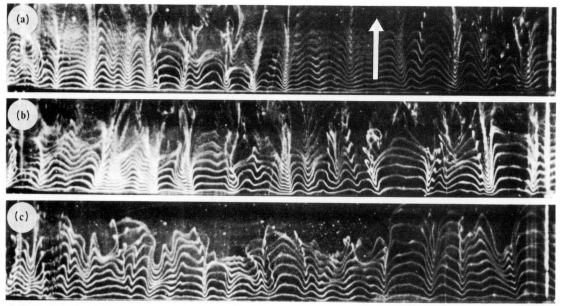

**Figure 6.8** Boundary-layer streaks produced in turbulent flow in a water channel as visualised by hydrogen-bubble time lines released into a wall-parallel plane (a) half-way out in the viscous sublayer, (b) at the edge of the viscous sublayer, and (c) in the inner flow one-and-a-half sublayer-thicknesses outwards. Each photograph represents a flow width of 0.3 m. Photographs courtesy of Dr I. Nezu, Kyoto University.

density and fluid viscosity (Fig. 6.9). For each flow condition, the range of streak spacings is considerable and varies little with distance from the bed (Smith & Metzler 1983). The average spacing increases noticeably with outward distance, however, while in harmony the spacing frequency distribution becomes increasingly symmetrical. In the sublayer, large numbers of closely spaced streaks occur, whereas a larger proportion of widely separated streaks is present low down in the logarithmic region. Smith and Metzler (1983) found it visually difficult to detect streaks at distances from the boundary exceeding about $Y = 40$. The frequently reported variability in transverse spacing is related to the behaviour of the streaks, which as noted from our experiments are seen to waver, divide and coalesce. Smith and Metzler (1983) also explored this aspect of the streaks, by measuring the lengths of time (persistence) at which low-momentum fluid could be detected at a fixed point in the near-bed flow. The streaks proved to be as variable in persistence as in spanwise spacing, but the largest persistences were equivalent to downstream distances of two to three boundary-layer thicknesses, had specific fluid masses been tracked.

What flow pattern do the streaks represent? Many workers consider that they record contra-rotating

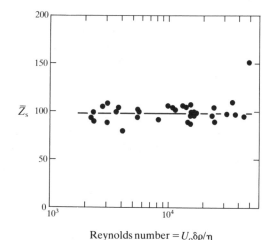

Reynolds number $= U_o \delta \rho / \eta$

**Figure 6.9** Non-dimensional transverse streak spacing as an experimental function of Reynolds number. Data of Smith and Metzler (1983).

streamwise vortices in the viscous and lower logarithmic regions (Richardson & Beatty 1959, Bakewell & Lumley 1967, Blackwelder & Eckelmann 1979). At their simplest (Fig. 6.10a), such vortices are uniform in size and regularly spaced across the flow. In order to explain observed streaks, however, the vortices must undergo

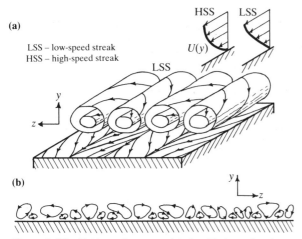

**Figure 6.10** Fluid motion associated with boundary-layer streaks in (a) the idealized case and (b) the real case (speculative). The tallest vortices in (b) are about eight sublayer-thicknesses high.

**Figure 6.11** Parting lineation formed by wave swash on a beach, Norfolk, England. Trowel is 0.28 m long and points up beach.

**Figure 6.12** Parting lineations (also parting-step lineations) in a laminated sandstone, Old Red Sandstone, Forest of Dean, England. Scale bar is 0.05 m. Inset shows idealized grain long-axis fabric (plane of bedding) associated with parting lineations.

time-dependent changes in all dimensions. Figure 6.10b is an instantaneous transverse section through a possible series of vortices, and suggests an explanation for the observed increase (1) in mean spacing away from the boundary, and (2) in the symmetry of the spacing distribution with distance from the bed. One important implication of the time-dependent behaviour of the vortices is that they are constantly exchanging fluid amongst themselves, so that, at any instant, some are growing in volume while other are contracting.

### 6.4.2 Implications for deformable boundaries

How will a deformable sand or mud surface respond to boundary-layer streaks? The streaky structure should be expressed in terms of streamwise features related to differential erosion and/or deposition, depending on general regime, for the boundary shear stress beneath the low-speed streaks is less than below the high-speed ones where the profile of mean velocity is steeper (Fig. 6.10a). But as the streaks are in a state of continuous motion and change, whereas the engendered sedimentary structures may be stationary for either a part or all of the time, the transverse spacing of the bed features may not precisely equal that of the parent streaks.

Parting lineation is the sand-bed structure most likely to be created by boundary-layer streaks, and is widely recorded from parallel-laminated sands and sandstones of very fine to medium grade (Allen 1964). Such deposits form on upper-stage plane beds (Sec. 4.8) under wave (Fig. 6.11) and fluvial (Fig. 6.12) conditions and, together with the lineations, are known

experimentally (Allen 1964, Mantz 1978). The lineations occur on the tops of millimetre-scale laminae as streamwise ridges and hollows measuring a few grain diameters in height, a few millimetres to a centimetre or so transversely, and up to several decimetres in length. In the ridges are concentrated the coarser grains. The particles forming a lamina are aligned roughly parallel with the lineations on its surface. In detail, their long axes are distributed in two modes, one on each side of the lineation trend (Fig. 6.12), recalling the 'herring-bone' pattern of bottom currents and forces inferred to explain Grass's (1971) streamwise visualizations (Fig. 6.7), as well as the proposals summarized in Figure 6.10.

Does the spacing of parting lineations conform to Equation 6.2, remembering that the formula describes an average? To see how $\bar{\lambda}_{z,s}$ varies with flow conditions we recall that $U_\tau = (\tau/\varrho)^{1/2}$, and that by the quadratic stress law (Eqn 1.27)

$$\tau = \frac{f}{8} \varrho U_m^2 \quad \text{N m}^{-2} \tag{6.3}$$

where $f$ is the Darcy–Weisbach friction coefficient, and $U_m$ the mean flow velocity. Expressing $U_\tau$ in terms of $f$ and $U_m$, and substituting into Equation 6.2, the streak spacing becomes

$$\bar{\lambda}_{z,s} \simeq 100 \frac{\eta}{\varrho} \left( \frac{8}{fU_m^2} \right)^{1/2} \quad \text{m} \tag{6.4}$$

varying in a given fluid inversely as the mean flow velocity and the square root of the bed friction coefficient. Figure 6.13 shows the average transverse spacing for friction coefficients between 0.01 and 0.05, approximately the range for upper-stage plane beds in sands of medium grade and finer (Guy et al. 1966). The spacings

are of the order of 0.001–0.01 m, which is roughly what is observed of parting lineations. We saw in Section 1.8 that upper-stage plane sand beds exist only when

$$\frac{\tau}{(\sigma - \varrho)gD} \geqslant \theta_{cr} \tag{6.5}$$

where $\theta_{cr}$ is the non-dimensional boundary shear stress, $\sigma$ the solids density, $g$ the acceleration of gravity, and $D$ the sediment diameter. Eliminating the stress between Equations 6.3 and 6.5 and rearranging, lineations on upper-stage plane sand beds can occur only for

$$U_m \geqslant \left( \frac{8\theta_{cr}(\sigma - \varrho)g}{\varrho} \right)^{1/2} \left( \frac{D}{f} \right)^{1/2} \quad \text{m s}^{-1} \tag{6.6}$$

where $\theta_{cr} \simeq 0.58$. Using this inequality, thresholds corresponding to constant $D$ have been plotted in Figure

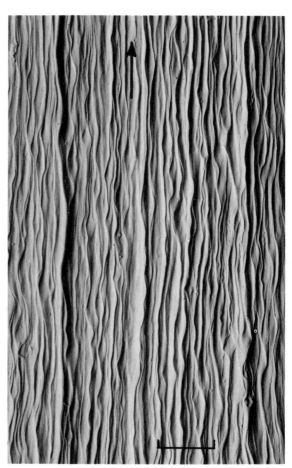

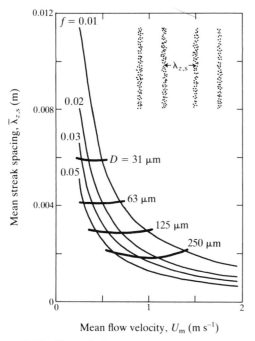

**Figure 6.13** Theoretical average transverse spacing of parting lineations as a function of mean flow velocity and grain diameter.

**Figure 6.14** Mould of streamwise structures formed during the erosion of a mud bed by a turbulent water stream. Mean velocity is 0.38 m s$^{-1}$. Scale bar is 0.05 m long.

6.13, showing that observed lineation spacings exceed by a factor of 2 or 3 the permissible spacings. It may be that only the larger streaks leave a discernible lineation, perhaps because they are the most powerful on account of their greater outward range.

Streamwise bed features resembling boundary-layer streaks in transverse spacing arise experimentally during mud erosion (Allen 1969) and are known to occur beneath turbidity-current sandstones. The grooves have V-shaped profiles, and meander, branch and rejoin in a complicated way (Fig. 6.14). Their plan appearance is remarkably like an aluminium-powder visualization of boundary-layer streaks (Cantwell *et al.* 1978). The mud grooves are of course stationary, whereas the streaks move beneath the flow. If the grooves do record the activities of streaks, it is puzzling how the moving flow pattern becomes imprinted in stationary form on the bed.

## 6.5 Streak bursting

The way in which boundary-layer streaks waver, divide and rejoin suggests that the near-wall longitudinal vortices are metastable under the influence of the fast outer flow. Kline *et al.* (1967) noticed during their experiments that low-speed streaks were locally and intermittently subject to a cyclic process described as bursting. Subse-

quent visualization (Corino & Brodkey 1969, Grass 1971, Kim *et al.* 1971, Head & Bandyopadhyay 1981) and hot-wire (Kim *et al.* 1971, Wallace *et al.* 1972, Brodkey *et al.* 1974) studies have given us a fairly clear idea of what happens during the bursting cycle. Essentially, it involves the creation of a horseshoe or hairpin vortex, which moves outwards and undergoes a partial breakdown into turbulence. The vortex has transverse dimensions similar to the associated boundary-layer streaks, and seems to retain these general dimensions throughout its life.

Figure 6.15 summarizes the cycle as it would be seen by an observer travelling with a vortex. The diagram shows the succession of flow structures within the cycle, and also the changing pattern of the instantaneous velocity profile $u(y, t)$ in the axial plane of the flow structure as compared to the profile of mean velocity $U(y)$. Vertical distances appear as before in terms of $Y = \varrho y U_\tau / \eta$ and spanwise distances in the corresponding non-dimensional form $Z = \varrho z U_\tau / \eta$. Notice that the vortex develops at the site of a low-speed streak, its limbs straddling the streak.

Because of the rotation implied by the streamwise velocity gradient, the vortex may be imagined to start life as a transverse cylindrical structure lying in the viscous sublayer and lower logarithmic region. As the vortex travels down stream, it begins to bend forwards and move upwards above the streak, which decelerates

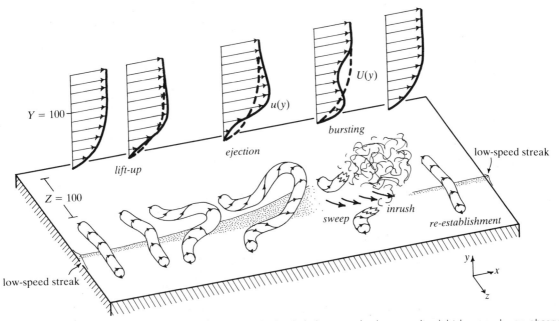

**Figure 6.15** Model of the streak-bursting cycle. The developing hairpin vortex is shown as it might be seen by an observer travelling with it. The velocity profiles on the left are plotted with respect to the ground.

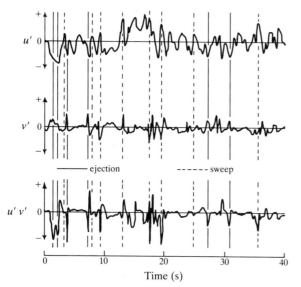

**Figure 6.16** Simultaneous time sequences of $u'$, $v'$ and $u'v'$ values measured near the wall in a turbulent boundary layer, to show the signals characteristic of ejections and sweeps. After Wallace *et al.* (1972).

and commences to rise. This is the stage called lift-up, the instantaneous velocity profile $u(y, t)$ differing hardly at all from the mean profile.

The continued bending and upward movement of the vortex creates a horseshoe- or hairpin-shaped structure and leads to the ejection stage. The instantaneous profile of velocity shows an increasing and eventually marked inflection, the fluctuating quantity measured close to the bed consisting of $u' < 0$ and $v' > 0$, giving a negative product $u'v'$ contributing to the Reynolds stresses. Figure 6.16 (Wallace *et al.* 1972) shows several ejections. The development of the vortex to the point of ejection is marked visually in ways depending on the relationship of the vortex to the plane in which it may have been visualized. Figure 6.17a, a hydrogen-bubble visualization in the *xy*-plane, shows a corkscrew motion suggestive of one of the limbs. Figures 6.17b and c illustrate the circulation within the vortex head, as seen in or near the plane of symmetry (see also Fig. 6.3).

Two actions occurring almost simultaneously but in different parts of the flow, are involved in the final stage of bursting. The sweep and inrush see the downward movement of a parcel of high-speed fluid and its subsequent spread down stream and sideways over the bed. The velocity signal measured near the boundary (Fig. 6.16) now consists of $u' > 0$ and $v' < 0$, continuing to give a negative product contributing to the Reynolds stresses. Grass (1971) illustrated a sweep at the site of a

low-speed streak (Fig. 6.18). What is not yet clear is whether the sweep demolishes the lower limbs of the vortex, or if high-speed fluid is simply thrust forwards between them. Certainly the upper part of the vortex remains coherent far out into the flow, as may be seen from Head and Bandyopadhyay's (1981) remarkable smoke visualizations (Fig. 6.19). These also show that the vortex becomes longer-limbed with increasing Reynolds number, changing from a horseshoe to a hairpin form. More or less simultaneously with the sweep and inrush, the head of the vortex 'bursts', that is, the motion within it becomes disordered and rapidly fluctuating (Fig. 6.17d). The Reynolds stress $-\varrho u'v'$, however, is contributed very largely by the ejections and sweeps, making these events the main creators of turbulence. Ejections seem to grow in importance relative to sweeps with increasing boundary roughness (Grass 1971, Raupach 1981).

As the sweeps and ejections are the main creators of turbulence, it is important to know the time interval separating bursts, that is, the local mean burst period $\bar{t}_b$. Most workers who have measured this quantity agree with Rao *et al.* (1971) that the period is a constant when scaled on outer flow variables. For a boundary layer flow

$$\bar{T}_b = \frac{\bar{t}_b U_o}{\delta} \simeq 5 \qquad (6.7)$$

where $\bar{t}_b$ is the measured mean burst period, $U_o$ the flow velocity outside the boundary layer, and $\delta$ the boundary-layer thickness ($U = 0.99U_o$). A variety of laboratory measurements assembled partly by Rao *et al.* (1971) appear in Figure 6.20, but note the limited range of Reynolds numbers represented. Equation 6.7 becomes applicable to free-surface flows when we substitute the surface velocity $U_s$ for $U_o$ and the flow depth $h$ for $\delta$. Figure 6.20 also shows some data for tidal currents (e.g. Heathershaw 1979), in fair agreement with the laboratory measurements, and at substantially larger Reynolds numbers. In their open-channel study, Nakagawa and Nezu (1981) found that $\bar{T}_b$ ranged between approximately 1.5 and 3.0. These values are low compared to most laboratory data, but are perhaps due to differences of flow conditions and evaluation technique.

How might bursting streaks affect deformable beds? Aside from the advection of bedload grains into the outer flow – an important process of suspension transport (Jackson 1976, Sumer & Deigaard 1981) – bursting streaks may influence the bed primarily through the inrush and sweep events. The fluid involved

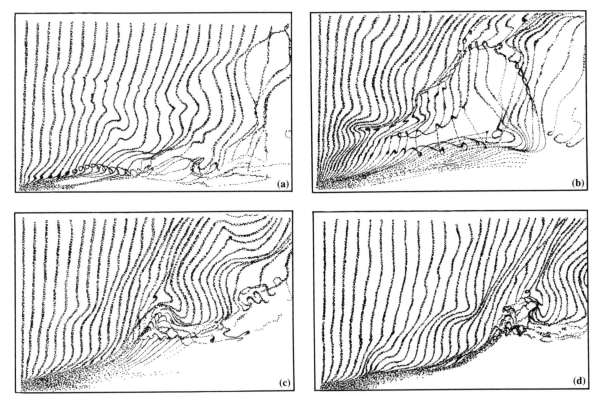

**Figure 6.17** Hydrogen-bubble time lines (flow from left) in the streamwise plane to illustrate features of bursting streaks. (a) A near-bed streamwise vortex. (b) Early development of inflected velocity profile as vortex lifts away from bed. (c) Inflected velocity profile in advanced stage of lift-up of vortex from bed. (d) Break-up of vortex (zone of confused and tangled bubble lines). Drawn from photographs by Kim *et al.* (1971).

in an inrush and sweep exerts a comparatively large boundary shear stress. As Figure 6.18 suggests, inrushes and sweeps may therefore modify or even obliterate any structures shaped by the apparently gentler streaks.

## 6.6 Large eddies (macroturbulence)

### 6.6.1 Fluid motion

Turbulent boundary-layer and channelized flows contain many eddies similar in scale to the flow itself. To be convinced of this, you have only to examine the surface of a flooded river, either from the bank or, better still, from the vantage point of a bridge. The water surface is in a state of constant agitation, under the influence of a sequence of large disturbances apparently thrust up from below (Fig. 6.21), called 'kolks' or 'boils' by river engineers (Coleman 1969, Jackson 1976). Each boil lasts a definite period, in a large river up to several tens of seconds, moving down stream all the while. Typically, a boil starts life as

a raised circular to oval patch on the water surface, which in the course of time widens and becomes lower, until eventually it subsides completely and merges with the surrounding less agitated parts. In a larger river, a boil may grow to several metres across before fading away. All is motion within the growing boil. There is a general sense of a radially outward movement of water, but superimposed on this are many local motions in a variety of directions, the surface being disturbed by cauliflower-head eddies resembling except for size the patch itself. The margin of the boil is a sharp convergence, marked by wavelets and commonly by an accumulation of foam or other debris. Thus the motion within a boil is just as if one of our ideal horseshoe eddies (Fig. 6.5a) had reached up to touch the water surface for a period.

Flow visualization provides the most accessible evidence for large eddies in laboratory-scale turbulent flows. Nychas *et al.* (1973) obtained some visual indications of their presence, but it was Falco (1977) who first revealed large eddies. Falco released oil-droplet

**Figure 6.18** A sweep (down stream of arrow) affecting a largely obliterated low-speed streak, as visualized by Grass (1971) using sand grains. Scale bar is 0.01 m. From Grass (1971) *J. Fluid Mech.* **50**, 233–55, with permission of Cambridge University Press. Photograph courtesy of Dr A. J. Grass, University College London.

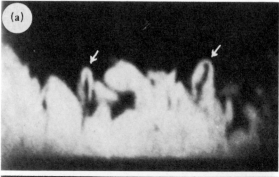

**Figure 6.19** Ejection-related vortices (arrowed) in a smoke-filled turbulent boundary layer, as visualized in a transverse plane tilted down at 45° in the upstream direction: (a) moderate Reynolds number; (b) high Reynolds number. Vortex width is virtually identical in each case with the average spacing of the streaks developed on the bed. From Head and Bandyopadhyay (1981) *J. Fluid Mech.* **107**, 297–338, with permission of Cambridge University Press. Photographs courtesy of Dr P. R. Bandyopadhyay, Systems and Applied Sciences Corporation.

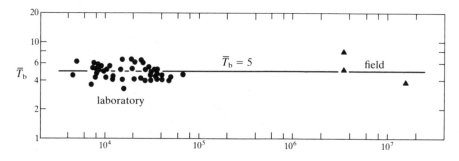

Reynolds number $= U_o \delta \rho / \eta$

**Figure 6.20** Non-dimensional local mean burst period as a function of Reynolds number at the laboratory and field scales. Data of Rao *et al.* (1971) with additional laboratory observations and three results from tidal flows.

116

**Figure 6.21** Boils on the surface of the flood tidal current, Severn Estuary. Flow from right to left at 1–2 m s$^{-1}$, with larger boils reaching 2 m across.

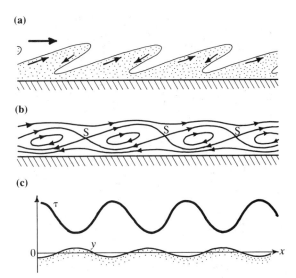

**Figure 6.23** Idealized features of larger coherent structures present in the turbulent boundary layer. (a) Bulges in the outer margin of the layer. (b) Streamlines associated with the bulges, as seen by an observer travelling with the bulges. (c) Distribution of mean shear stress beneath the bulges in (a) and (b) and the expected form of a deformable granular bed adjusted to the flow.

smoke into a boundary layer formed in a wind tunnel and illuminated using a sheet of light in the streamwise plane (Fig. 6.22). He noticed that the boundary between the smoky and unmarked air became folded into flat quasi-periodic bulges, separated by narrow upstream-sloping zones containing clear air drawn down almost to the bed (Fig. 6.23a). Simultaneous hot-wire measurements showed that the fluid within the bulges was more turbulent than that above and between them. The spacing of the bulges was rather larger than the boundary-layer thickness, and their convection velocity was similar to the mean flow velocity. Turbulence could therefore be observed only intermittently near the outer edge of the boundary layer, but at a characteristic mean period. Visual studies by Fiedler and Head (1966), and hot-wire work by Kovasznay *et al.* (1970) and Antonia (1972), also demonstrated the uneven thickness of turbulent boundary layers and the intermittency of turbulence in their outer reaches.

The zones between the bulges in Figures 6.22 and 6.23a are shear layers. Observed at a fixed station, these layers show up as sharp quasi-periodic 'velocity steps' in a continuous hot-wire signal (Blackwelder & Kaplan

1976, Thomas & Bull 1983). Similar 'temperature steps' become detectable on introducing heat as a contaminant into a boundary layer, run beneath a roof so as to preserve the gravitational stability of its density profile (Chen & Blackwelder 1978, Antonia *et al.* 1982). Figure 6.24 shows the simultaneous temperature records obtained by Chen and Blackwelder at five positions vertically above each other within the lowermost two-thirds of a heated boundary layer. As each bulge passes at the level of a sensor, the warm fluid carried from the boundary downwards along its back is quickly replaced across the shear zone by the cooler fluid being returned along the front of the following eddy. Notice how in Figure 6.24 the temperature steps appear at successively later times as we go progressively deeper into the boundary layer. The locus of the steps therefore slopes up stream, like the smokeless zones between visualized

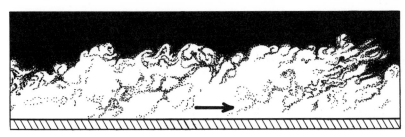

**Figure 6.22** Large-scale flow structures (note evidence of smaller-scale horseshoe or hairpin vortices) in a turbulent boundary layer filled with smoke (white). Redrawn from a photograph by Falco (1977).

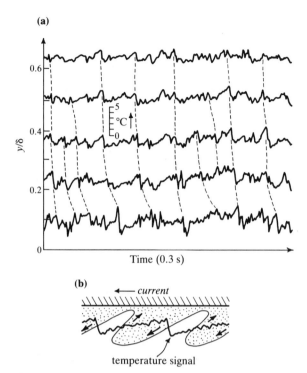

(a)

$y/\delta$

°C

Time (0.3 s)

(b) ← current

temperature signal

**Figure 6.24** Sequence of temperature values obtained at five levels within a turbulent boundary layer formed beneath a heated wall. The tie-lines between the different records correlate individual coherent structures. After Chen and Blackwelder (1978).

bulges (Fig. 6.22). Further evidence for large eddies in boundary layers comes from wall-pressure spectra (Bull 1967). As well as rapid fluctuations, consistent with boundary-layer streaks, there are slower variations on a timescale similar to bulges detectable visually.

These distinctive velocity, temperature and pressure patterns can tell us (1) the convection velocity and timescale characteristic of the bulges, (2) the slope of the shear zones between bulges, and (3) the transverse scale and downstream coherence of the eddies. The bulge convection velocity $U_c$ varies within broad limits but was found by Thomas and Bull (1983) to average $0.67U_o$, where $U_o$ is the velocity of the undisturbed stream outside the boundary layer. As the mean velocity $U_m$ of a channelized turbulent flow of depth $h$ is approximately $0.85U_s$, where $U_s = U_o$ is the velocity at the free surface, we obtain the corresponding convection velocity $U_c \simeq 0.79U_m$. Experimentally, the average time between the passage of bulges at a fixed station is the same as the mean burst period discussed in the preceding section. Non-dimensional bulge periods range

from 2.6, in the upper part of Nakagawa and Nezu's (1981) range for bursts, up to very nearly the Rao *et al.* (1971) figure of approximately 5 (Chen & Blackwelder 1978, Antonia *et al.* 1982, Thomas & Bull 1983). Jackson (1976) measured the mean non-dimensional period of boils in a large river as 7.6, the relatively high value possibly reflecting the fact that some large eddies failed to reach the surface. Multiplying the bulge period by the convection velocity suggests that the eddies range in streamwise length between approximately $1.6h$ and $3.35h$, according as the smaller or larger non-dimensional period is used. The transverse scale of bulges is poorly known, but the data of Brown and Thomas (1977) and Thomas and Bull (1983) suggest a greatest width of $(0.5-1.6)h$. A bulge can probably retain its identity over a downstream distance of several times its own length (Willmarth & Wooldridge 1962, Bull 1967). The bulges slope upward at about $20°$ from the horizontal.

Brown and Thomas (1977) and Thomas and Bull (1983) thought the bulges were large horseshoe vortices (see Fig. 6.5). Together with Falco (1977), they concluded that the structures could be represented as 'cat's eye' eddies when viewed at the convection velocity in the plane of the flow (Fig. 6.23b). Each shear zone contains a saddle-point marked S, and it will be noticed that the near-bed flow is compressed at or just up stream of its foot, suggesting that here is a local maximum of bed shear stress.

A strong periodic fluctuation of bed shear stress is in fact detectable as the bulges pass a fixed station (Thomas & Bull 1983), such that

$$\tau = \bar{\tau} + \tau_{sv}(t) + \tau' \quad N\,m^{-2} \qquad (6.8)$$

where $\tau$ is the instantaneous value of the stress, $\bar{\tau}$ the long-term mean, $\tau_{sv}(t)$ the slowly varying part changing with time $t$ on the same scale as the bulges, and $\tau'$ the high-frequency fluctuating component. The stress is a maximum when the foot of one of the shear zones lies at or near the station, and a minimum when the expanded flow between and below two cat's eyes is above. Ignoring the high-frequency component, Figures 6.23c shows how the stress $\tau = (\bar{\tau} + \tau_{sv})$ varies as a function of downstream distance beneath a sequence of bulges. Thomas and Bull found that the amplitude $a_{sv}$ of the slowly varying part roughly equalled the root-mean-square value of the total stress. As the root-mean-square stress is approximately 0.4 times the long-term mean in a turbulent boundary layer (Bridge 1981), we can write $a_{sv} \simeq 0.4\bar{\tau}$.

### 6.6.2 Implications for deformable boundaries

Could these large eddies affect a sand bed so as to form some kind of bedding? In answering this question, note first their large (but not too large) temporal and spatial scales: the dimensions are of the order of the flow depth, and their period in river and tidal currents measures from a few to many tens of seconds, depending on flow speed and depth. Secondly, if the laboratory results on bed shear stress (Brown & Thomas 1977, Thomas & Bull 1983) are applicable at the field scale, the passage of each eddy must see an almost two-fold variation in the local bed stress.

Given these properties, it seems inevitable that the local bedload transport rate will respond to the convection of the eddies. Bedload grains saltating in water rise no higher than 2–3 diameters above the static bed. If flow strength were reduced, such grains should sink one diameter closer to the bed in a time of order $D/V$, where $D$ is the particle diameter and $V$ the terminal fall velocity. As $D$ for sand is even smaller than $V$, this characteristic time is very small compared to the eddy scale and corresponding transport rate fluctuations. Hence there should be no significant delay between the bedload transport rate and the eddy-induced stress variations. Now if the transport rate varies with the passage of large eddies, promoting alternate erosion and deposition, some kind of bed wave must consequently be present. As the eddies travel at something like the mean flow velocity, the postulated bed waves cannot be either current ripples or dunes, which advance slowly compared to the current. That leaves only plane beds as the bedform with which eddy-related bed waves might be associated.

Consider in Figure 6.25a how the bed shear stress changes at a point on a plane sand bed during the passage of an eddy within a steady uniform two-dimensional turbulent aqueous flow. The stress is a maximum when the foot of the shear zone reaches the point, and a minimum when the expanded flow between two cat's eyes in Figure 6.23b lies above. The bedload transport rate may be written, neglecting the threshold stress as small,

$$J(t) = k[\bar{\tau} + \tau_{sv}(t)]^{3/2} \quad \text{kg m}^{-1}\text{s}^{-1} \qquad (6.9)$$

where $J(t)$, with $t$ as time, is a periodic function describing the dry-mass rate, $k$ a dimensional transport coefficient to be determined empirically, and the stress components are as in Equation 6.8. This formula is compatible with earlier proposals (Ch. 4), for as the square root of the shear stress is proportional to the mean flow velocity, the term under the exponent in

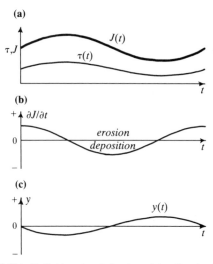

**Figure 6.25** Definition sketch for the origin of laminations on an upper stage plane bed due to the passage of macroturbulence.

Equation 6.9. corresponds to the stream power. Evidently, the transport rate varies on the same period and phase as the stress, as sketched in figure 6.25a.

But as the transport rate is changing, by Equation 4.19 the bed must experience alternate deposition and erosion, and so vary in level. Flow width makes no contribution to changes of bed elevation, because we have already assumed the flow to be two-dimensional. A contribution comes from the unsteady transport (Eqn 6.9) and also from its implied non-uniformity, for time and streamwise distance are interchangeable through $x = -tU_c$. The non-uniformity of the transport rate can be safely ignored, however, as the eddies are very large compared to the fluctuations of bed level permitted by the stress variation. The reduced formula (Eqn 4.19) giving us the rate of erosion or deposition on the bed is therefore

$$R = -\frac{1}{U_s}\frac{\partial J}{\partial t} \quad \text{kg m}^{-2}\text{s}^{-1} \qquad (6.10)$$

where $R$ is the transfer rate between bed and flow (positive for deposition), and $U_s$ is now the mean grain transport velocity, equal to $U_m/9$ in a deep turbulent current (Bagnold 1966). Figure 6.25b shows how $\partial J/\partial t$ varies with time. Multiplying Equation 6.10 by the reciprocal of the sediment dry bulk density $\gamma$, the time rate of change of bed elevation $y$ beneath the eddy is

$$\frac{dy}{dt} = -\frac{1}{\gamma U_s}\frac{\partial J}{\partial t} \quad \text{m s}^{-1} \qquad (6.11)$$

The bed elevation itself follows an appropriate integration of this equation (Fig. 6.25c). Notice the changes in phase between $J(t)$, $(\partial J/\partial t)(t)$, and $y(t)$ in Fig. 6.25. The bed elevation is a minimum when the bed shear stress is a maximum and vice versa.

What integration of Equation 6.11 proves is that the elevation of a plane sand bed at a fixed station varies in harmony with the convection of large eddies over the bed. The bed is only nominally plane, and a series of bed waves, as well as of eddies, pass over it. The streamwise profile of the waves follows by substituting distance for time in Equation 6.9, as sketched in Figure 6.23c. The bed waves are forced by the eddies, and share their time and length scales. The bed-wave height $H$ can be calculated by integrating Equation 6.11 for stated bed and flow conditions and using Equation 6.3 for the mean stress. This has been done in Figure 6.26 for an upper-stage plane bed of $D = 2.5 \times 10^{-4}$ m, $k = 0.0435$ kg$^{-1/2}$ m$^{1/2}$ s$^2$, $\gamma = 1600$ kg m$^{-3}$, $h = 2$ m, $f = 0.028$, $a_{sv} = 0.4\bar{\tau}$, and a range of values for $U_m$ above the critical value for an upper-stage plane bed. The experimental data of Guy et al. (1966) gave the friction and transport coefficients. At the given depth of 2 m, the calculated wavelength is $L = 6.7$ m assuming the Rao et al. (1971) value for the eddy period. Hence the predicted bed waves are extremely flat. The flatness

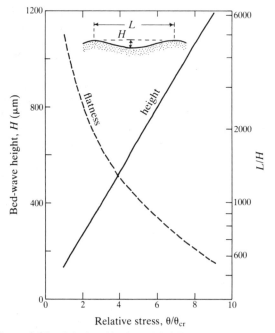

**Figure 6.26** Calculated bed-wave height (maximum lamina thickness) and relative height, as a function of the relative non-dimensional time-averaged bed shear stress.

index $L/H$ is of order $5 \times 10^3$ at the threshold of an upper-stage plane bed, and only one order of magnitude less at the upper limit of the chosen velocity range. It is hardly surprising that such long flat bed waves have not yet been detected beneath bedload layers, even under laboratory conditions.

As each bed wave is horizontally extensive but extremely flat, what our arguments amount to is the suggestion that widely extensive but millimetre-scale laminae are being continually generated on a plane sand bed as the result of the convection of large eddies by the overlying turbulent flow. Parts of some of these laminae (i.e. bed waves) would be preserved if the overall regime were such as to allow deposition in the long term. Aggradation on plane beds should therefore take place in the form of discrete and laterally persistent but fine-scale laminations. Just such a bedding style abounds in fluvial and turbidity-current deposits (Fig. 6.12), particularly on upper-stage plane beds in association with streak-related parting lineations (Allen 1964).

Finally, we can make an inference concerning the textural grading of the laminae. As was noted, the amplitude of the slowly varying stress in Equation 6.8 approaches one-half the long-term mean value, generating an approximately two-fold range in the local boundary shear stress. Such an extreme stress variation should lead to a detectable textural sorting in the deposited sediment, probably through its effect on the calibre of the near-bed suspended load. As discussed in Chapter 7, the suspension threshold increases with particle size and terminal fall velocity, and the relative amount of grains of a given diameter that can be transported in suspension rises steeply with increasing stress. The sediment in the bedload layer, from which deposition is largely expected, should therefore be coarsest on the average in the trough where the shear stress is greatest (Fig. 6.23c), on account of the relatively greater loss of fine grains to the suspended load. Hence the laminae will be graded from coarse at the base (bed-wave trough) to fine at the top (bed-wave crest). Many if not most laminae in parallel-laminated fluvial sandstones seem to be graded in this way (normal grading).

## 6.7 Relation of small to large coherent structures

Like Jonathan Swift's fleas, the large and small eddies of turbulent flows seem to form an inescapable association. However rough the flow, small eddies associated with ejections, bursts and sweeps are combined with large flow-filling vortices. Moreover, the large and small eddies share the same timescale, although at large Reynolds numbers differing greatly in size. Only the streaks are restricted in their occurrence, to smooth and transitional flows in which a viscous sublayer can be formed. What causes the streaks, and are the ejection–burst–sweep events and large eddies genetically related?

The streaks of the sublayer and innermost logarithmic region are possibly an expression of Taylor–Görtler instability, a variety of hydrodynamic instability affecting curved flows in which the velocity increases inwards towards the centre of flow curvature. Thomas and Bull (1983) found experimentally that new streaks are generated where streamlines are compressed beneath each cat's eye eddy (Fig. 6.27a). The streamlines in this region are curved and concave in the direction of increasing flow velocity.

To see how the instability is caused, let us imagine that the fluid in the region of curved streamline is rotating steadily about a fixed transverse axis and that its velocity $U(r)$ parallel to streamlines is a declining function of the radial distance $r$ from the common centre of streamline curvature (Fig. 6.27b). As was demonstrated in the case of motion in a channel bend (Ch. 5), the forces acting on a fluid element are the centrifugal force $\varrho U^2/r \, \mathrm{N\,m^{-3}}$ and an opposing pressure gradient. If the flow is steady, the forces are in balance, and the magnitude of the pressure gradient can also be stated as $\varrho U^2/r \, \mathrm{N\,m^{-3}}$. Consider a ring of fluid of

velocity $U_1$ at a radial distance $r_1$. The centrifugal force acting on an element in this ring is $\varrho U_1^2/r_1$. Introducing $r_1$ into both numerator and denominator, the centrifugal force can be put in the equivalent form $\varrho(U_1 r_1)^2/r_1^3$. Now if this ring of fluid is displaced outwards to $r_2 > r_1$, the centrifugal force on the element becomes $\varrho(U_1 r_1)^2/r_2^3$. But the pressure gradient at $r_2$ is $\varrho(U_2 r_2)^2/r_2^3 = \varrho U_2^2/r_2$. Hence if $(U_1 r_1)^2 > (U_2 r_2)^2$ the ring continues its outward movement, and the fluid motion is unstable. Since this movement is perpendicular to the shearing motion of the fluid, the instability will be expressed as paired streamwise vortices, as in Figure 6.10.

Whether or not instability is developed evidently depends on the way $U$ and $r$ change together, that is, on steepness of the profile of mean velocity where the streamlines are curved. A necessary condition for instability is that the velocity should decline in the direction towards which the streamlines are convex. However, for the instability actually to arise, the velocity gradient must be sufficiently steep. In turbulent boundary-layer and channel flows, this condition can be found only near the wall, and particularly within and just outside the viscous sublayer.

By all the evidence, streak bursting has the same timescale as the large flow-filling coherent structures. The two scales of eddy must therefore be connected in some way, but is one necessarily the cause of the other? Thomas and Bull (1983) found experimentally that streak lift-up began towards the upstream side of each bulge (Fig. 6.27a), at a time when the streamwise pressure gradient relative to the wall was not such as could have caused the associated local fluid deceleration. Hence they concluded that bursting was not due to the large eddies, at least as expressed through flow-parallel effects. However, the eddies are large rotating structures, and it is possible that bursting is triggered by pressure gradients acting normal to the boundary, which should experience an alternate pulling and pushing as the large eddies pass by. The relationship of the stages of the burst cycle to the pattern in Figure 6.27a suggests that lift-up might result from a pull, and the burst and sweep from the subsequent push.

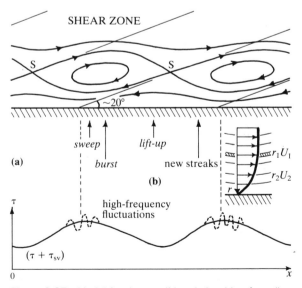

**Figure 6.27** Model for the possible relationship of small to large coherent structures in the turbulent boundary layer.

## Readings

Allen, J. R. L. 1964. Primary current lineation in the Lower Old Red Sandstone (Devonian), Anglo-Welsh Basin. *Sedimentology* **3**, 89–108.

Allen, J. R. L. 1969. Erosional current marks of weakly cohesive mud beds. *J. Sed. Petrol.* **39**, 607–23.

Antonia, R. A. 1972. Conditionally sampled measurements near the outer edge of a turbulent boundary layer. *J. Fluid Mech.* **56**, 1–18.

Antonia, R. A., S. Rajagopalan, C. S. Subramanian and A. J. Chambers 1982. Reynolds number dependence of the structures of a turbulent boundary layer. *J. Fluid Mech.* **121**, 123–40.

Bagnold, R. A. 1966. *An approach to the sediment transport problem from general physics.* Prof. Pap. US Geol. Surv., no. 422-I.

Bakewell, H. P. and J. L. Lumley 1967. Viscous sublayer and adjacent wall region in turbulent pipe flow. *Phys. Fluids* **10**, 1880–9.

Blackwelder, R. F. and H. Eckelmann 1979. Streamwise vortices associated with bursting phenomena. *J. Fluid Mech.* **94**, 577–94.

Blackwelder, R. F. and R. E. Kaplan 1976. On the wall structure of the turbulent boundary layer. *J. Fluid Mech.* **76**, 89–112.

Bradshaw, P. 1971. *An introduction to turbulence and its measurement.* Oxford: Pergamon.

Bridge, J. S. 1981. Hydraulic interpretation of grain-size distributions using a physical model for bedload transport. *J. Sed. Petrol.* **51**, 1109–24.

Brodkey, R. S., J. M. Wallace and H. Eckelmann 1974. Some properties of truncated turbulence signals in bounded shear flows. *J. Fluid Mech.* **63**, 209–24.

Brown, G. L. and A. S. W. Thomas 1977. Large structure in turbulent boundary layers. *Phys. Fluids* **20**, (10, II), S243–52.

Bull, M. K. 1967. Wall-pressure fluctuations associated with subsonic turbulent boundary layer flows. *J. Fluid Mech.* **28**, 719–54.

Cantwell, B. J. 1981. Organized motion in turbulent flow. *Annu. Rev. Fluid Mech.* **13**, 457–515.

Cantwell, B., D. Coles and P. Dimotakis 1978. Structure and entrainment in the plane of symmetry of a turbulent spot. *J. Fluid Mech.* **87**, 641–72.

Chen, C.-H. P. and R. F. Blackwelder 1978. Large-scale motion in a turbulent boundary layer in a study using temperature contamination. *J. Fluid Mech.* **89**, 1–31.

Coleman, J. M. 1969. Brahmaputra River: channel processes and sedimentation. *Sed. Geol.* **3**, 129–239.

Corino, E. R. and R. S. Brodkey 1969. A visual investigation of the wall region in turbulent flow. *J. Fluid Mech.* **37**, 1–30.

Falco, R. E. 1977. Coherent motions in the outer regions of turbulent boundary layers. *Phys. Fluids* **20** (10, II), S124–32.

Fiedler, H. and M. R. Head 1966. Intermittency measurements in the turbulent boundary layer. *J. Fluid Mech.* **25**, 719–35.

Grass, A. J. 1971. Structural features of turbulent flow over smooth and rough boundaries. *J. Fluid Mech.* **50**, 233–55.

Guy, H. P., D. B. Simons and E. V. Richardson 1966. *Summary of alluvial channel data from flume experiments, 1956–61.* Prof. Pap. US Geol. Surv., no. 462-I.

Head, M. R. and P. Bandyopadhyay 1981. New aspects of turbulent boundary-layer structure. *J. Fluid Mech.* **107**, 297–338.

Heathershaw, A. D. 1979. The turbulent structure of the bottom boundary layer in a tidal current. *Geophys. J. R. Astron. Soc.* **58**, 395–430.

Jackson, R. G. 1976. Sedimentological and fluid-dynamic implications of the turbulent bursting phenomena in geophysical flows. *J. Fluid Mech.* **77**, 531–60.

Kim, H. T., S. J. Kline and W. C. Reynolds 1971. The production of turbulence near a smooth wall in a turbulent boundary layer. *J. Fluid Mech.* **50**, 133–60.

Kline, S. J., W. C. Reynolds, F. A. Schraub and P. W. Runstadler 1967. The structure of turbulent boundary layers. *J. Fluid Mech.* **30**, 741–73.

Kovasznay, L. S. G., V. Kibbens and R. F. Blackwelder 1970. Large scale motion in the intermittent region of a turbulent boundary layer. *J. Fluid Mech.* **41**, 283–325.

Laufer, J. 1975. New trends in experimental turbulence research. *Annu. Rev. Fluid Mech.* **7**, 307–26.

Mantz, P. A. 1978. Bedforms produced by fine, cohesionless, granular and flaky sediments under subcritical water flows. *Sedimentology* **25**, 83–103.

Merzkirsch, W. 1974. *Flow visualization.* New York: Academic Press.

Nakagawa, H. and I. Nezu 1981. Structure of space-time correlations of bursting phenomena in an open-channel flow. *J. Fluid Mech.* **104**, 1–43.

Nychas, S. G., H. G. Hershey and R. S. Brodkey 1973. A visual study of turbulent shear flow. *J. Fluid Mech.* **61**, 513–40.

Rao, K. N., R. Narasimha and M. A. B. Narayanan 1971. The 'bursting' phenomenon in a turbulent boundary layer. *J. Fluid Mech.* **48**, 339–52.

Raupach, M. R. 1981. Conditional statistics of Reynolds stress in rough-wall and smooth-wall turbulent boundary layers. *J. Fluid Mech.* **108**, 363–82.

Richardson, F. M. and K. O. Beatty 1959. Patterns in turbulent flow in the wall-adjacent region. *Phys. Fluids* **2**, 718–19.

Smith, C. R. and S. P. Metzler 1983. The characteristics of low-speed streaks in the near-wall region of a turbulent boundary layer. *J. Fluid Mech.* **129**, 27–54.

Sumer, B. M. and R. Deigaard 1981. Particle motions near the bottom in turbulent flow in an open channel. Part 2. *J. Fluid Mech.* **109**, 311–37.

Thomas, A. S. W. and M. K. Bull 1983. On the role of wall-pressure fluctuations in deterministic motions in the turbulent boundary layer. *J. Fluid Mech.* **128**, 283–322.

Wallace, J. M., H. Eckelmann and R. S. Brodkey 1972. The wall region in turbulent shear flow. *J. Fluid Mech.* **54**, 39–48.

Willmarth, W. W. and C. E. Wooldridge 1962. Measurements of the fluctuating pressure at the wall beneath a thick turbulent boundary layer. *J. Fluid Mech.* **14**, 187–210.

# 7 A matter of turbidity

Suspended sediment profile – eddy diffusion model of suspension transport – suspension transport across drowned river floodplains – suspension transport and turbulence asymmetry.

## 7.1 Introduction

Have you ever stood by a river swollen after rain? If the river lies in the mountains it is quite likely that pebbles and cobbles are being carried over the bed. Above the rush of the invigorated current, you may hear a rapid succession of hollow-sounding crashes and bangs, caused by frequent impacts between stones in motion and between these and the stationary bed. You would be justified in concluding that you were listening to the stream's bedload, but it is most unlikely that you would be able to see that load, on account of the turbidity of the water. What makes it turbid? Dip a jar into the stream and let the contents settle. You will find that the turbidity is due to the presence of particles of mainly silt and clay size, perhaps with a little relatively fine sand. The contrast in texture with the bedload is striking. You will also notice that the settled grains form only a thin layer at the bottom of the jar. The fine particles were therefore widely dispersed and at a very low concentration when in the river, in sharp difference from the coarse bedload debris. Finally, your direct observation of the bedload was prevented because, as you will have noticed, the fine sediment was dispersed throughout the whole body of the stream. It constitutes the suspended load of the river.

Why did these dispersed grains settle out once the moving river water had been trapped as a sample in the jar? From our previous consideration of particle settling and fluidization (Ch. 3), we can suggest that the grains settled through the stagnant water contained in the jar because of the lack of upward currents able to balance their excess weight. What then 'fluidized' the same grains when in the river and part of its suspended load? Observing the river again, the water not only exhibits a translational motion, but is also highly turbulent. Large internal eddies are continually driving upwards to the free surface, there to spread out as interfering, short-lived boils (Fig. 6.21). The stagnant water in the jar was neither turbulent nor in translational motion. Could the eddies in the river therefore in some manner provide the upward-acting fluid force necessary to maintain against gravity a suspended load composed of grains more dense than the fluid? Make a stirrer by firmly gluing a disc to the end of a rod, so that the rod is normal to the plane of the disc. Try stirring the sample of water and sediment from the river, using a vigorous vertical motion of the stirrer. The resulting disorganized turbulent motion will quickly resuspend the sediment and, if maintained, in turn maintain the new suspension.

So far as rivers are concerned, it would seem that a suspended sediment load involving grains more dense than the fluid can exist only in the presence of fluid turbulence. Indeed, one widely accepted definition of the suspended load of a stream is that it is that part of the total load supported by fluid turbulence (Bagnold 1966). However, rivers are not the only agents capable of suspending fine sediment. Tidal currents in estuaries and near coasts also are turbulent. They are responsible for the dispersal in suspension of very large amounts of muddy sediment, as is testified by the brown turbid waters that ebb and flood over the mud-flats so extensively developed along estuarine and other protected shores. Turbulence may also accompany the currents associated with water waves generated by strong winds. The wind itself is turbulent, its ability to suspend and transport dust and other light detritus being a matter of common experience. Mineral dust from the Sahara has been detected in the West Indies more than 4000 km to the west. The loess-blanketed plains drained by the Huang Ho in central China bear witness even more dramatically to the ability of the wind to carry fine sediment in suspension.

## 7.2 A diffusion model for transport in suspension

### 7.2.1 Eddy transport

We saw in Chapter 6 that fluid turbulence has a mixed stochastic and deterministic character. It is just as fair to treat a turbulent flow statistically, in terms of averages, ranges and correlations, as deterministically, in terms of eddies of a definite size, shape, dynamical character and persistence in time. Turbulent eddies with closely definable characteristics are not just convenient fictions helpful in theoretical work.

The fluid in a turbulent eddy viewed at some instant is flowing in many different and partly opposed directions. A fluid particle on being tracked, however, is found to circulate around an eddy, although in the process the eddy itself may have experienced some translation. The continual circulation of fluid around eddies makes it possible for a property acquired by an element of fluid when the element was in one part of an eddy to be carried to some other part, where the property is released to a neighbouring eddy. This process is called eddy transport. The most important properties redistributed by eddy transport are heat and fluid momentum, but matter can also be transported as the result of exchanges between eddies. Why for example do we stir sugar into a cup of tea or coffee rather than simply wait for molecular diffusion to take its course? By stirring one creates a turbulent environment in which the sugar dissolves and disperses rapidly and completely while the beverage remains hot. The dissolution is rapid because eddies low in sugar from distant parts of the cup are continually being swept past the undissolved crystals. Dispersal proceeds rapidly because exchanges between the eddies quickly spread the sugar molecules through the fluid as a whole. The eddies thereby act so as to even out the distribution of dissolved sugar. Once an even spread is attained, no amount of further stirring will make the distribution uneven again. Although the eddies continue to exchange fluid and sugar molecules, there is no further net transport of sugar because the directed fluxes cancel out over the cup as a whole. Evidently, there can be a net eddy transport of matter

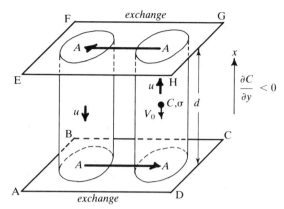

**Figure 7.1** Definition sketch for sediment diffusion in turbulent eddies.

or of a property only if there exists in the fluid a gradient of the matter or property in question. We can also infer from what happened in the cup that the transport takes place down the property gradient, and with the effect, in a closed system, of reducing and finally destroying the gradient. Let us develop these ideas when the matter transported is fine sediment in suspension.

Figure 7.1 shows a highly idealized eddy in a turbulent stream containing dispersed uniform fine sediment. The grains are of density $\sigma$ and occur at a local average fractional volume concentration $C$ which decreases with increasing distance $x$, where the $x$-direction may be parallel to any one of the three principal flow directions. It is assumed that the sediment concentration is low, so that the grains may be characterized by $V_0$, their terminal fall velocity in an unbounded fluid (Ch. 3). We also assume that the grains are very much smaller in size than the eddies through which they are dispersed, so that the grains have no difficulty in following the movement of fluid within the eddy. The eddy has a characteristic size $d$ and internal velocity $u$, where $d$ is the average size measured in the $x$-direction of the eddies in the stream and $u$ is the average magnitude of the fluctuating turbulent velocity component, also measured in the $x$-direction (i.e. $\bar{u}'$, in the notation of Chapter 6).

Consider what happens between the two planes ABCD and EFGH which just contain the eddy and lie normal to the $x$-direction, and imagine that the limbs of the eddy parallel with this direction are contained in cylindrical tubes of uniform cross-sectional area $A$. The portion of the eddy flowing along the plane ABCD lies towards the high end of the concentration gradient and can exchange fluid with elements of the flow bearing relatively large quantities of sediment. The discharge of grains through the right-hand of our imaginary cylinders therefore equals

$$\sigma u A C - \tfrac{1}{2} \sigma u d A \frac{\partial C}{\partial x} \quad \text{kg s}^{-1}$$

where $C$ is the average sediment concentration at the midplane of the eddy. Note that the negative sign is used here because the concentration gradient is defined as decreasing in the positive $x$-direction, making $\partial C / \partial x$ negative. When the fluid reaches the plane EFGH, however, it encounters and exchanges grains with elements of the flow relatively poor in dispersed sediment, towards the low end of the concentration gradient. The sediment discharge in the opposite direction along the left-hand of the imaginary cylinders is therefore

$$\sigma u A C + \tfrac{1}{2} \sigma u d A \frac{\partial C}{\partial x} \quad \text{kg s}^{-1}$$

Hence the net transport through the midplane of the eddy is

$$\text{total net discharge} = \left( \sigma u A C - \tfrac{1}{2} \sigma u d A \frac{\partial C}{\partial x} \right)$$

$$- \left( \sigma u A C + \tfrac{1}{2} \sigma u d A \frac{\partial C}{\partial x} \right) \quad (7.1)$$

$$= - \sigma u d A \frac{\partial C}{\partial x} \quad \text{kg s}^{-1} \quad (7.2)$$

or, dividing through by the cross-sectional area $A$,

$$F = - \sigma u d \frac{\partial C}{\partial x} \quad \text{kg m}^{-2} \text{s}^{-1} \quad (7.3)$$

where $F$ is the dry-mass flux measured normal to the concentrated gradient.

Equation 7.3 shows that the transport rate of dispersed sediment through a plane normal to the concentration gradient increases linearly with the magnitude of the gradient. According to this equation, there is no directed transport when the gradient is zero, for the local transports along the different limbs of the eddies then balance out, as was surmised from our consideration of sweetening a cup of coffee or tea. The transport rate is also seen to increase with the length and velocity scales characteristic of the eddies. As these scales may be expected to bear some deterministic relationship to the overall properties of the flow (Ch. 6), they combine to form what amounts to a coefficient describing the transport of suspended sediment by eddy exchange. This coefficient is conveniently denoted by the Greek letter $\epsilon$ (epsilon) and may be defined as

$$\epsilon = kUL \quad \text{m}^2 \text{s}^{-1} \quad (7.4)$$

where $U$ and $L$ are respectively velocity and length scales characteristic of the flow overall, and $k$ is a numerical constant expected to be of order unity or somewhat less. The quantity $\epsilon$ is called the sediment eddy diffusion coefficient, and by its use Equation 7.3 can be simplified to

$$F = - \epsilon \sigma \frac{\partial C}{\partial x} \quad \text{kg m}^{-2} \text{s}^{-1} \quad (7.5)$$

Equation 7.5 may be employed to solve problems of sediment transport in suspension, provided that we know how $\epsilon$ varies with direction in the flows under consideration. As Equation 7.5 was derived for a single direction, the value of $\epsilon$ may change somewhat from one direction to another in the same flow. For rivers, the diffusion coefficients in the streamwise, vertical and spanwise directions differ slightly in value.

### 7.2.2 Vertical profile of suspended sediment in channelized flows

In establishing the rates at which rivers discharge sediment, engineers have made countless measurements of the way in which the local concentration of suspended grains varies vertically through the water column contained in the channel. As an example, Figure 7.2 shows the upward variation of total sediment concentration in the Mississippi River at St Louis on 24 April 1956 (Colby 1963). Grains of both silt and sand sizes were present, the finer particles being more evenly distributed than the coarser ones, which occurred chiefly in the lowermost few metres of the flow. The second example (Fig. 7.3) is representative of concentration profiles measured from laboratory channels, in this case a uniform steady flow 0.057 m deep and of average velocity $1.4 \text{ m s}^{-1}$ carrying in suspension a narrowly graded quartz sand of diameter 130 $\mu$m (Montes & Ippen 1973). Again the concentration decreases upwards and at a faster rate near the bed than the free surface. What law or laws describe such concentration profiles?

One point needs to be settled before we can attempt to derive these laws. How does it come about that a concentration gradient of suspended sediment can be steadily maintained in the case of a steady uniform channelized flow, when Equation 7.5 applied in the

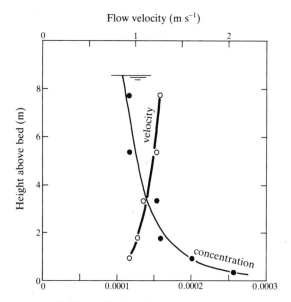

Figure 7.2 Vertical variation of local mean flow velocity and total sediment fractional volume concentration in the Mississippi River at St Louis. Data of Colby (1963).

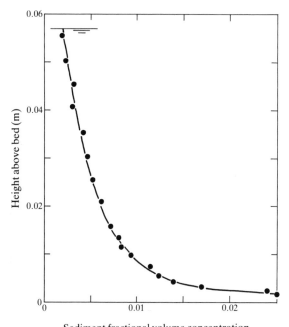

Figure 7.3 Vertical distribution of the total sediment fractional volume concentration in a laboratory channel. Data of Montes and Ippen (1973).

vertical $y$-direction states that there is a continual upward flux of grains? We are bound to conclude that another mechanism must operate to provide a downward flux of particles just balancing the upward flux due to eddy transport. Only one mechanism is available, namely, the settling of the grains under gravity arising from their excess of density over the fluid. Ignoring concentration effects, the magnitude of the downward flux due to settling is $\sigma C V_0 \text{ kg m}^{-2}\text{s}^{-1}$. In the steady state, the upward flux due to eddy transport is just balanced overall by the downward flux due to settling, so taking Equation 7.5 we can write

$$\sigma C V_0 = -\sigma \epsilon(y) \frac{\partial C}{\partial y} \quad \text{kg m}^{-2}\text{s}^{-1} \qquad (7.6)$$

in which $\epsilon(y)$ represents as yet unknown functions describing the vertical variation of the sediment diffusion coefficient.

What form will the concentration profile take distant from the bed in a channelized flow? Since the turbulence itself results from the shearing of the fluid in the channel, it is logical in settling the form of $\epsilon(y)$ to choose the shear velocity as the velocity scale in Equation 7.4. It will be remembered from Chapter 5 that the shear velocity $U_\tau = (ghS)^{1/2}$, where $g$ is the acceleration due to gravity, $h$ the flow depth and $S$ the water-surface slope. As we are considering those parts of the flow distant from the bed, it is reasonable to suppose that proximity to the bed no longer has any further influence on the size of the eddies. An appropriate length scale in Equation 7.4 is therefore the flow depth $h$. Combining these suggestions, we may write

$$\epsilon(y) = khU_\tau = \text{constant} \quad \text{m}^2\text{ s}^{-1} \qquad (7.7)$$

for the region beyond the influence of the bed. Substituting Equation 7.7 into Equation 7.6, and rearranging the resulting differential equation so that like variables appear on the same side of the equals sign (i.e. separating the variables), we then obtain

$$\frac{1}{\sigma C} \partial C = -\frac{V_0}{\sigma khU_\tau} \partial y \qquad (7.8)$$

By introducing a suitable boundary condition, Equation 7.8 may be integrated to give the concentration $C$ as a function of the height $y$ above the stream bed. In view of the variables so far introduced, the only possible choice is to suppose that a reference sediment concentration $C_{\text{ref}}$ has already been measured at some arbitrary height $y_{\text{ref}}$ near the bed. Integration of the right-hand side of Equation 7.8 from $y_{ref}$ to $y = h$ and of the left-hand side from $C = C_{ref}$ leads to

$$\ln\left(\frac{C}{C_{ref}}\right) = -\frac{V_0}{khU_\tau}(y - y_{ref}) \qquad (7.9)$$

where ln denotes natural logarithms. Equation 7.9 is equivalent to

$$C = C_{ref}\exp[-V_0(y - y_{ref})/khU_\tau] \qquad (7.10)$$

where exp stands for the base of natural logarithms.

Equation 7.10 states that, in the region beyond the influence of the bed, the sediment concentration decreases exponentially upwards, and at a rate increasing with the terminal fall velocity of the grains. In a river, all grain sizes should decrease in amount away from the bed, but the smaller particles should be more evenly spread over the depth than the larger grains. Equation 7.10 describes quite well the distribution of sediment in the outermost 70–80% of a channelized flow, as may be seen from Figure 7.4 in which the data of Figure 7.3 are plotted on a semilogarithmic scale. In the case of this experiment, the numerical constant $k = \epsilon/hU_\tau$ takes the value 0.11.

How does the concentration vary in the near-bed region, where Equation 7.10 no longer satisfactorily describes the distribution? The shear velocity $U_\tau$ is again the correct choice for the velocity scale in Equation 7.4 but the flow depth $h$ is no longer appropriate. As there can be no eddies at the bed itself, it is reasonable that the characteristic eddy size should change with distance

$y$ from the bed. Suppose that the size increases linearly with the distance. The expression corresponding to Equation 7.7 then becomes

$$\epsilon(y) = kyU_\tau \quad m^2\,s^{-1} \qquad (7.11)$$

where $k$ is a new numerical constant of order unity. Substituting into Equation 7.6, and separating the variables and integrating as before, gives

$$\ln C = -\frac{V_0}{kU_\tau}(\ln y - \ln y_{ref}) + \ln C_{ref} \qquad (7.12)$$

and its equivalent

$$C = C_{ref}\left(\frac{y}{y_{ref}}\right)^{-V_0/kU_\tau} \qquad (7.13)$$

This equation states that the concentration in the near-bed region declines upwards according to a power function. As in the outer region of the flow, the rate of decrease grows larger as the terminal fall velocity of the sediment increases. Figure 7.5 with double-logarithmic scales compares Equation 7.13 with the experimental data of Montes and Ippen (1973) earlier shown in Figure 7.3. The agreement is good over a substantial proportion of the flow thickness, $k$ taking the value 0.35.

Our analysis therefore shows that the vertical distribution of suspended sediment in a steady uniform channelized flow can be accurately described by means of a 'two-layer' model. The inner layer, where the bed

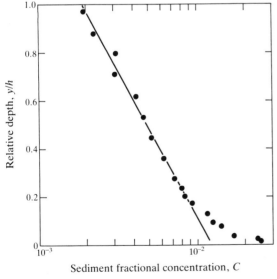

**Figure 7.4** Semilogarithmic plot of the data of Figure 7.3.

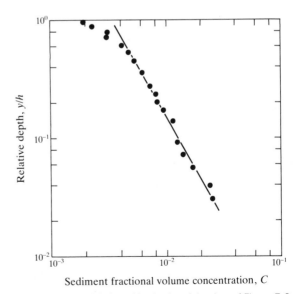

**Figure 7.5** Double-logarithmic plot of the data of Figure 7.3.

is the dominant influence, is modelled by a power function (Eqn. 7.13) and the outer layer, dominated by the flow thickness, is well described by an exponential relationship (Eqn. 7.10). It is not difficult to devise a procedure allowing the two equations to be matched; together they are superior to the single widely used equation advanced by Rouse (1937), which attempts to incorporate both bed and depth influences. It only remains to decide whether the numerical constants appearing in Equations 7.10 and 7.13 are universal and whether they are in any way influenced by grain size. Each constant is in effect a non-dimensional sediment diffusion coefficient, as may be seen by rearranging Equations 7.5 and 7.11 (e.g. $k = \epsilon/hU_\tau$).

Coleman (1969, 1970) explored the universality of the constants appearing in Equations 7.10 and 7.13 by making a careful analysis of extensive laboratory and field data. Figure 7.6 shows a selection of his results from laboratory channels. In the outer 70–80% of the flow, $k = \epsilon/hU_\tau$ is approximately a constant which grows larger with increasing terminal fall velocity (increasing grain size for quartz-density solids). Coleman obtained results similar in value and trend upon analysing Anderson's (1942) observations from the Enoree River in South Carolina (Fig. 7.7). The data collected from several river and laboratory channels are combined in Figure 7.8 into a graph of $\epsilon/V_0$ against $y$. The rising limbs of the curves representing the various channels lie very close to a single line. Evidently the dispersal of sediment in the inner region is independent

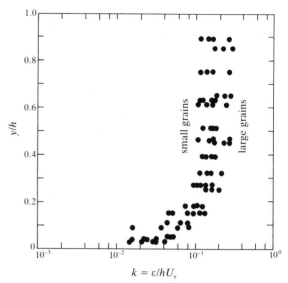

$$k = \epsilon/hU_\tau$$

**Figure 7.7** Non-dimensional sediment transfer functions as measured from the Enoree River. Data of Anderson (1942) and Coleman (1969).

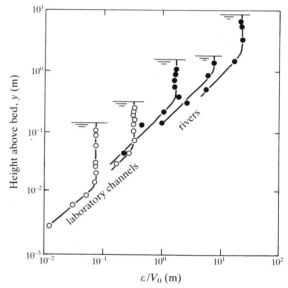

$$\epsilon/V_0 \text{ (m)}$$

**Figure 7.8** Sediment transfer ratio as a function of height above the bed in two laboratory channels and three rivers (Enoree, Snake, Mississippi). Data of Coleman (1970).

of the scale of the flow, and the linear assumption about eddy size is justified.

### 7.2.3 The suspended load and its transport rate

Now that expressions for the vertical distribution of suspended sediment have been obtained, we can proceed

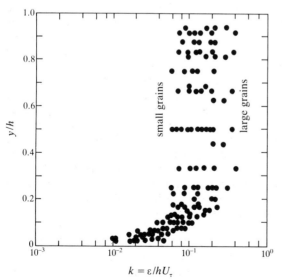

$$k = \epsilon/hU_\tau$$

**Figure 7.6** Non-dimensional sediment transfer functions as measured from laboratory channels. Data of Coleman (1969).

128

to calculate the suspended sediment load of a fluid stream and the transport rate of that load. Proceeding as in Section 4.4.1, the suspended load may be defined as the quantity of sediment above unit area of the bed whose excess weight is supported by fluid turbulence. The rate of transport follows as the product of the suspended load with the vertically averaged tangential velocity of the suspended grains. As before, the load may be expressed in terms of either dry mass or immersed weight, according as the sediment transport is being treated.

The suspended load in terms of dry mass per unit bed area is the integral over the depth $h$ of the product of the solids density (assumed constant) with the local sediment concentration. That is

$$m_s = \sigma \int^h C(y)\,\mathrm{d}y \quad \mathrm{kg\,m}^{-2} \qquad (7.14)$$

where $m_s$ is the dry-mass load and the other symbols carry their previous meanings. This measure of the load becomes the load in terms of immersed weight on multiplying $m_s$ by $g(\sigma - \varrho)/\sigma$, where $\varrho$ is the fluid density. Here $C(y)$ is the vertical variation in the local sediment concentration, which may have been either measured, as in Figure 7.2, or calculated using Equations 7.10 and 7.13 on the basis of a single reference measurement. The calculation of the transport rate is completed by introducing the grain transport velocity. The transport velocity can be estimated for each level in the velocity profile of the flow by applying the settling laws explained in Chapter 3, but this elaborate procedure is hard to justify in the face of the simplifying assumption suggested by the general conditions under which suspension occurs. Experience shows that suspension transport is significant only for relatively fine particles whose terminal fall velocity is small or very small compared to the mean translation velocity of the fluid. Hence the local grain transport velocity will be sensibly the same as the local mean fluid velocity, except very close to the bed where the fluid velocity is itself small. To a good approximation we can therefore calculate the suspended load transport rate (dry mass) as

$$J_s = m_s U_m \quad \mathrm{kg\,m}^{-1}\mathrm{s}^{-1} \qquad (7.15)$$

where $U_m$ is the vertically averaged flow velocity applicable at the same site as $C(y)$. Empirically, the suspended load transport rate is proportional to the mean flow velocity raised to approximately the fourth power, which means that the suspended load increases as roughly the cube of the mean flow velocity. Of

course, to obtain a river's total discharge of suspended sediment ($\mathrm{kg\,s}^{-1}$) through a cross section, Equation 7.15 must be integrated over the entire flow, since both $m_s$ and $U_m$ vary with position.

## 7.3 Transport in suspension across river floodplains

### 7.3.1 Some field observations

Rivers that water alluvial valleys and plains have the widely recognized habit of building themselves up within broad sediment ridges rising above the frequently waterlogged lowlands to either side. The best descriptions of these features, called alluvial ridges, come from the alluvial valley and delta of the Mississippi, where two varieties are recognized (Fisk 1944, 1947, 1952). In one kind, called a meander-belt ridge, the active meanders of the river, together with channel loops that had been cut off, lie atop a low sedimentary ridge elevated several metres above the adjacent floodbasins. The second variety, called a levée ridge, is particularly associated with the relatively straight bayous of the lower valley and delta. These ridges are many times wider than the channels they border and rise as much as 5 m above the adjoining swamps.

How does such alluvial relief arise? The answer is only partly that the relatively coarse-grained sediments deposited in and near the channels consolidate more slowly and to a lesser extent than the thick silts and clays typical of the floodbasins. The remaining and perhaps more important part of the explanation seems to be that there is a gradient of sediment deposition rate transversely across the floodplain of a river, the largest rates prevailing closest to the channel. It is otherwise difficult to explain why, for example, the sand layers present within levées become less frequent and individually thinner as they are traced away from the channels. Successive floods, each depositing a gradually outward-tapering layer of sediment, would therefore seem capable over the years of slowly elevating the active channel by building up its more immediate surroundings. The process of elevating the channel cannot proceed indefinitely, however, for the more the channel and its borders are raised, the easier it is for the river to abandon its elevated position in order to construct a new course at a lower level in another location. River avulsion − the process of abandoning an alluvial ridge in favour of constructing a new one − is one of the most important but least understood of all fluvial processes. The Mississippi River, for example, has avulsed every 1000 years or so over the past 5000 years. Yet an understanding of avulsion is vital to the appreciation of alluvial morphology and stratigraphy. Our ignorance stems from the long and unjustifiable neglect of what happens on river floodplains in times of flood.

### 7.3.2 Outline of a model for floodplain deposition

Using ideas introduced earlier in this chapter, we are now going to look into the reasons for the creation of alluvial ridges. The insight will not be complete, however, because of the lack even of essential empirical data, particularly from natural rivers. Fortunately, some help can be gained from the laboratory studies involving flows in a channel combined with a floodplain.

Consider in Figure 7.9 the steady flow of a river that has drowned its surrounding broad floodplain. The river occupies a straight symmetrical channel of depth $H$ and slope $S$, and has inundated the surrounding plains, of the same slope $S$, to a uniform depth $h$, where $h$ is a moderate or small fraction of $H$. The water depth above the main channel is therefore $(H + h)$. Although the mean currents above both the main channel and the floodplain are everywhere directed parallel to the slope, because the flow is turbulent there could be an eddy transport of suspended grains away from the channel and across the floodplain in the $z$-direction. However, Equation 7.3 shows that transport will occur only if there is a decline in the concentration of suspended sediment transversely away from the river-flooded channel. It may also be surmised from the same relationship that the transport, particularly of individual grain sizes, will be strongly influenced by the transverse variation shown by the sediment diffusion coefficient.

Will there be a cross-floodplain gradient in the suspended sediment concentration? Let us argue this question as follows, recognizing that it is no longer the point concentration that is of interest, but the concen-

tration averaged over the flow depth at a site. The channel and floodplain flows in Figure 7.9 are interacting, but if a tall enough barrier were to exist along the channel margins, their respective uniform depth-averaged velocities $U_{ch}$ and $U_{fl}$ could be written (Eqn. 5.8) as

$$U_{ch} = \left(\frac{8}{f_{ch}}gS\right)^{1/2}(H+h)^{1/2} \quad m\,s^{-1} \quad (7.16a)$$

$$U_{fl} = \left(\frac{8}{f_{fl}}gS\right)^{1/2}h^{1/2} \quad m\,s^{-1} \quad (7.16b)$$

in which $f_{ch}$ and $f_{fl}$ are the Darcy–Weisbach friction coefficients for respectively the channel and the floodplain. Now from Equation 7.15, and the empirical observation that the suspended-load transport rate varies as the fourth power of the mean flow velocity, it is clear that the suspended sediment loads $m_{s,ch}$ and $m_{s,fl}$ present in respectively the undisturbed channel and floodplain are given by

$$m_{s,ch} \propto U_{ch}^3 \quad (7.17a)$$

$$m_{s,fl} \propto U_{fl}^3 \quad (7.17b)$$

assuming that $f_{ch}$ and $f_{fl}$ and all other constants and coefficients are sufficiently alike between the two cases that they may be ignored. Substituting the velocities from Equation 7.16, we discover that

$$m_{s,ch} \propto (H+h)^{3/2} \quad (7.18a)$$

$$m_{s,fl} \propto h^{3/2} \quad (7.18b)$$

Equation 7.14 allows the sediment loads also to be written as

$$m_{s,ch} = \sigma(H+h)C_{ch} \quad kg\,m^{-2} \quad (7.19a)$$

$$m_{s,fl} = \sigma h C_{fl} \quad kg\,m^{-2} \quad (7.19b)$$

where $C_{ch}$ and $C_{fl}$ are the depth-averaged concentrations for the undisturbed channel and floodplain respectively. Eliminating the loads between Equations 7.18 and 7.19, and expressing the result as a ratio, we find

$$\frac{C_{ch}}{C_{fl}} = \left(\frac{H+h}{h}\right)^{1/2} \quad (7.20)$$

This equation states that the depth-averaged suspended sediment concentrations for the undisturbed channel

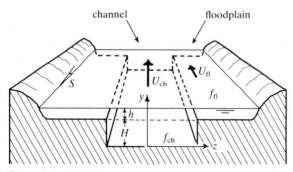

**Figure 7.9** Definition sketch for a river showing combined channel and floodplain flow.

and floodplain are in the ratio of the square roots of the flow depths. For example, when $H = 5$ m and $h = 1$ m, the concentration in the channel is approximately 2.5 times that on the floodplain. Note that the concentration ratio increases as $h$ declines relative to $H$.

The result expressed by Equation 7.20 therefore justified the view that there is a continuous transverse gradient of suspended sediment concentration within interacting channel and floodplain flows. Putting $C$ as the local depth-averaged concentration, what is the likely form of $C(z)$? It would be logical to expect the concentration to decrease continuously from $C = C_{ch}$ at the channel midline outwards to $C = C_{fl}$ at some station on the floodplain so remote that the influence of the channel was no longer felt. Rajaratnam and Ahmadi (1979, 1981) measured experimentally the transverse variation of flow properties in interacting channel and floodplain flows. So long as $h$ is not too similar in value to $H$, their data on the tranverse variation in the local mean flow velocity and local mean boundary shear stress can be fitted to a good approximation by an error function. The only disturbing influence is a small jump in the value of the velocity and stress at the drowned edge of the channel. Therefore for the concentration of suspended sediment we can suggest

$$\frac{C - C_{fl}}{C_{ch} - C_{fl}} = \exp\left(-k_C Z_C^2\right) \qquad (7.21)$$

where $k_C$ is a numerical constant and $Z_C = z/z_{ref,C}$ is a non-dimensional distance, $z$ being the distance from the channel midline, and $z_{ref,C}$ the distance to the place where $(C - C_{fl})/(C_{ch} - C_{fl})$ attains some chosen value. The form of Equation 7.21 is sketched in Figure 7.10a; unfortunately, there are no means as yet of predicting $k_C$ and $z_{ref,C}$ for natural rivers. The range of concentration values of course depends on the depth ratio according to Equation 7.20.

Our discovery that the suspended sediment concentration declines outwards across the floodplain implies that grains are being transported transversely away from the channel. The channel is behaving as a linear source of particles, which, by the action of eddies of turbulence, become diffused therefrom and spread over the adjoining surface. The turbulence is also diffusing outwards the momentum of the vigorous channel flow, as is evident from the outwardly declining flow velocity measured experimentally by Sellin (1964), Toebes and Sooky (1967), and Rajaratnam and Ahmadi (1979, 1981). Some of the eddies involved in the transport of flow properties are quite large and well organized, but, while not strictly turbulence, would not differ in effect.

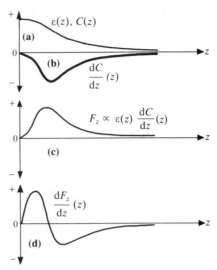

**Figure 7.10** General features of a model for floodplain deposition in response to eddy diffusion away from a channel. In order to accommodate all of the formulae involved, the dependent quantities are plotted in terms of the actual transverse distance $z$.

Sellin (1964) found in his experiments that a train of regularly spaced vortices with vertical axes lay above each drowned channel margin (Fig. 7.11). This observation should not occasion surprise, however, because the interplay between a vigorous channel flow and a neighbouring sluggish floodplain current has many similarities to that between a jet and its calm surroundings (see Sec. 11.3).

What are the implications of the inferred spanwise sediment transport? Consider in Figure 7.12 a vertical stationary reference column within the steady floodplain flow and appeal to sediment continuity. The column is of height $h$, length $\delta z$ in the transverse $z$-direction, and length $\delta x$ parallel to the streamwise $x$-direction. It is assumed that the depth-averaged suspended sediment concentration varies with $z$ but is independent of $x$. Therefore the current discharges sediment into the column through its upstream face at a rate equal to the product of the area of the face with the unit flux, that is, $F_x h\,\delta z \; \text{kg s}^{-1}$, where $F_x \; \text{kg m}^{-2}\text{s}^{-1}$ is the depth-averaged unit flux in the $x$-direction. But as $C$ is independent of $x$, Equation 7.3 tells us that an equal discharge leaves the column through the downstream side. Hence the quantity of sediment present in the reference column is unchanged by the streamwise transport of suspended particles. Next consider the effect of the transverse eddy transport of sediment, and let $F_z$ be the depth-averaged unit flux in the $z$-direction.

131

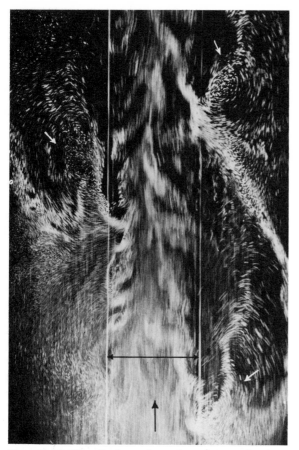

**Figure 7.11** Visualization of vortices (arrowed) along margins of a channel (between arrows, 0.115 m wide) in laboratory combined channel–floodplain flow. Camera travelling at mean speed of vortex cores. Photograph courtesy Dr R. H. J. Sellin, University of Bristol.

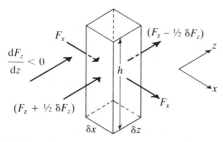

**Figure 7.12** Definition sketch for floodplain deposition by tranverse eddy diffusion.

Because the sediment concentration declines with increasing $z$, Equation 7.3 now states that the unit flux will also fall in the $z$-direction. Let the change of unit flux between the faces be the small quantity $\delta F_z$.

The total flux entering the column is therefore $(F_z + \frac{1}{2} \delta F_z)h\ \delta x$ while that leaving is $(F_z - \frac{1}{2} \delta F_z)h\ \delta x$. Clearly the eddy transport causes the column to gain sediment. Introducing the area of the base of the column, we can write

rate of gain of suspended sediment

$$= \frac{h(F_z + \frac{1}{2} \delta F_z)\ \delta x - h(F_z - \frac{1}{2} \delta F_z)\ \delta x}{\delta x\ \delta z} \quad (7.22)$$

which in the limit becomes

rate of gain of suspended sediment

$$= -h \frac{dF_z}{dz} \quad \text{kg m}^{-2}\text{s}^{-1} \quad (7.23)$$

where $dF_z/dz$ is the gradient of the depth-averaged unit flux. A unit column therefore gains sediment at a rate proportional to its height and the flux gradient, the negative sign being introduced to make the gain a positive quantity.

But the reference column cannot go on receiving sediment indefinitely, without becoming so choked that the flow as a whole becomes halted. As with the above case of the vertical sediment profile, we are bound to conclude that some mechanism operates to balance the sediment inflow and so maintain the transverse concentration gradient. The present mechanism must be deposition at the base of the column, at a rate just equal to the rate of gain. Hence Equation 7.23 is also a statement of how the local rate of deposition from suspension varies spanwise over the floodplain. Referring to Equation 7.5, its form can be derived once we know $\epsilon(z)$, the depth-averaged diffusion coefficient, and also $C(z)$, a suggestion for the latter appearing in Equation 7.21.

The following argument suggests the form of $\epsilon(z)$. By Equation 7.7 the depth-averaged diffusion coefficient increases with the flow depth, whence the coefficient should be larger in the channel than on the floodplain for the same shear velocity. Other things being equal, Equations 7.7 and 7.11 state that the coefficient is largest where the shear velocity is greatest. As the shear velocity characteristic of the undisturbed channel and floodplain flows varies with the square root of the corresponding depths, the coefficient in the case of interacting flows may be expected to decline continuously from a maximum at the channel midline outwards to a minimum at some remote station on the floodplain. It again seems appropriate to suggest an error function

$$\frac{\epsilon - \epsilon_{fl}}{\epsilon_{ch} - \epsilon_{fl}} = \exp(-k_\epsilon Z_\epsilon^2) \qquad (7.24)$$

where $k_\epsilon$ and $Z_\epsilon = z/z_{ref,\epsilon}$ are a new numerical constant and non-dimensional distance respectively.

The ideas sketched above are assembled in Figure 7.10. Equations 7.21 and 7.24 appear in Figure 7.10a and in Figure 7.10b is shown the slope of the curve of Equation 7.21, that is, the graph representing $dC/dz$. The equation represented by this curve is multiplied by Equation 7.24 to give the curve for $F_z$ plotted in Figure 7.10c. The graph of the slope of $F_z(z)$ shown in Figure 7.10d is therefore the required form of Equation 7.23 for $dF_z/dz$.

The graph of $dF_z/dz$ shows two distinct regions. In a narrow zone adjoining the channel midline, the flux gradient is positive because the flux is increasing outwards, and the deposition of outward-diffusing sediment cannot occur. The much wider outer region is typified by a negative flux gradient and by deposition at a rate declining with increasing $z$. Deposition from successive floods according to the pattern shown in Figure 7.10d should therefore result in the channel gradually becoming elevated at the axis of a broad alluvial ridge.

The relationships summarized in Figure 7.10 refer to the suspended sediment as a whole, without distinction as to grain size. The increase in the sediment diffusion coefficient with increasing terminal fall velocity noted by Coleman (1969, 1970) may partly explain the decreasing coarseness of floodplain sediments with increasing distance from the channel. The outward decline in floodplain velocity, however, may be of greater significance.

## 7.4 Limitations of diffusion models

The idea of suspension as expressing a diffusion process is the classical approach to the transport of fine sediment in turbulent currents. Models based on this approach are rather successful and have considerable practical value, as may be judged from the quality of the fit of theory to observation in Figures 7.4 and 7.5. The classical approach therefore continues to claim wide adherence.

Yet diffusion models have one very serious limitation. The sediment suspended by a stream, because the grains are more dense than the fluid, represents a force equal to the immersed weight of the sediment acting downwards on the stream bed. Diffusion models give us no hint as to the nature and origin of the upward-acting force that must necessarily exist under Newton's first law in order to balance this load, and so permit a steady transport in suspension. As reference to Equations 7.10 and 7.13 will show, the suspended load is predictable using a diffusion model only when we already have empirical information about the suspended sediment concentration at some known level in the flow. While there may be little practical difficulty in providing this information, it is rather unsatisfying to be left without a rigorous dynamical explanation for suspension transport. Referring again to Equations 7.10 and 7.13, we find that the sediment particles are merely characterized kinematically, in terms of their terminal fall velocity, and not by their immersed weight. Several workers have therefore sought dynamical alternatives to the diffusion model of suspension transport.

McTigue (1981) is the most recent worker to approach the problem from the continuum theory of mixtures. Under this theory, each component of the mixture of grains and fluid is treated as a continuum, the interaction of the components then being explored in the context of a turbulent channel flow. The resulting equations point to an equilibrium between the immersed weight of the suspended grains and a correlation associated with the fluid drag due to vertical velocity fluctuations. A specific result is an equation for the flux balance which has the same form as Equation 7.6. Eddy transport in McTigue's view is therefore a model of the correlation associated with the fluid drag. The fact that empirical data support the diffusion model of course means that McTigue's theory is equally well substantiated.

We are now going to examine in somewhat greater detail Bagnold's (1966) rather more accessible dynamical theory of suspension transport.

## 7.5 A dynamical theory of suspension

Bagnold (1966) acknowledged that only dynamical theories of sediment transport can have any pretence to validity. He recognized that the sediment load carried by a stream represents a force acting downwards on the stream bed and that this force must be balanced by another acting upwards from the bed. As the transported sediment can travel in a number of different modes, it is possible for the total upward-acting force to be made up of more than one component. We have already seen in Chapter 4 that, in the case of the bedload, the upward-acting force arises from impacts between closely arrayed shearing grains and between these grains and the static bed. Bagnold associated transport of the suspended load strictly with the turbulence of channelized flows at large Reynolds numbers, and explicitly defined the suspended load as that part of the total load supported by fluid turbulence. On the basis of some early but highly suggestive visualizations of turbulent eddies being formed at the boundary of a stream, he proposed that the eddies are able to support the suspended grains because, in the thin region near to the stream bed, the turbulence is asymmetrical in the direction normal to the bed. He postulated that the upward-acting eddy velocities are on the average larger in magnitude than the downward-acting ones. Consequently, the upward flux of fluid momentum through a plane parallel to the bed must exceed the downward momentum flow. By the principle of momentum conservation (Sec. 1.4), a net upward momentum flux is tantamount to an upward-acting force, and it is this force, dependent on the asymmetry of the turbulence, which is to be regarded as lifting the suspended load against gravity. Bagnold developed his idea as follows.

Consider in Figure 7.13 a representative eddy of unit volume whose vertical limbs intersect a plane ABCD lying close to and parallel to the stream bed. Putting $\varrho$ as the fluid density, a mass of fluid $\varrho(\frac{1}{2} - a)$ is moving upwards in the left-hand limb at the root-mean-square velocity $v_{up}$ characteristic of the turbulence (i.e. $(v'^2_{up})^{1/2}$ in the notation of Chapter 6). Here $a$ is a measure of the asymmetry of the turbulence, defined as the volume fraction by which the mass of fluid in the limb deviates from one-half of the unit volume. The remaining larger mass $\varrho(\frac{1}{2} + a)$ is travelling downwards in the right-hand limb at the smaller root-mean-square velocity $v_{dn}$. The momentum parallel to $y$ over ABCD must sum to zero, whence

$$\varrho(\tfrac{1}{2} - a)v_{up} = \varrho(\tfrac{1}{2} + a)v_{dn} \quad \text{N s m}^{-3} \quad (7.25)$$

or, rearranging,

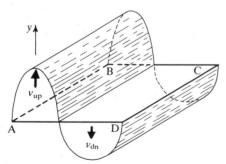

**Figure 7.13** Definition sketch for suspension transport in response to vertical turbulence asymmetry.

$$v_{dn} = v_{up} \frac{(\tfrac{1}{2} - a)}{(\tfrac{1}{2} + a)} \quad \text{m s}^{-1} \quad (7.26)$$

Now the momentum $\varrho(\frac{1}{2} - a)v_{up}$ of the fluid in the left-hand limb is being transported upwards through ABCD at the rate $v_{up}$. Momentum $\varrho(\frac{1}{2} + a)v_{dn}$ is similarly being carried down in the other limb at the rate $v_{dn}$. The fluxes of momentum are therefore unequal, and there is consequently an impulsive force $F$ directed upwards away from the stream bed, of value

$$F = \varrho(\tfrac{1}{2} - a)v^2_{up} - \varrho(\tfrac{1}{2} + a)v^2_{dn} \quad \text{N m}^{-2} \quad (7.27)$$

On substituting from Equation 7.26 for $v_{dn}$, Equation 7.27 simplifies to

$$F = \varrho v^2_{up} \frac{a(1 - 2a)}{(\tfrac{1}{2} + a)} \quad (7.28)$$

Differentiating Equation 7.28 with respect to $a$, and setting $dF/da = 0$, we find that the flux at the level of ABCD attains an upper limiting value when $a$ is approximately equal to 0.207. Because fluid turbulence is an instability phenomenon, Bagnold expected the turbulence asymmetry in rivers to be such as always to yield the limiting flux.

Equation 7.28 is not particularly useful as it stands, because it is far from straightforward to obtain reliable measurements of upward and downward eddy velocities. However, remembering how a root-mean-square value is calculated, we can relate $v_{up}$ to the conventional root-mean-square velocity $v$ (i.e. $(v'^2)^{1/2}$ in the notation of Chapter 6) describing the vertical turbulence at ABCD through

$$v^2 = v^2_{up}(\tfrac{1}{2} - a) + v^2_{dn}(\tfrac{1}{2} + a) \quad \text{m}^2 \text{ s}^{-2} \quad (7.29)$$

Substituting from Equation 7.26 for $v_{dn}$, Equation 7.29 becomes

$$v^2 = v^2_{up} \frac{(\tfrac{1}{2} - a)}{(\tfrac{1}{2} + a)} \quad (7.30)$$

Elimination of $v^2_{up}$ between this equation and Equation 7.28 leads to

$$F = 2a\varrho v^2 \quad \text{N m}^{-2}$$
$$= 0.414\varrho v^2 \quad (7.31)$$

on the supposition that the flux at the level of ABCD is maximized.

We now want to relate the flux as given by Equation

7.31 to the overall properties of the flow. Many experimenters, for example, Grass (1971), have measured the variation of $v$ with distance from the bed in turbulent channel flows. The root-mean-square velocity increases steeply from zero at the bed to a practically constant value at a height of between one-tenth and one-fifth of the flow depth. The ratio of the maximum value of $v$ to the shear velocity $U_\tau$ is close to unity in all of these studies. Hence the impulsive force acting upwards from the bed is, to a good approximation,

$$F = 0.414\tau \quad \text{N m}^{-2} \qquad (7.32)$$

where $\tau$ is the mean boundary shear stress, since $U_\tau = (\tau/\varrho)^{1/2}$.

Equation 7.32 states that the sediment-suspending force generated by a turbulent stream is of the same order of magnitude as the mean boundary shear stress itself. Do rivers appear to comply with this picture? Consider the suspended load measured in the Mississippi River (Fig. 7.2). The channel is 8.6 m deep and the overall mean concentration of sediment is approximately $2 \times 10^{-4}$. Referring to Equation 7.14 and its discussion, the suspended load in terms of immersed weight is approximately $2.8$ N m$^{-2}$, assuming a solids density of 2650 kg m$^{-3}$. Now the mean flow velocity is approximately $1.05$ m s$^{-1}$, so that by the quadratic stress law (Eqn 7.27) the mean boundary shear stress is $5.5$ N m$^{-2}$, assigning the plausible value of 0.04 to the Darcy–Weisbach friction coefficient. The sediment load therefore turns out to be similar in magnitude to the bed shear stress, as is required.

Another test, advocated by Leeder (1983), is to compare from flume observations the measured immersed-weight suspended load with the load as predicted by Equation 7.32 given the mean boundary shear stress. Figure 7.14 shows the relative load measured from above an upper-stage plane bed in flume experiments by Guy et al. (1966), Jopling and Forbes (1979) and Mantz (1983), who worked on a range of quartz silts and sands. The values of the relative load vary widely, from almost an order of magnitude smaller than the expected figure of unity, to over two orders of magnitude larger. Hence either the theory is incorrect, or laboratory flumes provide an inappropriate environment in which to test the theory, or effects not accounted for are capable of increasing suspension transport, particularly for fine grains, at the steep slopes and shallow depths encountered on a laboratory scale.

A third way of testing Bagnold's (1966) model is to seek direct evidence from boundary-layer flows for the

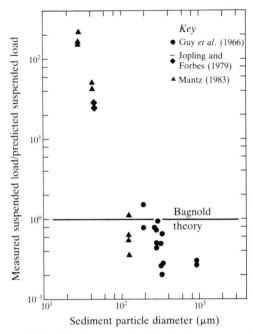

**Figure 7.14** Relative suspended load above upper-stage plane beds in laboratory flume channels, as a function of grain size.

asymmetry of the instantaneous vertical component of the fluctuating eddy velocity, that is, $v'$ in the notation of Chapter 6. Grass (1971) gave some laboratory measurements suggesting that the turbulence has the required asymmetry, but they are suspect because of the likelihood that additional vertical velocity components due to secondary flow (Ch. 11) occurred in his flume channel. The later work of Brodkey et al. (1974) appears to lack this disadvantage. By ingenious signal processing, they were able to distinguish up from down vertical turbulent velocities, and so could plot frequency diagrams for $v'$ as a vector. These diagrams are almost symmetrical far out in the boundary-layer flow. Near the bed they are distinctly asymmetrical and show a long tail of large upward velocity values, attributable to ejections during streak bursting (Ch. 6). Figure 7.15a shows one such distribution, for a station near the outer edge of the laminar sublayer, where $v'$ has been normalized by the velocity $U$ outside the boundary layer. The two halves of this distribution are practically of equal area, as required by conservation of both mass and momentum. If we square the values of $v'/U$, however, we obtain a quantity proportional to the momentum flux. The new frequency distribution (Fig. 7.15b) has patently unequal halves, the flux upwards into the boundary layer exceeding that downwards towards the wall. This

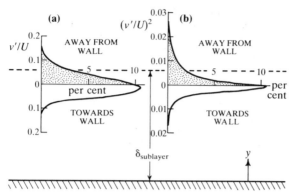

**Figure 7.15** Experimental distribution at a level just inside the viscous sublayer of (a) the vertical fluctuating component of the eddy velocity, and (b) its square (proportional to momentum flux). Data of Brodkey *et al.* (1974).

is the imbalance required if a turbulence-related outward-acting force is to support the suspended load.

## 7.6 A criterion for suspension

Bagnold (1966) and Middleton (1977) asked under what flow conditions would grains of a given size and excess density be transported in suspension? From our considerations of particle settling and fluidization (Ch. 3), it is reasonable to suggest that the coarsest particles that can be taken into suspension by a turbulent current are those whose terminal fall velocity $V_0$ is of the same order of magnitude as the root-mean-square value of the vertical component of the eddy velocity. Since empirically the root-mean-square value on the average is closely comparable with the shear velocity $U_\tau$, it follows that an approximate criterion for the onset of suspension transport is

$$\frac{V_0}{U_\tau} = 1 \tag{7.33}$$

Now remembering that $U_\tau = (\tau/\varrho)^{1/2}$, where $\tau$ is the mean boundary shear stress and $\varrho$ the fluid density, let us define for the onset of suspension a critical non-dimensional stress $\theta_s = \tau/(\sigma - \varrho)gD$, where $\sigma$ is the solids density, $g$ the acceleration due to gravity, and $D$ the particle diameter. Introduced into Equation 7.33, the preceding relationships lead to the alternative suspension criterion

$$\theta_s = \frac{\varrho V_0^2}{(\sigma - \varrho)gD} \tag{7.34}$$

This criterion is shown in Figure 4.4 on the assumption that the particles are quartz spheres in water. The criterion increases steeply with increasing grain diameter but assumes a constant value at large particle sizes, when the terminal fall velocity is proportional to $D^{1/2}$ rather than to $D^2$ (Ch. 3). The curves for the suspension criterion and for the threshold of bedload motion therefore define three fields of sediment transport behaviour. Fine grains become suspended at all flow stages up from the motion threshold, whereas coarse ones travel first as bedload and in suspension only at the highest stages. Middleton (1977) has exploited the suspension criterion to create an interesting explanation for the grain-size distributions of sandy sediments.

## Readings

Anderson, A. G. 1942. Distribution of suspended sediment in a natural stream. *Trans. Am. Geophys. Union* **23**, 678–83.

Bagnold, R. A. 1966. *An approach to the sediment transport problem from general physics.* Prof. Pap. US Geol. Surv., no. 422-I.

Brodkey, R. S., J. M. Wallace and H. Eckelmann 1974. Some properties of truncated turbulence signals in bounded shear flows. *J. Fluid Mech.* **63**, 209–24.

Colby, B. R. 1963. *Fluvial sediments – a summary of source, transportation, deposition and measurement of sediment discharge.* Bull. US Geol. Surv., no. 1181-A.

Coleman, N. L. 1969. A new examination of sediment suspension in open channels. *J. Hydraul. Res.* **7**, 67–82.

Coleman, N. L. 1970. Flume studies of the sediment transfer coefficient. *Water Resources Res.* **6**, 801–9.

Fisk, H. N. 1944. *Geological investigations of the alluvial valley of the lower Mississippi River.* Vicksburg, Mississippi: Mississippi River Commission.

Fisk, H. N. 1947. *Fine grained alluvial deposits and their effects on Mississippi River activity.* Vicksburg, Mississippi: Mississippi River Commission.

Fisk, H. N. 1952. *Geological investigation of the Atchafalaya Basin and the problem of Mississippi River diversion.* Vicksburg, Mississippi: Mississippi River Commission.

Grass, A. J. 1971. Structural features of turbulent flow over smooth and rough boundaries. *J. Fluid Mech.* **50**, 233–55.

Guy, H. P., D. B. Simons and E. V. Richardson 1966. *Summary of alluvial channel data from flume experiments, 1956–61.* Prof. Pap. US Geol. Surv., no. 462-I.

Jopling, A. V. and D. L. Forbes 1979. Flume study of silt transportation and deposition. *Geogr. Annlr* A **61**, 67–85.

Leeder, M. R. 1983. On the dynamics of sediment suspension by residual Reynolds stresses – confirmation of Bagnold's theory. *Sedimentology* **30**, 485–91.

McTigue, D. F. 1981. Mixture theory for suspended sediment transport. *J. Hydraul. Div. Am. Soc. Civ. Engrs* **107**, 659–73.

Mantz, P. A. 1983. Semi-empirical correlations for fine and coarse cohesionless sediment transport. *Proc. Instn Civ. Engrs* (2) **75**, 1–33.

Montes, J. S. and A. T. Ippen 1973. *Interaction of two-dimensional turbulent flow with suspended particles.* Rep. Ralph M. Parsons Lab. Water Resources and Hydrodyn., no. 164.

Middleton, G. V. 1977. Hydraulic interpretation of sand size distributions. *J. Geol.* **84**, 405–26.

Rajaratnam, N. and R. M. Ahmadi 1979. Interaction between main channel and floodplain flows. *J. Hydraul. Div. Am. Soc. Engrs* **105**, 573–88.

Rajaratnam, N. and R. Ahmadi 1981. Hydraulics of channels with flood-plains. *J. Hydraul. Res.* **19**, 43–60.

Rouse, H. 1937. Modern conceptions of the mechanics of fluid turbulence. *Trans. Am. Soc. Civ. Engrs* **102**, 463–543.

Sellin, R. H. J. 1964. A laboratory investigation into the interaction between the flow in the channel of a river and that over its floodplain. *Houille Blanche* **19**, 793–801.

Toebes, G. H. and A. A. Sooky 1967. Hydraulics of meandering rivers with floodplains. *J. Waterways Harbours Div. Am. Soc. Civ. Engrs* **93**, 213–36.

# 8 The banks of the Limpopo River

Clay minerals – deposition, packing and erosion of muddy sediments – stress and strain – strength and rupture of muddy sediments – tool marks – desiccation cracks.

## 8.1 Introduction

The curious fate that befell the Elephant's Child on the slimy banks of the Limpopo River in Rudyard Kipling's story reminds us that muddy sediments are as remarkable in their properties and behaviour as they are commonplace in a wide range of sedimentary environments. What is a muddy sediment? It must be one containing enough clay minerals to render it cohesive. In practice, the majority of muddy sediments are dominated by particles smaller in size than sand (Table 2.2). River and tidal muds contain numerous small fragments of rock-forming minerals other than the clays, and have an average particle size typically in the silt grade. It is only in caves, the oceans and deep lakes that true clays with an average particle size less than $2\,\mu m$ are to be found. Try chewing pieces of clay from these environments; the harsh gritty feel of the river and tidal deposits will soon convince you of their relative coarseness. Some muddy sediments even contain significant amounts of sand-sized and coarser debris, for example, many glacial tills and debris-flow deposits, and texturally may be neither silts nor clays.

The most immediately striking property of muddy sediments is their cohesion, the tendency for the constituent particles to stick together. Some freshly deposited muds are liquid, resembling cream, whereas others are jelly-like, until disturbed. On losing water, muddy sediments change into a material that flows plastically at moderate stresses, as potters for thousands of years have known, but which retains virtually any acquired shape when the stress is removed. At very high stresses, such as those existing within the Earth's crust, muddy sediments change into hard strong rocks of negligible water content which are impermeable to fluids. How are muddy sediments formed, and why are their properties so diverse? What governs the response of muddy sediments to applied forces? These are some of the questions to be tackled in this chapter.

## 8.2 Clay minerals

### 8.2.1 Crystal chemistry and structure

The essence of a mud is its component clay minerals, of which usually more than one species is represented. Although very many clay mineral species are known (Brindley & Brown 1980), only four species (or species complexes) are important naturally, namely, kaolinite, illite, chlorite and smectite (Table 2.1). These minerals occur in platy, lath-shaped or needle-like crystals (Beutelspacher & van der Marel 1968), that characteristically are less than $10\,\mu m$ in size, making them of subcolloidal to colloidal dimensions. Because of their relatively large surface area compared to mass, the crystals have unusual electrical properties and, when dispersed in a liquid, exhibit Brownian motion.

Clay minerals are hydrous silicates belonging to the larger family of silicate minerals known as the phyllosilicates, that is, the silicates with a sheet-like atomic structure. In the clay minerals generally, two kinds of crystallographic sheet structure are recognized. A tetrahedral sheet consists of a layer of cations (Si, Al, or $Fe^{3+}$) each of which is co-ordinated by four oxygens arranged at the corners of a tetrahedron. In each such sheet the tetrahedra share three corners in the same plane, so forming an open regular pattern of hexagonal rings, while the apical oxygens all point in the same direction. An octahedral sheet consists of three linked octahedra, in which cations are co-ordinated by hydroxyl anions. When all three octahedral positions are occupied by cations (divalent), the clay mineral is classified as trioctahedral, but the description dioctahedral is applied when every third position is vacant and there are just two trivalent cations. The two kinds of sheet can be articulated in three main ways to form layers. A 1:1 layer consists of one tetrahedral sheet linked to one octahedral sheet, one plane of hydroxyls in the octahedral sheet remaining unshared. A 2:1 layer sandwiches a single octahedral sheet between two tetrahedral ones, the apical oxygens of the latter articulating with the hydroxyls of the former. A 2:1:1 layer consists of the 2:1 arrangement just described to which is added externally an additional octahedral sheet. Any excess electrical charge in these various kinds of layers can be neutralized by adding interlayer cations.

These layer structures are represented as follows amongst the four main species or species complexes of Table 2.1. Kaolinite is a 1:1 dioctahedral layer silicate. Illite is somewhat variable in character but is predominantly a 2:1 dioctahedral layer silicate closely allied to the true micas (e.g. muscovite). Also rather variable is chlorite, a 2:1:1 generally trioctahedral layer silicate. Smectite has a 2:1 layer structure and may be either dioctahedral or trioctahedral. It has the interesting property of allowing water and organic liquids to penetrate in substantial amounts between the layers. Hence muddy sediments with a significant smectite content swell and shrink considerably – much more so than muds dominated by other clays – as they are alternately wetted and dried.

### 8.2.2 Dispersions of colloidal particles

Muddy water encountered in the field is a dispersion of generally very small clay mineral and other mineral grains in an aqueous solution which, even in the case of rivers, contains small amounts of a surprising variety of organic as well as inorganic substances. Such a subcolloidal or colloidal dispersion shows unusual properties which depend on the size, composition and relative amount of the solids, and on the chemistry and concentration of the solutes. In particular, the dispersion may be either stable or unstable.

The following experiment illustrates the circumstances under which a colloidal dispersion becomes unstable. Dissolve a tiny fragment of sodium hexametaphosphate in roughly one litre of distilled water placed in each of two beakers, and stir into each con-

**Figure 8.1** Experimental effect of a strong electrolyte on settling of an aqueous dispersion of kaolinite: (a) weak sodium hexametaphosphate; (b) solution with added common salt (settling started at same time as in (a)).

tainer about 0.025 kg of fine-grade china clay (mainly kaolinite). Any lumps should be carefully broken up and smoothed out. Almost immediately the clay becomes uniformly dispersed throughout the extremely weak sodium hexametaphosphate solution. The dispersion is milky and the smallest crystals remain suspended for hours and even days (Fig. 8.1a). In this stable colloidal dispersion, the interparticle forces must be repulsive, for how else can we explain the lack of particle agglomeration and the uniformity of particle concentration (except for effects due purely to gravitational settling)? Now gently stir into one of the beakers a little saturated aqueous sodium chloride solution and carefully observe the ensuing changes. The dispersion is now unstable and looks granular, with swirls of particle-rich and particle-poor fluid appearing against the glass. It will be noticed that the clay is settling out (Fig. 8.1b), and much more rapidly than in the previous case. Examine the settling mass with a reading glass. It will be found to consist of clay floccules, the largest perhaps 500–1000 $\mu$m across. Such floccules (Fig. 8.2) comprise very many loosely aggregated clay mineral crystals and other particles (Zabawa 1978). They abound particularly in salty natural waters, as is easily proved by dipping up a jar of turbid water from a tidal current or flooding river. Evidently the effect of adding the strong electrolyte sodium chloride to the clay dispersion is now to make attractive the forces acting between the crystals. The same effect occurs where muddy rivers meet salt water in estuaries.

Theoretical and experimental studies show that whether a dispersion of clay minerals in an electrolyte

140

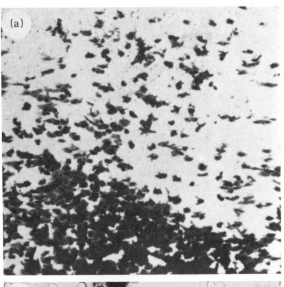

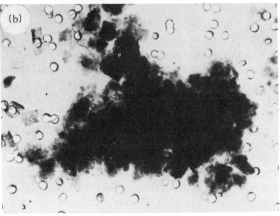

**Figure 8.2** Clay-mineral floccules. (a) Macrofloccules 1000–3000 μm across formed in a laboratory experiment. Photograph courtesy of Dr K. Kranck, Bedford Institute of Oceanography. (b) Floccule about 8 μm long from the Bristol Channel, caught on a nucleopore filter. Photograph courtesy of Dr D. J. A. Williams, University College of Swansea.

is stable or not depends on the relative strength and distribution of two kinds of forces acting on the particles (Van Olphen 1963). One is the weakly attractive van der Waals' force, to which all matter is subject. The other is the repulsive electrical charge known as the Coulomb force. Clay minerals have a net negative charge which results from three main mechanisms: (1) vacancies or unbalanced ionic substitutions in the crystal structure; (2) the physical or chemical adsorption of ions where there are broken bonds on particularly the faces of the crystals; and (3) the adsorption of ions of the same type as those in the structure of the mineral.

The net negative charge on each particle is balanced by electrolyte ions of positive charge, known as counterions, which lie in the solution immediately around the particle. The concentration of these counter-ions decreases exponentially with increasing distance outwards from the particle, so that the negative charges on the particle surface, combined with the positively charged counter-ions close by, form an electrical double layer sheathing the particle like an atmosphere. As the particles of any one clay mineral species in a given electrolyte have the same double layer, they repel each other on randomly making contact under the influence of fluid turbulence and/or Brownian motion, but only so long as the double layer is sufficiently thick. The dispersion is then stable. At high electrolyte concentrations, the double layer becomes thinned, allowing particles to be influenced by the weak but attractive van der Waals' force. Particle agglomeration and floccule growth ensue, as in the beaker to which salt was added.

### 8.2.3 Ionic exchange reactions

The charge on and thickness of the electrical double layer depends on the clay mineral species and particle size and on the electrolyte composition and concentration. If the electrolyte is modified chemically, then the ions forming the double layer correspondingly alter, by a rapid process of ion exchange. The ions chiefly exchanged are the cations, but some anions on the particle surface (chiefly edges) can be either added to or replaced. The organic compounds found in natural waters and in soils can also become involved in exchange reactions.

The ability of a clay mineral species to engage in cation exchange is its cation exchange capacity, a distinctive and measurable property. Kaolinite, illite and chlorite have similar small to moderate cation exchange capacities. That of smectite, however, is large and about an order of magnitude greater than the capacity of illite.

The significance of cation exchange is that any change in the pore fluid chemistry of a muddy sediment is likely to cause an alteration in the physical properties of the sediment, simply because the extent and strength of the electrical double layer surrounding the particles depend partly on the electrolyte. For example, the flocculant structure to be expected in a mud deposited intertidally could be weakened on leaching the sediment with fresh water in response to rainfall or a lowering of sea level. Conversely, a lacustrine mud may shrink and crack on being inundated by the sea, as the result of the increased electrolyte concentration in the pore fluid and the consequent thinning of the double layer.

## 8.3 Deposition of muddy sediments

In our experiments with dispersions of china clay, deposition occurred from a stagnant fluid, rapidly in the case of the clay in salt solution but much more slowly from the stable mixture containing only sodium hexametaphosphate. In each case, the rate of transfer or flux of clay in dry mass per unit area and time to the deposit forming over the bottom of the beaker may be written

$$R = \sigma C_b V_b \quad \text{kg m}^{-2}\text{s}^{-1} \qquad (8.1)$$

where $\sigma$ is the solids density of the clay mineral, $C_b$ the near-bed fractional volume concentration of the clay, and $V_b$ the terminal fall velocity of the near-bed particles measured relative to the ground. Strictly, the velocity term should be the fall velocity relative to the ground plus the velocity also relative to the ground at which the bed is building up. For slow deposition, however, the latter is sufficiently small as to be ignored and we shall not include it. But what is the expression corresponding to Equation 8.1 for a turbulent current, in which clay mineral particles, perhaps agglomerated in floccules, contribute to a suspended load?

McCave and Swift (1976) suggest a model that plausibly describes the deposition of mud from turbulent currents. They reason that there is a flow zone close to the bed from which all dispersed particles can be deposited, and specifically propose that the viscous sublayer (Ch. 6) is this zone. Consider mud deposition from a uniform steady turbulent current flowing over a smooth bed (Fig. 8.3). In the figure a rectangular volume element is shown, with one large face ABCD of unit area contained in the surface marking the outer limit of the sublayer, and the other EFGH of unit area contained in the plane of the bed. Consider the possible mud fluxes, averaged over space and time, into and out

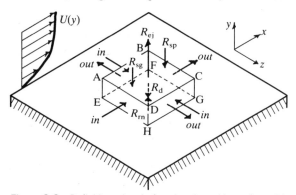

**Figure 8.3** Definition sketch for the deposition of muddy sediments.

of this volume. As the current is uniform and steady, we expect the streamwise discharges of mud entering and leaving the volume to be the same, and should find that the transverse fluxes are also balanced. Net deposition can therefore depend only on the vertical fluxes. Those through ABCD are: (1) the inward flux $R_{sg}$ due to the gravitational settling of particles fed from above; (2) the inward flux $R_{sp}$ due to sweeps within the wall region of the turbulent flow; and (3) the outward flux $R_{ej}$ due to ejections associated with streak bursting (Ch. 6). Acting through EFGH are (1) the outward rate of deposition $R_d$, and (2) the inward rate of resuspension $R_{rn}$. The principle of sediment continuity demands that

$$R_d + R_{ej} = R_{sg} + R_{sp} + R_{rn} \quad \text{kg m}^{-2}\text{s}^{-1} \qquad (8.2)$$

or, alternatively,

$$R_d - R_{rn} = R_{sg} + R_{sp} - R_{ej} \qquad (8.3)$$

In the last equation, $(R_d - R_{rn}) = R$ is the net rate of bed deposition that would actually be measured by comparing levels over a period of time.

We now need to explore each of the terms on the right-hand side of Equation 8.3. As in the case of the stagnant system, $R_{sg} = \sigma C_b V_b$, where 'near-bed' now means just outside the viscous sublayer. The other two terms refer to gains or losses due to the coherent fluid motions in the near-bed turbulent flow. It is reasonable to suppose that $(R_{sp} - R_{ej})$ is negative. This is because experiments show that transported particles become concentrated into the low-speed streaks from which the ejections arise (Grass 1971, Mantz 1978), whereas the sweeps come from a region of lower particle concentration. Now two factors seem to determine the likelihood that a particle can be deposited on the bed after it has in some manner entered the viscous sublayer but before it is ejected from a low-speed streak. First, there is the chance given to the particle of touching and sticking to the mud already forming the bed. This chance should decline as the sweeps and ejections become shorter in duration and more vigorous, that is, as the boundary shear stress increases. Secondly, progressively fewer particles should stick permanently to the bed as the fluid shear approaches closer to the maximum possible strength of the bond between bed and recently alighted clay floccules. These ideas suggest that Equation 8.3 takes the form

$$R = \sigma C_b V_b (1 - \tau/\tau_{cr}) \qquad (8.4)$$

where $\tau$ is the boundary shear stress due to the fluid motion, and $\tau_{cr}$ the critical stress for mud deposition implied by the idea of a maximum bond strength.

Equation 8.4 has yet to be tested rigorously, but it is appealing insofar as it gives sensible results. The formula states that the deposition rate of mud from a turbulent current increases with near-bed mud concentration and with the terminal fall velocity of the near-bed particles, but decreases with increasing boundary shear stress. Deposition ceases when the fluid force reaches the threshold stress appropriate to the particular clay–electrolyte system. The threshold stress is hard to measure reliably, and few clay–electrolyte systems have so far been investigated. However, it appears to be similar to the corresponding erosion threshold and to lie in the range $0.01$–$0.1\,\mathrm{N\,m}^{-2}$, implying mean flow velocities of at most a few decimetres per second. Particle volumetric concentrations vary considerably in natural environments. Near-bed fractional values of $10^{-5}$–$10^{-3}$ typify estuaries, with local concentrations as high as $10^{-2}$ and even more. More open seas contain less dispersed clay. Thus the near-bed particle concentration may be the dominant control on the rate of mud deposition in natural environments, since there is experimental evidence that $V_b$ increases linearly to steeply as the clay content increases, at least in stagnant systems. This means that floccule size is a steeply to very steeply increasing function of clay concentration for a given electrolyte. But it should be remembered that the forces holding floccules together are relatively weak. The stresses due to turbulence may therefore limit the growth of floccules, by causing those which have become too large to be torn apart.

## 8.4 Packing of muddy sediments

The obviously very watery nature of freshly deposited muddy sediments means that we cannot expect their particles to be arranged much like the packings of equal spheres discussed in Section 2.6. Instead, we should expect bridge-like, cardhouse and chain arrangements to predominate, for it is only such as these which can explain the water contents of 60–80% measurable from fresh muds. Only the packings of equal spheroids discussed in connection with cohesionless sediments begin to approach the required structure.

There is an important theoretical reason for expecting open clay packings. Van Olphen (1963) pointed out that the net negative charge of clay mineral particles is associated chiefly with the crystal faces. It is therefore

only adjacent to faces that the counter-ions of the double layer are cations. As their ability to be exchanged shows, anions can form the counter-ions in the portion of the double layer adjacent to crystal edges, where cations are exposed on the mineral surface. Hence when particles are momentarily brought together under the influence of turbulence and/or Brownian motion, crystal edges and faces should tend to adhere on account of their opposite charge.

Because the particles are so tiny and sensitive to changes in the chemical environment, it is no easy matter to establish the mode of packing of a muddy sediment. The main techniques are the study of thin or ultrathin sections and the examination by scanning electron microscope of surfaces exposed by fracturing variously dried specimens (Smart & Tovey 1981a). Figure 8.4 illustrates actual packings, further examples of which are described by Smart and Tovey (1981b). In Figure 8.5 appear schematic representations of the more important modes of packing which have either been proposed or claimed for muddy sediments. They are basically of three kinds: (1) arrangements of single flakes, (2) assemblages of crystal aggregates, and (3) chain or honeycomb arrangements. When the porosity is extremely high, single-flake arrangements are probably unco-ordinated (SU type). They may otherwise be edge-to-edge (SEE type) or edge-to-face (SEF type) arrangements. In aggregate or turbostratic arrangements, the basic unit is an ordered packet of platy crystals, or a bundle of lath-shaped or needle-like ones, set face-to-face (there may be some stepping from one crystal to another, so that some edges may adjoin faces). These packets or bundles may be unco-ordinated at low clay concentrations (AU type) or, at higher particle concentrations, arranged edge-to-face (AFFEF type), edge-to-edge (AFFEE type) or in a mixed manner (AFFM type). Otherwise crystals can be linked partly edge-to-edge and partly edge-to-face to form long honeycomb-like chains. Aggregate and chain arrangements seem to be more characteristic of muddy sediments formed in shallow marine environments, where clay and other colloidal debris are in generous supply, than in either the deep ocean, where particulate matter is highly dispersed, or in freshwater systems of low electrolyte concentration.

With progressive burial in the Earth's crust, muddy sediments become increasingly consolidated or compacted. Consolidation is accompanied by two main physical changes: (1) the slow expulsion of pore water, indicated by a progressive decrease in porosity, and (2) the rotation of clay crystals and crystal aggregates towards the horizontal (Fig. 8.5). If the burial is deep enough, the elevated temperature and pressure may chemically change the original clay mineral species, as well as causing new minerals to be precipitated as a cement within voids.

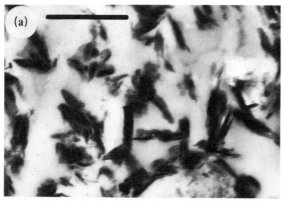

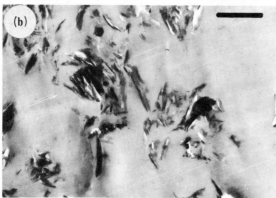

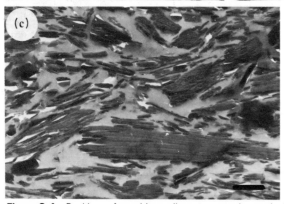

**Figure 8.4** Packing of muddy sediments seen in vertical ultra-thin sections. (a) Very open structure with local single-flake edge-to-face arrangement, recently deposited mud, Severn Estuary, England. (b) Very open aggregate structure, recently deposited mud, Severn Estuary. (c) Subhorizontal fabric in artificially consolidated kaolinite. Scale bar is 1 μm long in each case. Photographs courtesy of Dr P. Smart, University of Glasgow.

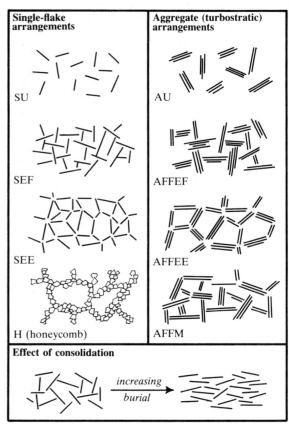

**Figure 8.5** Schematic modes of packing of muddy sediments, and the general effect of consolidation.

## 8.5 Coming unstuck

### 8.5.1 An experiment

Muddy sediments like other materials are responsive to external forces. Just how they behave can be summed up by the following deceptively simple experiment.

The principal apparatus you will need consists of two large glass beakers, a 0.15 m diameter glass filter funnel, some sodium hexametaphosphate, and a quantity of fine-grade china clay. Dissolve a small fragment of sodium hexametaphosphate into approximately 0.5 litres of distilled water in one of the beakers, and gradually mix into the solution enough china clay to make a thick cream (consistency while stirred). Carefully smooth out all lumps. Set up the funnel in a retort stand, sealing its outlet with a piece of Plasticine or similar material. Pour in the clay dispersion, shake the funnel gently so that a flat surface forms on the dispersion, and let all stand for a while.

**Figure 8.6** Coin supported by an aqueous slurry of kaolinite.

For the first part of the experiment you will need a medium-sized coin, a piece of cotton thread about 0.15 m long, and a tiny piece of Plasticine or putty. Using the Plasticine, hang the coin from the thread so that its faces lie horizontally. Now slowly lower the coin onto the surface of the clay mixture in the funnel. The coin sinks perhaps its own thickness into the dispersion, a strongly convex-outward meniscus appearing around it, but surprisingly it does not sink out of sight, even though the metal is much denser than the dispersion (Fig. 8.6). The surface tension of the water cannot be supporting the coin, for the same coin lowered into a weak aqueous solution of hexametaphosphate without added clay at once sinks. Evidently the clay dispersion has sufficient strength to support the weight of the coin, but not without undergoing a modest deformation in the process, as shown by the meniscus. The same conclusion follows on lifting out the coin. As can also be observed from whipped cream, the surface of the dispersion stands in a little peak, and similar peaks can be formed by dipping a glass rod in and out of the mixture. These peaks could not arise if the dispersion lacked the inherent strength to resist being flattened under gravity, as would have been the case if aqueous sodium hexametaphosphate without clay had been substituted.

The second part of the experiment is performed by removing the plug from the outlet of the funnel. The dispersion, under its own weight, flows continuously down the neck of the funnel (Fig. 8.7a). Evidently there is a deforming force above which the mixture is constrained to flow plastically. Instead of emerging in a

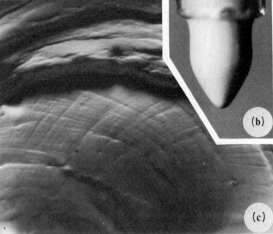

**Figure 8.7** Behaviour of an aqueous slurry of kaolinite flowing under gravity from a funnel. (a) Experimental apparatus (note sloping sides of slurry accumulating in dish). (b) Detail of clay cylinder emerging from funnel. (c) Rhombohedral fractures present on surface of slurry retained in funnel (largest fractures about 0.01 m long; long diagonal of rhomb tangent to rim of funnel).

steady stream, however, the dispersion emerges from the funnel in large, uniformly sized dollops (Fig. 8.7b). Each dollop takes the form of a cylindrical mass which, under its own weight, ruptures at the upper end once a critical length has appeared. The dollops are much too large to be explained as drops whose size is governed only by surface tension. Instead, they express in a further way the strength of the mixture, and specifically its ability to resist tension. Each dollop will hang freely from the funnel under its own weight, but only up to the extrusion of a limiting weight. The constancy of this weight can easily be checked by collecting separate dollops on weighed watch glasses.

For the third part of the experiment, attend to the surface of the dispersion that remains in the funnel. Examination with a reading glass will show that the surface is broken into tiny rhomb-shaped elements elongated parallel to the rim of the funnel (Fig. 8.7c). These elements are defined by oppositely spiralling step-like features originating close to the axis of the funnel. The steps and rhombs are constantly changing as the dispersion empties from the funnel, although preserving the general features described. In particular, the steps will be observed to vary in height, and some rhombs will be seen to become smaller as new steps appear. It is impossible to avoid the conclusion that the steps represent steeply inclined internal rupture surfaces, and that the rhombs are elements which, without being further deformed internally, are moving relative to similar neighbours. The deforming force is arguably compressive, for as the dispersion flows through the neck of the funnel, the area of its free surface steadily declines, bringing elements increasingly together. Paradoxically, rupture is a mode of behaviour associated with brittle solids, yet the dispersion flows plastically through the neck of the funnel.

### 8.5.2 Stress and strain

The preceding experiment illustrates many of the possible responses of a muddy sediment to deforming forces. How can we further analyse a deforming force? Consider within a muddy sediment under an applied force an arbitrarily oriented cubical element, the edges of which define an orthogonal co-ordinate system $(x, y, z)$ (Fig. 8.8). Let equilibrium and a homogeneous state of stress be assumed. The force acting on each face can be resolved into two parts, each of which is a stress, that is, a force per unit area with the units $N\,m^{-2}$. One of these is a normal stress, denoted conventionally by the symbol $\sigma$ (sigma is used equally conventionally elsewhere in this book for the solids density, but the difference of usage is always made clear). The normal stresses acting on the cube are therefore $\sigma_x$, $\sigma_y$ and $\sigma_z$, the subscripts denoting the faces affected; these stresses may be either tensile (negative) or compressive (positive). Both sorts of normal stress operated in our previous experiment. The other component of the force is a traction or tangential stress. This component can be

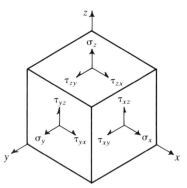

**Figure 8.8** Stresses acting on an arbitrarily oriented cubical element.

resolved on each face into two parts, one parallel to one of the edges, and the other parallel to the remaining edge. For example, using appropriate subscripts, the tangential force acting parallel to the $y$-direction on the face normal to the $x$-axis is $\tau_{xy}$, while that acting on the same face parallel to the $z$-direction is $\tau_{xz}$. Hence nine stresses in all act on the cubical element, three normal and six tangential. It can be shown that of these stress components only six are independent. A knowledge of these six is sufficient to define any stress system and state of stress at a point, provided that the assumptions about the homogeneity of the stress field and the equilibrium of the element are satisfied.

A suitable rotation of our arbitrarily oriented cubical element will clearly make the tangential component of the force on each face equal to zero, leaving only the normal stresses (Fig. 8.9). By definition, any stress acting on a surface along which the shear stress is zero is called a principal stress. As the element has three orthogonal faces, it is affected by three principal stresses. Speaking generally, these will not be equal but can be ranked by means of numerical subscripts in preference to co-ordinate directions. In Figure 8.9 the greatest principal stress is $\sigma_1$, the least principal stress is $\sigma_3$, while $\sigma_2$ is the intermediate stress, that is $\sigma_1 > \sigma_2 > \sigma_3$. The

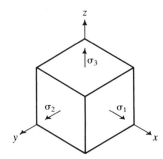

**Figure 8.9** Principal stresses acting on a cubical element.

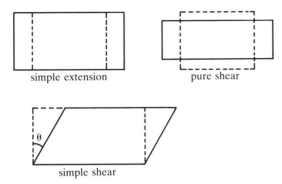

simple extension
pure shear

simple shear

**Figure 8.10** Three kinds of two-dimensional strain.

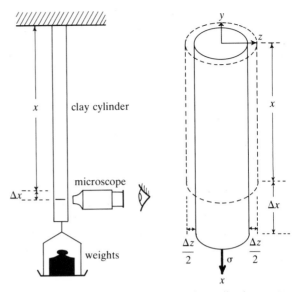

**Figure 8.11** Deformation of a solid under tension (apparatus and definition sketch).

difference between the greatest and least principal stresses is called the stress difference or deviatoric stress.

When a material body is acted on by a force, its elements experience a distortion, deformation or change in relative position called a strain. Three basic kinds of two-dimensional strain can be recognized (Fig. 8.10). The force may elongate the body in one direction only, affording the type of strain called simple extension. A change in the dimensions of the body in two perpendicular directions is described as pure shear. The third type of strain, called simple shear, is that met with in connection with the viscosity of liquids (Sec. 1.8). It will be seen that simple shear involves the translation of the elements of the body parallel to an axis, but in such a manner that the ordinate line of any point is rotated or sheared through an angle.

How are stress and strain related? This is a fundamental question, because the answer with respect to a given material gives us an essential insight into how that material behaves under stress.

Suppose we have a long cylinder of dried muddy sediment arranged in an apparatus in such a way that its upper end is fixed whereas weights can be attached to the lower end, the position of which can be measured (Fig. 8.11). As more weights are added, the cylinder lengthens progressively, that is, it becomes more strained. At a critical weight, however, the cylinder ruptures, snapping audibly. Up to this point, a graph of stress (added weight) against strain (elongation of cylinder) shows a linear proportionality between the two (Fig. 8.12). This kind of stress and strain relationship, combined with the failure by sudden rupture, is typical of a brittle substance. Brittle substances also display an almost exclusively elastic behaviour. In an ideally elastic material, strain is directly proportional to stress, and the deformation is completely reversible on removing the force, up to the elastic limit, or yield stress, at which the material ruptures. In the case of our dried muddy

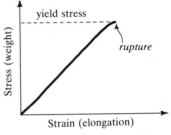

**Figure 8.12** Schematic stress-strain behaviour of a brittle substance.

sediment, the elastic limit is conveniently specified by the yield strength, calculated by dividing the weight added up to rupture by the cross-sectional area of the cylinder.

Suppose the cylinder had consisted of a muddy sediment with a moderate water content? To answer the question for yourself, slowly stretch a cylinder of Plasticine or modelling clay held between the hands. The stress–strain graph now takes the form shown in Figure 8.13. Up to a limiting stress (added weight) the specimen behaves elastically and the stress and strain vary linearly. The elastic limit is much smaller than in the case of the dried specimen, and for any stress above this limit the stress and strain vary non-linearly. At a sufficiently large stress, called the plastic limit, the specimen goes on deforming for no further increase in stress. One need only remove the weights to discover that deformation above the elastic limit is irreversible.

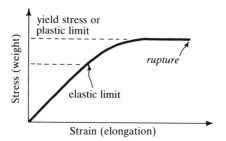

**Figure 8.13** Schematic stress-strain behaviour of a ductile substance.

The increase in deformation is not indefinite, however, and eventually rupture occurs. Like the china clay dispersion in the experiment (Fig. 8.7), the specimen above the elastic limit is likely to elongate markedly, then develop a neck, and finally rupture more or less abruptly. The behaviour described is that typical of a ductile material, which behaves elastically at low stresses but plastically at stresses above the elastic limit. An ideally plastic substance, when stressed beyond the elastic limit, deforms continuously under constant stress and becomes permanently changed in shape. In the case of our specimen, the yield strength is conveniently specified by dividing the added weight at which continuous deformation occurs by the cross-sectional area of the specimen.

It is important to recognize that the deformation behaviour of muddy sediments (as of materials generally) is never either simple or ideal. Water content, temperature and overburden, pore fluid pressure, and the rate of application of the load all affect deformation. A material which is brittle under one set of conditions can be ductile under a different set. Gillott (1968) and Bloor (1957) discuss many aspects of the deformation of muddy sediments and clay-based materials.

### 8.5.3 Elastic constants

An elastic material can be characterized using a number of elastic constants which, in the case of muddy sediments, are important for understanding the behaviour of sound in modern sea and lake deposits (Hamilton 1971, 1979).

The constant of proportionality between stress and strain for an elastic material, called Young's modulus, is defined as stress divided by strain. In the case of the clay cylinders, the measure of strain to be used is the non-dimensional form $\Delta x/x$, where $x$ is the unstretched length of the cylinder and $\Delta x$ the extent by which the length is increased by the stress (Fig. 8.11). Young's modulus takes a unique value, being constant in tension

and compression and in all directions within the substance. But the effect of loading the cylinder is to reduce its diameter by an amount $\Delta y = \Delta z$, where $y = z$ is the cylinder diameter, as well as to stretch it. The non-dimensional ratio $(\Delta z/z)/(\Delta x/x)$, that is, the ratio of the strains, is called Poisson's ratio, which for muddy sediments is a small fraction. Two other elastic constants, known as the modulus of rigidity and the bulk modulus, are also useful and can be calculated given Young's modulus and Poisson's ratio (Jaeger 1956).

### 8.5.4 Mohr's circle and the Coulomb failure criterion

The experiments with cylinders of muddy sediment give us many insights into the properties and behaviour of such materials under stress, but are rather artificial, inasmuch as the force field is one-dimensional.

Let us consider the response of a body of muddy sediment acted on by a two-dimensional force field composed of a greatest principal stress $\sigma_1$ acting in the x-direction and a least principal stress $\sigma_3$ directed parallel to z (Fig. 8.14). Consider the forces acting on a square plane ABCD of unit area, oriented so that two of its edges lie parallel with the y-axis along which an intermediate principal stress might act. The normal to ABCD makes an angle $\alpha$ with the direction of the greatest principal stress. Now AD is of unit length, since ABCD is by definition of unit area, whence OD $= \sin \alpha$ and OA $= \cos \alpha$. To determine the normal stress $\sigma$ and tangential stress $\tau$ acting on ABCD, we must resolve $\sigma_1$ and $\sigma_3$ in turn onto this plane (any intermediate principal stress can be ignored as making no contribution to forces in the xz-plane). When resolving the forces, remember that $\sigma_1$ acts on a plane normal to the x-direction, while $\sigma_3$ acts on a surface normal to the z-axis. With this in mind, the normal force on ABCD is

$$\sigma = \cos \alpha (\sigma_1 \cos \alpha) + \sin \alpha (\sigma_3 \sin \alpha) \quad \text{N m}^{-2} \quad (8.5)$$

$$= \sigma_1 \cos^2 \alpha + \sigma_3 \sin^2 \alpha \quad (8.6)$$

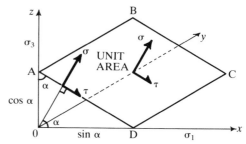

**Figure 8.14** Definition sketch for the calculation of Mohr's circle.

Since the components differ in their direction of action, the tangential force becomes

$$\tau = \sin\alpha(\sigma_1 \cos\alpha) - \cos\alpha(\sigma_3 \sin\alpha) \quad \text{N m}^{-2} \quad (8.7)$$

$$= (\sigma_1 - \sigma_3) \sin\alpha \cos\alpha \qquad (8.8)$$

where $(\sigma_1 - \sigma_3) = \Delta\sigma$, the stress difference.

Making use of the trigonometric rules that $\sin\alpha \cos\alpha = \frac{1}{2}\sin 2\alpha$ and $\sin^2\alpha + \cos^2\alpha = 1$, we can substitute the double angle $2\alpha$ into Equations 8.6 and 8.8, to yield

$$\sigma = \tfrac{1}{2}(\sigma_1 + \sigma_3) + \tfrac{1}{2}(\sigma_1 - \sigma_3)\cos 2\alpha \qquad (8.9)$$

and

$$\tau = \tfrac{1}{2}(\sigma_1 - \sigma_3)\sin 2\alpha \qquad (8.10)$$

Inspection of Equations 8.9 and 8.10 shows that they describe points on a circle in a graph with $\sigma$ as the abscissa and $\tau$ as the ordinate (Fig. 8.15). This convenient curve, called Mohr's circle, is of centre $\tau = 0$, $\sigma = \frac{1}{2}(\sigma_1 + \sigma_3)$ and has a diameter equal to the stress difference $(\sigma_1 - \sigma_3)$. By geometry, at the position $\tau = 0$ and $\sigma = \sigma_3$, any point on the circle subtends the original angle $\alpha$.

Suppose that $\sigma_1$ and $\sigma_3$ are just large enough that rupture occurs along ABCD. The Mohr's circle is then limiting. Now it is possible to estimate in a variety of apparatus the value of $\sigma_1$ necessary to cause the rupture of a specimen for fixed $\sigma_3$. Let us suppose that the failure conditions have been measured for as wide as possible a range of $\sigma_3$ values. We then have a series of Mohr's circles, each a limiting curve (Fig. 8.16). Clearly, an infinite number of limiting circles is possible, which together define an envelope indicating the general condition for the rupture of the material. In

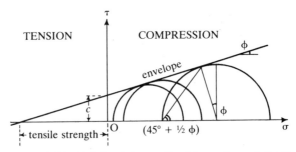

**Figure 8.16** Mohr's circles at the failure of a substance under load.

practice, the envelope over most of its range turns out to be a straight line, of equation

$$\tau = c + \sigma \tan\phi \quad \text{N m}^{-2} \qquad (8.11)$$

where $c$ is the intercept on the ordinate and $\tan\phi$ is the slope of the envelope. This formula, known as Coulomb's equation, has the following physical interpretation. The quantity $\tau$, called the shear strength, is the tangential stress on the failure surface at the moment of failure; it is not a constant, but depends on the normal stress and the slope of the envelope, that is, on the conditions of measurement. The intercept $c$, called the cohesion, occurs for $\sigma = 0$ and represents the inherent strength of the material. The cohesion is therefore a tangential stress, with the units of force per unit area. Another intercept (negative under the conventions adopted) is with the abscissa. This occurs for $\tau = 0$ and is the nominal tensile strength of the material, a quantity which is a normal stress. As the envelope to Mohr's circles tends in practice to bend downwards for $\sigma < 0$, observed tensile strengths are substantially smaller than the nominal values. The constant quantity $\phi$ is called the angle of internal friction and $\tan\phi$ the coefficient of internal friction. Referring to the geometry of Figures 8.14 and 8.16, the angle between the least principal stress $\sigma_3$ and the failure surface ABCD is $(45° + \phi/2)$. Hence the greatest principal stress $\sigma_1$ makes an angle $(45° - \phi/2)$ with the failure surface. Should failure surfaces appear in conjugate sets, the smallest angle between the surfaces will take twice this last value.

Failure or slip surfaces may appear in conjugate or complementary sets, and one such set is represented by the rhomb pattern observed from experiment (Fig. 8.7c). It is clear from Mohr's circle that the direction of the greatest principal stress bisects the dihedral or acute angle between any pair of conjugate shear surfaces. Our clay, for which $(45° - \phi/2) \simeq 30°$, and therefore

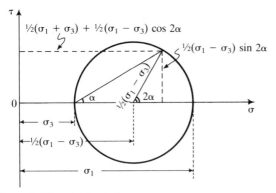

**Figure 8.15** Mohr's circle.

$\phi \simeq 30°$, was therefore under a circumferential compression, as was inferred on independent grounds.

### 8.5.5 Controls on shear strength

Many factors affect the shear strength of water-saturated muddy sediments under specified loading conditions, but the most important are (1) water content and (2) whether or not the sediment has recently been disturbed.

Measurement shows that the shear strength of undisturbed water-saturated muddy sediments decreases very steeply with increasing water content. As the water content, or porosity, represents the amount of potential void space in the sediment, it may therefore be surmised that strength is intimately related to particle packing. At a high porosity, individual particles and particle aggregates are relatively far apart and only weakly bonded. With decreasing porosity, stronger and more numerous bonds develop as the mineral grains approach closer together; additional strength may arise through particle interlocking and, perhaps, the commencement of cementation, as new minerals begin to crystallize at particle contacts. At a low porosity, particle interlocking and cementation contribute most of the strength.

Muddy sediments of moderate to high water content seem to have a metastable fabric. An undisturbed sample, showing a relatively high shear strength, when vibrated or stirred changes into a material of much lower strength, and in some instances even into a liquid. Such materials are described as sensitive and their 'remoulded' strength is much less than the 'unmoulded' value. The china clay dispersion used in the experiment described above is likely to prove markedly sensitive. It seems probable that disturbance has the effect of breaking many of the interparticle bonds. Some naturally occurring muddy sediments are extremely sensitive and present serious problems to civil engineers (Tavenas et al. 1971, Mitchell & Markell 1974, Smalley 1976). Sensitivity is one of the factors that promotes the flow of muddy debris on land and in the oceans.

## 8.6 Erosion of muddy sediments

Muddy sediments are commonly exposed to erosion by wind or water. The affected mud is likely in the first case to be strong and either dry or merely damp, whereas in the second we can expect it to be fully water-saturated and of any strength upwards from liquid or jelly-like. What mechanisms of erosion operate in such a wide range of cases? Broadly, the two most important

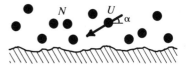

**Figure 8.17** Definition sketch for corrasion.

mechanisms are corrasion and what might be called either fluid stressing or fluid stripping (Allen 1982).

Corrasion is merely sand blasting under natural conditions. In this mechanism, transported sand and gravel particles serve as tools, releasing a fragment from the mud bed at each contact with the bed. The transported particles may be either sliding, rolling, saltating, or, best of all, in suspension. The bed must, of course, be unobscured by stationary grains. What factors affect the rate of erosion by corrasion? Consider a homogeneous cloud of uniform sediment particles, of number concentration $N$ per unit volume, impelled by a current towards a mud bed at an angle $\alpha$ measured from the bed (angle of attack) and at a uniform velocity $U$ (Fig. 8.17). The number of contacts between these particles and the bed per unit area and time is therefore $NU \sin \alpha$. If each particle on contact releases a quantity of mud of average mass $m_c$ from the bed, then the unit rate of bed erosion is

$$\frac{\mathrm{d}m}{\mathrm{d}t} = -m_c NU \sin \alpha \quad \mathrm{kg\ m^{-2}\,s^{-1}} \quad (8.12)$$

where $m$ is the mass lost per unit bed area and $t$ is time. In this simple model the erosion rate increases with the average mass released per contact, the concentration and speed of approach of the tools in the current, and the angle of their attack.

Experiments made by mechanical engineers show that $m_c$ is not a constant, as implied by Equation 8.12, but varies with the properties of the eroded material and with the momentum and angle of attack of the approaching particles. Two varieties of corrasion are recognized. In deformation wear, typical of brittle materials, such as dried mud affected by wind-blown sand, particle impact produces local fractures in the bed and consequent loss of mass as fragments are knocked or fall out. This type of corrasion is most effective at steep angles of attack and ceases below a critical angle set by the properties of the tools and the material eroded (Fig. 8.18). The other mechanism – cutting wear – is typical of ductile materials, such as plastic muds on the sea floor. The impacting tools cut slivers from the bed, acting like woodcarver's chisels, and are most effective at a low to moderate angle of attack (Fig. 8.18). The

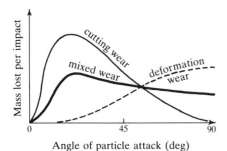

Mass lost per impact

cutting wear

mixed wear

deformation wear

0          45          90

Angle of particle attack (deg)

**Figure 8.18** Erosion rate as a function of angle of attack in varieties of corrasion.

mechanism is ineffective for glancing impacts, as the particles have little or no momentum normal to the bed. Cutting wear is also ineffective at large angles of attack, largely because the grains are then liable to imbed themselves in the surface. A sure sign that a mud bed has suffered cutting wear is the presence of a myriad of cross-cutting but subparallel striae on the surface, accompanied by perhaps a few larger grooves

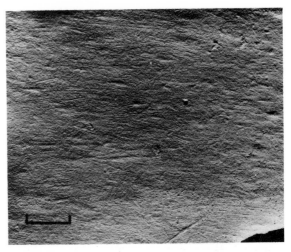

**Figure 8.19** Mud surface (dried sample) corraded by sand-laden tidal currents, Severn Estuary, England. Scale bar is 0.01 m.

**Figure 8.20** Large tool marks preserved as moulds beneath a Jurassic–Cretaceous deep-water sandstone, South Georgia. Scale bar is 0.05 m. Photograph courtesy of Dr P. W. G. Tanner, University of Glasgow.

(Fig. 8.19). Such structures represent a variety of tool marks. A larger kind of tool mark (Fig 8.20), typified by an internal ornament of strictly parallel longitudinal grooves, represents the ploughing of a comparatively large flow-impelled object through a muddy sediment, in a manner resembling the machining of metal. Some other sedimentary structures depending on corrasion are described in section 11.2.4.

The other mechanism of mud erosion, fluid stressing or fluid stripping, depends not on the mediation of transported particles, but on the direct action of the drag force exerted by the current on its bed. It is effective only in aqueous flows, and what happens is largely governed by the strength of the mud. Migniot (1968) formed mud beds by allowing various mixtures to settle in a flume channel, and then applied turbulent currents of plain water to them. For each muddy sediment, a fairly definite critical boundary shear stress $\tau_{cr}$ had to be exerted before erosion began. The value of this erosion threshold increased with the plastic yield strength $\tau$ of the mud, but in two regimes separated by a strength of approximately $3 \, \mathrm{N m^{-2}}$ (Fig 8.21). In the lower regime

$$\tau_{cr} = 0.32\tau^{1/2} \tag{8.13}$$

whereas in the upper regime the relationship is linear and

$$\tau_{cr} = 0.17\tau \tag{8.14}$$

The boundary between these regimes may mark some fundamental change in either particle packing or bed response.

What can be observed when a mud bed is eroded

**Figure 8.22** Mould of transverse waves eroded by the fluid stripping of a soft mud bed in a laboratory water channel. Mean flow velocity is $0.55 \, \mathrm{m s^{-1}}$. Scale bar is $0.01 \, \mathrm{m}$.

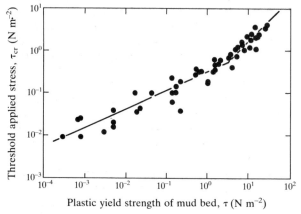

**Figure 8.21** Experimental threshold stress for the fluid stripping of a muddy sediment, as a function of plastic yield strength. Data of Migniot (1968).

through fluid stripping? Laboratory experiments by Migniot (1968) and Allen (1969, 1971) provide some answers.

The surface of a very weak mud bed is thrown by a turbulent aqueous current into travelling waves. Bursting low-speed streaks strip small masses of mud from their crests, as well as from other parts of the surface. Weak to moderately strong muds respond by forming into nearly stationary transverse waves (Fig. 8.22), whose spacing increases with flow depth. Small masses of mud are released from the bed by (1) bursting streaks, and (2) vortices carried to the bed in the mixing layer of the separated flow formed to leeward of each crest. These mud waves can easily be made by putting under a running tap a dinner plate coated with either the china clay dispersion used above or mayonnaise. From beds of strong mud large masses are ripped to give a

**Figure 8.23** Large cabbage-leaf (plumose) marking preserved beneath a Jurassic–Cretaceous deep-water sandstone, South Georgia. Current is from top to bottom. Scale bar is 0.05 m. Photograph courtesy of Dr D. I. M. McDonald (British Antarctic Survey) and Dr P. W. G. Tanner, University of Glasgow.

longtitudinally grooved surface. The mud torn from each groove is rolled up by the current like a carpet, and in places may remain at the downstream end of the furrow. As illustrated by the so-called cabbage-leaf markings preserved beneath turbidity current sandstones (Fig. 8.23), the floor of the furrow may bear a series of trumpet-shaped features which together form a plumose mark. This is a pattern of microfractures characteristic of the sudden rupture of a brittle or semibrittle material. These markings open out in the direction of propagation of the major fracture, like wind expanding in a trumpet bell. The major fracture is constrained in a plane parallel to the original mud surface, presumably in response to a comparatively marked bedding-parallel clay fabric.

The erosion of muddy sediments under natural conditions is clearly a complex phenomenon, and in aqueous environments corrasion and fluid stressing probably operate simultaneously. For these reasons it is impossible to describe the rate of erosion of mud by a unique erosion law. Each case should be treated on its merits.

## 8.7 Drying out

### 8.7.1 An experiment

Muddy sediments tend to shrink and crack on losing moisture beneath the wind and sun. These expressions of desiccation are commonplace where muds are exposed on river floodplains after floods, on lake margins during droughts, and on the higher parts of tidal flats, especially during neap tides and warm weather.

It is easy to produce for oneself the main stages in the desiccation process. Half-fill a pneumatic trough or shallow plastic bowl with water and sprinkle uniformly over the bottom a thin layer of clean sand. Now cautiously pour into the trough a weak aqueous dispersion of clay free from lumps. Let the mixture settle for a few days before carefully siphoning off the excess water with a pipette. Using a glass rod, make a short depression on the surface of the mud somewhere near its centre. Leave the trough in a draught-free place where it can dry out slowly and uniformly, observing the surface of the mud bed at frequent intervals.

Figure 8.24 illustrates some of the features that it is possible to observe during the desiccation of a muddy sediment. The bed at first is quite weak and watery, and the main early change is a progressive downward shrinkage or consolidation, indicating a gradual water loss. At the depression made on the surface there eventually appears a small crack, widest in the middle and with sharp tips. With further moisture loss, the tips of the crack advance until one of them meets the wall of the trough. By this stage or soon after, other fractures will have appeared, either at some new place on the bed, or as a second generation growing orthogonally from the first crack. A continuous crack is now likely to have appeared at the edge of the bed. Eventually, the still-moist bed becomes divided into a series of columns by connected straight or curved cracks. Possibly by this stage, and certainly by the time the mud has fully dried, the columns will have curled up and parted from the sand below.

What points emerge from this experiment? First, in the case of a fully saturated mud, a substantial amount of water must be lost before cracking begins. Secondly, the cracking mud is in tension and fails like a brittle or semibrittle material, as a succession of clean ruptures or cracks appear. Thirdly, the cracks propagate relatively

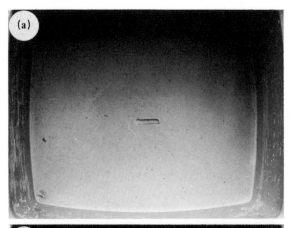

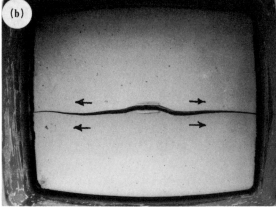

**Figure 8.24** Drying of a mud bed deposited in a rectangular basin. (a) Mud in water-saturated plastic state, with linear defect marked on surface. (b) Fully dried bed with continuous peripheral crack and a linear crack (arrows show growth) initiated at prior defect. Note narrowing of crack away from defect, indicating that growth was comparatively slow.

slowly, by the advance of their tips, and develop in generations. Fourthly, the cracks tend to be disposed orthogonally. Fifthly, irregularities in the bed can localize cracks.

### 8.7.2 Shrinkage

The mud in our experiment started out fully water-saturated and, indeed, quite watery, but ended up completely dry. In the course of losing its moisture, there was a substantial decline in the overall volume of the bed, that is, the volume of mineral particles plus the volume of voids filled with air and/or water. Since shrinkage and cracking is somehow connected with water loss, how does bed volume change with water content?

Haines (1923) explored this question experimentally

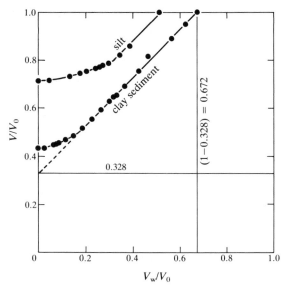

**Figure 8.25** Relative volume changes in two muddy sediments drying out. Data of Haines (1923).

for a clay sediment and a silt (Fig. 8.25). The graph shows the overall volume $V$ of the clay sediment as a fraction of its overall volume $V_0$ at the start of desiccation, plotted against the fractional volume of water $V_w/V_0$. Above a certain fractional water content, the overall sediment volume changes linearly and on a one-to-one basis with the water. Hence in this regime the shrinkage is a straightforward accommodation to water loss. The downward projection of this limb of the graph to meet the ordinate gives us the fractional volume of solids present in the original sample; the fraction indicated agrees with the value of 0.328 calculated from the measured solids density. The lower limb is weakly non-linear and represents substantially lower rates of shrinkage for a given water loss. In this regime, air as well as water occurs between the particles. The particle packing is now collapsing on itself, but the breakdown is at a lower rate than the water loss. The muddy sediment must now be under internal tension.

The relative importance of the two regimes varies with clay mineral content. Figure 8.25 compares the shrinkage of the clay just described with another of Haines' (1923) samples, a silt containing also some sand. As is to be expected, this relatively clay-poor sediment yields a graph with a much longer lower limb.

### 8.7.3 Forces causing shrinkage

It is understandable that a mud should shrink as water is lost while only mineral particles and pore water are present, but why should shrinkage continue once internal voids open to the air have appeared? There seem to be only two possible tensile stresses that can account for the continued shrinkage of the sediment and the increasing disruption and condensation of its fabric. One of these forces is related to the rearticulation within the clay mineral crystals of layer structures which had earlier been separated by the uptake of loosely bonded interlayer water. The other originates as surface tension films develop under the influence of air entering the pores. Surface tension will act whatever the mineralogy of the mud, but the rearticulation force cannot be significant unless substantial quantities of smectite are present. The following analysis is therefore restricted to surface tension.

What is surface tension? Surface tension is a property manifested only by liquid. A liquid poured into a wide container tends to form a flat surface, whereas a liquid released in small amounts from the end of a narrow tube tends to form spherical drops. In each case the free surface bounding the liquid has a minimum surface area for the mass involved. Independently of gravity, there must therefore exist a force tending to draw the elements of the liquid together. Now the molecules of a liquid are closely packed and strongly attracted to each other, but in violent relative motion. Within the interior of a body of liquid, the attractive force on a molecule is on the average uniform in all directions. A molecule at the surface, however, can be attracted only by molecules lying below. It therefore tends to move inwards. But as the molecule occupies a certain volume, its inward movement tends to shrink the surface area. By virtue of the inward-acting force, a molecule at the surface of a liquid possesses a certain amount of potential energy, the amount of which per unit area is called the surface tension or free surface energy. Surface tension has the units of newtons per metre, abbreviated to $N\ m^{-1}$, and acts along the local liquid surface.

The effect of surface tension on a solid in contact with a liquid depends to some extent on the degree to which the solid is wetted. Clean water spreads uniformly over chemically clean glass and in a clean glass capillary tube forms a spherical concave-up meniscus. On a greasy or dusty surface, however, water forms near-spherical globules. Similar drops are afforded by mercury, no matter how clean the surface, and in a capillary tube this liquid metal forms a convex-up meniscus. The wettability of a particular solid by a particular liquid can be described by means of the contact angle between the two, that is, the angle between the solid surface within the liquid and the tangent to the liquid surface where it meets the solid (Fig. 8.26). The angle for clean glass in clean water is zero, but for mercury and glass is approxi-

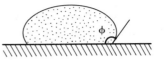

**Figure 8.26**  Definition sketch for contact angle.

mately $135°$. Glass and the common rock-forming minerals afford a small to moderate contact angle with water having a slightly contaminated surface.

Natural muds, composed of haphazardly packed grains of a range of sizes, are far too complicated for the tensile strength created within them by surface tension to be predicted mathematically. A simple two-dimensional model, however, gives us some insight into the likely magnitude of the tensile stress and its relationship to water content.

Consider in Figure 8.27 a sediment composed of straight cylindrical particles of uniform diameter $D$, arranged parallel to each other so as to touch in planes parallel to two of the planes of an orthogonal coordinate system. The unit cell for this two-dimensional packing comprises four touching circles of diameter $D$ centred at the corners of a square. Let the free water

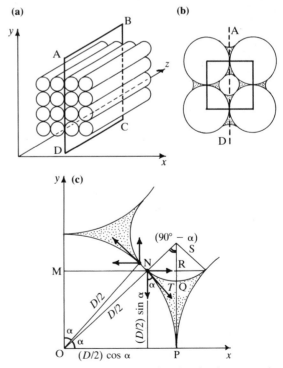

**Figure 8.27**  Definition sketch for the development of a surface tension force within a partly air-filled sediment composed of parallel cylindrical particles.

present be distributed uniformly between the particle contacts, and the contact angle between water and solids be zero. A plane ABCD dividing the particles is subject to a certain tensile stress $\sigma$ on account of surface tension along the internal air–water interfaces. Taking one side of ABCD, there are two meniscus edges per quadrant cut off by the borders of the unit cell. Let the lower meniscus edge subtend an angle $\alpha$ from the origin of the $x$-axis, and the upper meniscus edge subtend the same angle from the $y$-direction (Fig. 8.27c). The components normal to ABCD of the surface tension $T$ per unit cylinder length are $T\sin\alpha$ at the lower contact and $T\cos\alpha$ at the upper. But of these, only $T\sin\alpha$ acts on a normal plane between two particles, and so contributes to $\sigma$. As there are two such contacts within the half of the unit cell divided off by ABCD, and there are $1/D$ unit cells per unit distance parallel with the $y$-direction

$$\sigma = \frac{2T\sin\alpha}{D} \quad \text{N m}^{-2} \qquad (8.15)$$

that is, the normal stress is directly proportional to the surface tension and inversely proportional to grain size.

Equation 8.15 states that the normal stress also varies with the angle subtended by the meniscus edges at the particle centres, up to the value of $\alpha = \pi/4$ radians when the particles become wholly surrounded by water. Exploiting the geometry shown in Figure 8.27, and noting that the overall area of the unit cell is $D^2$, the fractional water content $P_w$ is found to be related to $\alpha$ through

$$P_w = 2\left[\sin\alpha - \tfrac{1}{2}(\alpha + \sin\alpha\cos\alpha)\right.$$

$$\left. - \frac{(1-\cos\alpha)^2}{2}\left(\frac{(\pi/2 - \alpha)}{\cos^2\alpha} - \frac{1}{\tan(\pi/2 - \alpha)}\right)\right] \qquad (8.16)$$

where $\alpha$ in the second and third terms is measured in radians. Referring to Figure 8.27c, the first term within the main bracket represents the area of the rectangle OPRM, the second term the areas of the sector OPN and the triangle ONM, and the third term the area of the sector QSN less the triangle RSN. The maximum fractional water content, or porosity, of the packing is equal to $(1 - \pi/4) \approx 0.2146$. The fractional water content when $\alpha = \pi/4$, that is, when the meniscus of a symmetrically placed cylinder of air just touches the particles, is approximately 0.0799.

Figure 8.28 shows Equation 8.15 plotted against

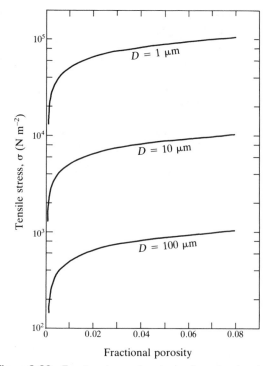

Figure 8.28 Tensile stress due to surface tension in a wetted sediment composed of cylindrical particles, as a function of fractional water content and cylinder diameter.

Equation 8.16 for three values of $D$, and on the assumption that the pore water is of zero contact angle and $T = 0.073 \, \text{N m}^{-1}$. The tensile stress due to surface tension is substantially constant over most of the permissible range of fractional water content. Its value for particle sizes representative of muds is very large, of the order of $10^4$–$10^5 \, \text{N m}^{-2}$. In a real mud the tensile stress would be expected to build up gradually to some maximum with progressive water loss, as menisci appeared in increasingly small pore spaces, before declining as complete dryness was approached and even the smallest pores emptied. The predicted stresses are much greater than the cohesion of muddy sediments, even very strong ones. Such deposits should inevitably rupture as they dry, on account of their comparatively weak mineral skeleton. Finally, Figure 8.28 reveals that the surface tension stress is quite large even in sediments as coarse as sand. It is presumably for this reason that moist sands form cliffs several decimetres high along stream banks. A drying sand, however, does not crack like a mud, on account of the great strength of quartz as compared to open-textured clay mineral aggregates. The quartz grains can accommodate the tensile stress wholly through elastic deformation.

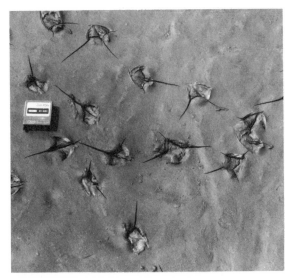

**Figure 8.29** Desiccation cracks initiated at the prints of birds' feet, Severn Estuary, England. Scale box is 0.05 m square.

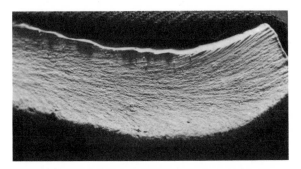

**Figure 8.30** Plumose marking on the vertical side of a desiccation crack polygon (0.025 m thick). Crack spread from left to right.

### 8.7.4 Development and scale of cracks

It was found experimentally (Fig. 8.24) that a mud cracks gradually on drying out, and that the ruptures can be localized by inhomogeneities in the material. Moreover, the ruptures appear in generations and tend to be disposed orthogonally. What are the mechanics underlying these observations? Lachenbruch (1962) developed a theory of cracks, applicable to muddy sediments, that depends on the slow growth of successive fractures, and their initiation at defects in the affected material.

Consider an extensive layer of muddy sediment that is drying out. Owing to evaporation at the surface, the rate of moisture loss declines downwards through the

**Figure 8.31** Large orthogonally related desiccation cracks, Severn Estuary, England. Spade gives scale.

layer, so that the tensile stress due to drying also decreases downwards. Below a certain depth the horizontal forces acting within the layer become compressive, on account of the weight of sediment above. As the mud dries out, strain energy accumulates within it, to the point where the tensile stress exceeds the tensile strength of the material at some place, and the layer cracks and consequently shortens. The body of muddy sediment can be likened to a stretched chain. There is one link weaker than all the others, and it is here, at what amounts to a defect in the chain, that rupture will occur. In the case of a mud layer, the defect could be a local thinning, a concealed pebble or shell, or a footprint or worm track on the surface. Now the fracture that appears in the mud will extend downwards to the level of zero horizontal stress and will relieve the stress in the adjacent sediment to a lateral extent increasing with its depth. Because the first crack is localized by the most pronounced defect in the layer, the remaining portions of muddy sediment will be stronger than the layer just before the first crack appeared. A further period of drying is therefore necessary before a second generation of cracks will form at the more pronounced of the remaining defects. As the first crack will have released the stress in its vicinity to a much greater extent normal to than parallel to itself, any new cracks departing from or approaching the fracture will tend to do so orthogonally, for cracks tend to form normal to the direction of greatest tensile stress. Thus with progressive drying out, successive generations of orthogonally related cracks will appear, at a spacing related to and increasing with the thickness of the mud layer. The locations of the cracks will tend to be determined by the magnitude and spatial distribution of the defects within the layer.

Lachenbruch's (1962) model gives a qualitatively accurate explanation of desiccation cracks in muddy sediments. Ruptures are observed to be initiated at such surface defects as bird's footprints (Fig. 8.29), gastropod trails (Baldwin 1974) and worm tracks (Soleilhavoup & Bertouille 1976), by such internal defects as bivalve shells (Kues & Siemers 1977), and by variations of layer thickness, such as result from the presence of ripple marks in underlying beds (Donovan & Foster 1972). The experiments of Corte and Higashi (1964) confirm that cracks grow slowly, from one or a few centres, and in generations. Plumose marks on the sides of the fractures (Fig. 8.30) allow the directions of crack propagation to be established (Corte & Higashi 1964). Typically, cracks are orthogonally related (Fig. 8.31), and their spacing is observed to increase with depth (Donovan & Archer 1975).

## Readings

Allen, J. R. L. 1969. Erosional current marks of weakly cohesive mud beds. *J. Sed. Petrol.* **39**, 607–23.

Allen, J. R. L. 1971. Transverse erosional marks of mud and rock. *Sed. Geol.* **5**, 167–385.

Allen, J. R. L. 1982. *Sedimentary structures*, Vol. I. Amsterdam: Elsevier.

Baldwin, C. T. 1974. The control of mud crack patterns by small gastropod trails. *J. Sed. Petrol.* **44**, 695–7.

Beutelspacher, H. and H. W. van der Marel 1968. *Atlas of electron microscopy of clay minerals and their admixtures*. Elsevier: Amsterdam.

Bloor, E. C. 1957. Plasticity – a critical survey. *Trans. Br. Ceram. Soc.* **56**, 423–81.

Brindley, G. W. and G. Brown 1980. *Crystal structure of clay minerals and their x-ray identification*. Monograph Mineral Soc., no. 5.

Corte, A. E. and A. Higashi 1964. *Experimental research on desiccation cracks in soil*. Rep US Snow, Ice and Permafrost Res. Establ., no. 66.

Donovan, R. N. and R. Archer 1975. Some sedimentological consequences of a fall in the level of Haweswater, Cumbria. *Proc. Yorks. Geol. Soc.* **40**, 547–62.

Donovan, R. N. and R. J. Foster 1972. Subaqueous shrinkage cracks from the Caithness Flagstone Series (Middle Devonian) of northeast Scotland. *J. Sed. Petrol.* **42**, 309–17.

Gillott, J. E. 1968. *Clay in engineering geology*. Amsterdam: Elsevier.

Grass, A. J. 1971. Structural features of turbulent flow over smooth and rough boundaries. *J. Fluid Mech.* **50**, 233–55.

Haines, W. B. 1923. The volume changes associated with variations of water content in soil. *J. Agr. Sci.* **13**, 296–310.

Hamilton, E. L. 1971. Elastic properties of marine sediments. *J. Geophys. Res.* **76**, 579–604.

Hamilton, E. L. 1979. $V_p/V_s$ and Poisson's ratios in marine sediments and rocks. *J. Acoust. Soc. Am.* **66**, 1093–101.

Jaeger, J. C. 1956. *Elasticity, fracture and flow*. London: Methuen.

Kues, B. S. and C. T. Siemers 1977. Control of mudcrack patterns by the infaunal bivalve, *Pseudocyrena*. *J. Sed. Petrol.* **47**, 844–8.

Lachenbruch, A. H. 1962. *Mechanics of thermal contraction cracks and ice-wedge polygons*. Spec. Pap. Geol. Soc. Am., no. 70.

McCave, I. N. and D. J. P. Swift 1976. A physical model for the rate of deposition of fine-grained sediments in the deep sea. *Bull. Geol. Soc. Am.* **87**, 541–6.

Mantz, P. A. 1978. Bedforms produced by fine, cohesionless, granular and flaky sediments under substantial water flows. *Sedimentology* **25**, 83–103.

Migniot, C. 1968. Études propriétés physique de différents sédiments très fins et de leur comportement sous des actions hydrodynamiques. *Houille Blanche* **23**, 591–620.

Mitchell, R. J. and A. R. Markell 1974. Flow-sliding in sensitive soils. *Can. Geotechn.* **11**, 11–31.

Smalley, I. 1976. Factors relating to the landslide process in Canadian quickclays. *Earth Surf. Processes* **1**, 163–72.

Smart, P. and N. K. Tovey 1981a. *Electron microscopy of soils and sediments: techniques*. Oxford: Clarendon Press.

Smart, P. and N. K. Tovey 1981b. *Electron microscopy of soils and sediments: examples*. Oxford: Clarendon Press.

Soleilhavoup, F. and H. Bertouille 1976. Figures de desiccation observées lors de crues de L'Oued à Laghouat (Sahara Septentrional). *Rev. Géomorph. Dyn.* **24** (3), 81–98.

Tavenas, F., J. Y. Chagnon and P. La Rochelle 1971. The Saint-Jean Vianney landslide: observations and eyewitness accounts. *Can. Geotechn.* **8**, 463–78.

Van Olphen, H. 1963. *An introduction to clay colloid chemistry*. New York: Interscience.

Zabawa, C. F. 1978. Microstructure of agglomerated suspended sediments in northern Chesapeake Bay estuary. *Science* **202**, 49–51.

# 9 Creeping, sliding and flowing

Kinds and distribution of mass movements — effective stress — soil creep — rotational and translational slides — debris flows — mass-movement associations.

## 9.1 Introduction

So far we have looked almost exclusively at situations in which sediment particles are transported as the result of the drag exerted on them by moving fluid. In these situations, in which the fluid drives the sediment, the concentration of the transported particles is very small overall, although locally it may be high. But as a sediment-laden flow becomes richer in debris, there comes a stage when the solids are no longer driven by the fluid. Instead, and normally under the influence of gravity, the solids and the fluid move in harmony together, combining into a new material which, although multi-phase, alone constitutes the flow. Such flows are called mass movements. They are characterized by a high solids concentration, movement of the solids *en masse* and, generally speaking, a laminar style of motion. Natural examples of mass movements abound, both on land and in the oceans, and some are of considerable importance in the earlier stages of denudation and sediment transport. The study of mass movements provides a major point of contact between sedimentologists, geomorphologists and engineers concerned with soil mechanics.

## 9.2 Mass movements in general

The coherent movement of debris under its own weight expresses itself in a huge variety of ways that are difficult to classify and are of unequal sedimentological interest. The difficulties of classification, and the diversity of available criteria, can be judged from the work of Ward (1945), Varnes (1958), Dott (1963), Skempton

and Hutchinson (1969) and Prior and Coleman (1979), spanning all of the disciplines given introductory mention. As sedimentologists, however, our concern is chiefly with the behaviour of soil materials (in the pedologists' sense) and of unlithified and partly lithified sediments. The mass movements generated from these materials (Fig. 9.1) are most tellingly considered in terms of (1) the cohesiveness of the debris involved, and (2) the overall geometry of the transported mass.

Much of the land is blanketed by soil (Fig. 9.1) which, on all but the lowest slopes, is moving down hill at a rate perceptible only over long time periods (creep). The soil blanket is extremely thin compared to its dimensions parallel to the ground surface. Most soils are moist, cohesive to a moderate degree, and granular on a number of scales, but exhibit marked seasonal variations particularly of moisture content.

The simplest avalanches (Fig. 9.1) involve cohesionless, granular materials, for example, sand on the leeward slope of a ripple or dune (Fig. 4.23), and angular stones on the talus slope below an eroding cliff. Avalanches are longest in the downslope direction and generally lobe-shaped in plan. Their thickness is of the order of 0.01–0.1 times the downslope length.

Creeping soils and avalanches differ in many ways, but share one property distinguishing them from slides. Whereas the components of avalanches and soils become fairly well mixed during downslope movement, the lithological elements of a slide tend (Fig. 9.1) to retain their integrity, remaining recognizably similar to the parent deposits in terms of stratigraphy and details of bedding.

A slide moves over a basal failure surface. As seen in

159

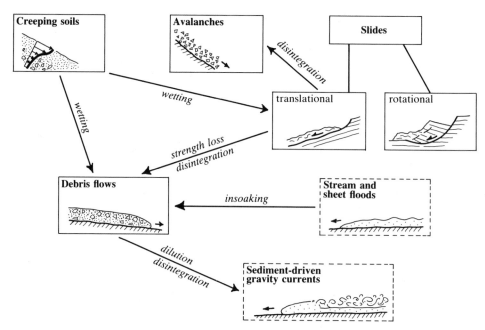

**Figure 9.1** Outline classification of mass movements.

a downslope slice, this underlying surface may be either virtually planar, any significant curvature being limited to the extreme upslope end, or noticeably curved and concave-up over its whole length, in which case it becomes a variety of listric fault. The term translational is applied to slides that are very long compared to their thickness and surmount an essentially planar, slope-parallel glide surface. Such slides can travel far. The ratio of thickness to downslope length for translational slides is between about 0.03 and 0.06 in the case of those on land (Skempton & Hutchinson 1969), and of the order of 0.01 or 0.02 for those beneath the sea (Prior & Coleman 1979). A rotational slide, lying on a concave-up failure surface, invariably represents a relatively small downslope movement. Their thickness : length ratio ranges on land between 0.15 and 0.33 and averages about 0.22 (Skempton & Hutchinson 1969). Slightly lower ratios (0.12–0.25) appear to characterize submarine examples (Prior & Coleman 1979).

In one special case, a translational slide can be transformed wholly or partially into an avalanche. The surface sand forming the leeward slope of an aeolian dune can become cohesive either because of wetting (dew, rainfall), drying after wetting, or freezing after moistening. Slope failure leads initially to a translational slide composed of a few large slabs of coherent sand rafted on a layer of flowing dry particles. With continued downslope movement, these slabs pro-

gressively disintegrate, and can end up either as numerous small masses or even as single grains, depending on the cause of their initial cohesion.

Creeping soils commonly evolve into translational slides. This occurs chiefly where a soil overlying impervious bedrock such as a shale formation becomes sufficiently wet.

A debris flow (Fig. 9.1) is a sluggish to fast movement of granular debris mixed with a roughly equal to subordinate quantity of water over a low to moderate slope, either on the land or beneath water. Most debris flows are channelized, so that they are long and narrow in plan, but can change into lobed sheets where they spread over open ground. Their thickness is generally less than 0.01 times the downslope length. Normally, the debris is an ill-sorted mixture of sand and rock, blended with a variable amount of silt and clay. The term mud flow is appropriate for debris flows in which the muddy component is abundant enough to form a continuous matrix supporting the coarser elements. Speaking generally, debris flows are well mixed, though some include rafts of as yet undisrupted source materials.

Debris flows arise in several ways. Water loss to the dry ground below can so concentrate the sediment in a stream or sheet flood that a debris flow is gradually created (Fig. 9.1). Saturation of dry or moist soils can lead to debris flows, as can the saturation of muddy sediments that have become extensively fissured through repeated wetting and drying. Some soils lose strength dramatically on being disturbed and so become debris flows. Many translational slides found on muddy sea beds appear to disintegrate to form debris flows as they move down slope.

A sediment-driven gravity current (see Ch. 12) is a flow which derives its motion from an excess of density conferred by the presence of dispersed solids, but which contains a concentration of particles intermediate between that typical of a debris flow and the value representative of ordinary rivers and the wind. Sediment-driven gravity currents (Fig. 9.1) can arise beneath the atmosphere as well as in lakes and oceans. Turbidity currents are the most important sediment-driven gravity currents of aqueous environments. They are surge-like flows which in perhaps most cases record the dilution and further disintegration of sub-aquatic debris flows (Hampton 1972). Sediment-driven gravity currents range in plan from long and tongue-like to sheet-like and lobed, depending on the form of the surface traversed. They appear to be extremely thin compared to their downstream length, more so even than debris flows.

## 9.3 Soil creep

### 9.3.1 Mechanisms and rates

How is it that soils can move down slope under gravity at rates that are perceptible only over long periods? Carson and Kirkby (1972) divide the mechanisms involved into two classes.

Continuous or rheological creep stems directly from the properties of the soil clay mineral particles. As we saw in Section 8.2, crystals of these minerals possess electrochemical surface charges which bond the particles together. The bonds are individually weak but, because of the small sizes of the particles involved, occur in very large numbers within a unit volume of soil. These bonds continually break and reform under the action of the downslope weight of soil, the resulting creep representing the cumulative effect of very many but individually insignificant ruptures.

The other and much more important creep mechanisms are diurnal to seasonal and either purely physical in character or related to organic activity in the soil.

The physical mechanisms include (1) the heating and cooling of soil particles, (2) the wetting and drying of soil clays, particularly the expandable species, and (3) the freezing (9% volume increase) and thawing of soil water. How do these mechanisms work? Each causes an

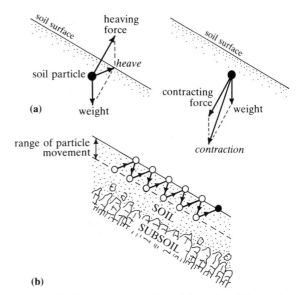

**Figure 9.2** Movements experienced by soil particles during soil creep.

expansion and contraction of the soil which, ideally, takes place wholly normal to the ground surface and declines downwards (Fig. 9.2a). The expansion or heave occurs along the resultant between the downward-acting weight of the soil above a selected soil element and the heaving force itself acting upwards normal to the ground surface. The return movement, or contraction, occurs along the resultant between the downward-acting weight and the downward-acting normal contraction force, assuming that the return is not due to gravity alone. Hence a soil particle tends to describe an irreversible zig-zag downslope path under the action of repeated expansion–contraction cycles (Fig. 9.2b).

Downslope movement also occurs in response to the activities of organisms as they rework the soil. For example, particles tend to spill down slope from an earthworm casting. Lastly, fine particles tend to be transported down slope in suspension within the soil, as rain water or snow-melt seeps through the internal voids.

Climate seems to be the main factor controlling the relative importance of creep mechanisms. Under tropical and subtropical conditions creep is probably related mainly to wetting and drying, organic activity, and heating and cooling. To these under temperate and maritime conditions can be added some winter contribution from freezing and thawing. Creep in montane and arctic climates is due mainly to freezing and thawing. This sporadic and relatively rapid form of creep (solifluction) combines the direct effects of heave due to freezing with enhanced rheological creep, and perhaps some internal slipping when the soil becomes saturated during the thaw.

How can such a slow process as soil creep be measured? Some of the more widely and frequently used of the many techniques proposed (Carson & Kirkby 1972) include the measurement of (1) the displacement of objects on the soil surface, (2) the progressive distortion of buried flexible pipes, and (3) the displacement of markers fixed to the sides of pits excavated in the soil, filled for a period, and then reopened. It is difficult with these methods to achieve any great accuracy and consistency.

What rates typify soil creep? They show some dependence on climate but are very small compared to other kinds of mass movement (Saunders & Young 1983). Solifluction-dominated creep under arctic and montane conditions proceeds at surface rates up to $0.3 \, \mathrm{m \, a^{-1}}$, although values between 0.01 and $0.1 \, \mathrm{m \, a^{-1}}$ are more representative. Temperate continental conditions afford surface rates between 0.002 and $0.01 \, \mathrm{m \, a^{-1}}$. Surface rates seldom exceed $0.002 \, \mathrm{m \, a^{-1}}$ in temperate maritime climates. Surface rates of the order of $0.005 \, \mathrm{m \, a^{-1}}$ have been measured from the rainforest environment, with its deep, moist soils rich in fauna and plant roots.

The rate of creep decreases with depth below the ground surface. Solifluction can give rise to detectable movements as deep as 5–10 m but is rarely significant at depths greater than 0.5–1 m. In other climatic zones, appreciable movement is restricted to the topmost few decimetres of the soil. Vertical profiles of the rate of creep (Fleming & Johnson 1975) vary from convex-up, through essentially rectilinear to inflected (Fig. 9.3), with perhaps the last two forms being the most common.

### 9.3.2 Models of soil creep

How can we create a mathematical model of soil creep? So many creep mechanisms combine in each case to determine the final soil motion that it seems inappropriate to single out any one as the sole basis for a model. Let us instead regard the soil material as possessing an apparent viscosity, where viscosity is the material constant that relates the shearing stress in a fluid to the rate of strain (Sec. 1.8). We are not of course giving the soil a true molecular viscosity, but rather a complex and only partly a continuum property that is analogous to viscosity. Using this analogy, we wish to calculate under steady conditions how the rate of downslope creep $U$ might vary with normal distance $y$ below the ground surface, that is, the function $U(y)$.

Consider in Figure 9.4 a column of soil material of uniform bulk density $\gamma$ arranged normal to the surface of the soil. The base of the column is of unit area and lies parallel to the ground at a normal depth $y$ below the surface, which slopes at an angle $\beta$. The weight of the column can be resolved into components parallel and normal to the slope. The slope-parallel component constitutes a shear stress acting on the base of the column

$$\tau = \gamma g y \sin \beta \quad \mathrm{N \, m^{-2}} \tag{9.1}$$

where $\tau$ is the stress. Introducing $\eta_a(y)$ as the apparent viscosity, another expression for the stess is given by Equation 1.31 for laminar (i.e. creep) conditions

$$\tau = \eta_a(y) \frac{\mathrm{d}U}{\mathrm{d}y} \tag{9.2}$$

where $\mathrm{d}U/\mathrm{d}y$ is the rate of change of the creep rate with depth, that is, the velocity gradient.

Suppose first of all that the apparent viscosity is independent of depth. Eliminating $\tau$ between Equations 9.1 and 9.2

$$\eta_a \frac{\mathrm{d}U}{\mathrm{d}y} = -(\gamma g \sin \beta) y \quad \mathrm{N \, m^{-2}} \tag{9.3}$$

which on rearrangement becomes

$$\frac{\mathrm{d}U}{\mathrm{d}y} = -\frac{(\gamma g \sin \beta)}{\eta_a} y \quad \mathrm{s^{-1}} \tag{9.4}$$

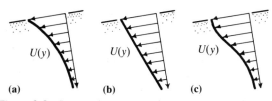

**Figure 9.3** Range of velocity profiles (schematic) observed from creeping soils.

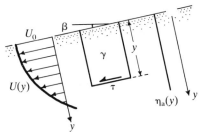

**Figure 9.4** Definition sketch for soil creep under depth-constant apparent viscosity.

where the minus sign indicates that the velocity declines downwards. Integrating Equation 9.4 leads to

$$U = -\frac{(\gamma g \sin \beta)}{2\eta_a} y^2 + C \quad \text{m s}^{-1} \qquad (9.5)$$

where $C$ is an integration constant. To find $C$ put $y = 0$ in Equation 9.5, whence $C$ is the creep rate $U_0$ measured at the ground surface. The velocity profile is therefore

$$U = U_0 - \frac{(\gamma g \sin \beta)}{2\eta_a} y^2 \qquad (9.6)$$

the rate of creep declining as the depth squared (Fig. 9.4). Equation 9.6 gives us a velocity gradient of zero at the ground surface, which is physically sound as the stress there is zero, and a velocity of zero at a finite depth below the surface, predictable only if we know $U_0$ and $\eta_a$. It is more interesting to calculate $\eta_a$ from observed rates of surface creep and limiting depths of movement. For example, an apparent viscosity of approximately $5.2 \times 10^{11} \text{ N s m}^{-2}$ is obtained for the representative values $\gamma = 1600 \text{ kg m}^{-3}$, $\sin \beta = 0.167$ ($\beta = 10°$), $y = 0.25$ m and $U_0 = 0.005 \text{ m a}^{-1}$.

The velocity profile given by Equation 9.6 is not especially like observed profiles (Fig. 9.3), so we should abandon our assumption that $\eta_a$ is a constant. We can instead regard the apparent viscosity as increasing with depth, since (1) the local soil moisture content and temperature fluctuate less downwards, (2) the average local temperature increases downwards, and (3) plant roots and the soil fauna decrease downwards in density. Let us try an exponentially increasing apparent viscosity

$$\eta_a = \eta_{a,0} \exp(ky) \quad \text{N s m}^{-2} \qquad (9.7)$$

where exp stands for the base of natural logarithms, $\eta_a$ is the apparent viscosity at depth $y$, $\eta_{a,0}$ the apparent viscosity at the surface $y = 0$, and $k$ a dimensional coefficient describing the rate of change of apparent viscosity with depth (Fig. 9.5). This gives us a finite apparent viscosity at the surface but an increasingly steep downward increase. Introducing Equation 9.7 into Equation 9.3 and rearranging gives

$$\frac{dU}{dy} = -\frac{(\gamma g \sin \beta)}{\eta_{a,0} \exp(ky)} y \quad \text{s}^{-1} \qquad (9.8)$$

which on integrating affords

$$U = \frac{(\gamma g \sin \beta)}{\eta_{a,0}} \frac{\exp(-ky)}{k^2} (1 + ky) + C \quad \text{m s}^{-1} \quad (9.9)$$

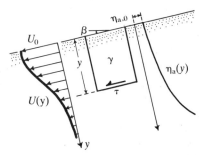

**Figure 9.5** Definition sketch for soil creep under apparent viscosity increasing exponentially downwards.

where $C$ is another integration constant. To find this constant we put $U = 0$ at $y = \infty$ in Equation 9.9, whence $C = 0$. Hence

$$U = \frac{(\gamma g \sin \beta)}{\eta_{a,0}} \frac{\exp(-ky)}{k^2} (1 + ky) \qquad (9.10)$$

making the surface velocity ($y = 0$)

$$U_0 = \frac{(\gamma g \sin \beta)}{\eta_{a,0}} \frac{1}{k^2} \qquad (9.11)$$

The velocity decreases downwards (Fig. 9.5), like the previous model, but the profile is inflected, as in many real examples (Fig. 9.3). At the ground surface the velocity gradient is again zero. The depth to the inflection point on the profile is determined by the value of $k$, and increases as this coefficient declines. Its position is given by differentiating Equation 9.10 and setting $d^2U/dy^2 = 0$. For sufficiently large values of $k$, the inflection point lies so close to the surface that the profile would in practice be difficult to distinguish from the convex-up sort (Fig. 9.3). The model with exponentially increasing apparent soil viscosity gives quite realistic results but suffers from the obvious defect of predicting some movement at all depths.

## 9.4 Effective stress and losses of strength

### 9.4.1 Effective stress

Before discussing the other forms of mass movement – slides and debris flows – it is useful to examine some ideas and findings concerned with (1) the forces actually effective in controlling the failure of sediments under load, and (2) the effects of disturbance on the shear strength of these materials.

In Chapter 8 we examined the conditions under which a muddy sediment fails and introduced the Coulomb

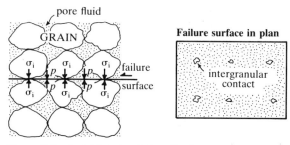

**Figure 9.6**  Definition sketch for effective stress.

equation (Eqn 8.11) as the failure criterion. That relationship includes the normal stress acting on the material, but this stress as it might arise under natural conditions was not interpreted further. It is now important to look at the contributions to the normal stress.

Imagine a failure surface cutting a sample of water-saturated sand (Fig. 9.6). Within an element of area $A$ on this surface, there are intergranular contacts whose total area is $A_{gr}$ and across which an intergranular normal stress $\sigma_i$ N m$^{-2}$ is exerted. The remaining area $(A - A_{gr})$ is taken up by the pore fluid, at a pressure $p$ N m$^{-2}$. Under conditions of static equilibrium, the total force acting on the failure plane therefore comprises a component arising from intergranular contacts and another due to the pore fluid pressure. Hence

$$\sigma A = \sigma_i A_{gr} + p(A - A_{gr}) \quad \text{N} \tag{9.12}$$

and dividing through by $A$

$$\sigma = \sigma_i \frac{A_{gr}}{A} + p\left(1 - \frac{A_{gr}}{A}\right) \quad \text{N m}^{-2} \tag{9.13}$$

where $\sigma$ is the applied or total unit normal stress. Now the static fluid cannot sustain a shear stress, so the only force preventing failure of the sample (other than cohesion) is that arising between the grains. Hence the intergranular force is the effective force, which may be defined as $\sigma_e = \sigma_i(A_{gr}/A)$. Making this substitution in Equation 9.13, and introducing the new coefficient $k = A_{gr}/A$, we arrive at

$$\sigma_e = \sigma - p(1 - k) \quad \text{N m}^{-2} \tag{9.14}$$

which reads approximately

$$\sigma_e = \sigma - p \tag{9.15}$$

as $k$ in practice is very small. The most general form of the Coulomb equation is therefore

$$\tau_s = c + (\sigma - p)\tan\phi \quad \text{N m}^{-2} \tag{9.16}$$

where $\tau_s$ is here used for the shear strength, c is the cohesion as before, and $\phi$ the angle of internal friction.

The Coulomb failure criterion in this form has interesting and important implications. When the pore pressure is negative, as in a partly saturated soil, with its numerous surface tension films, the applied stress and the pore pressure reinforce each other and so promote the stability of the sediment. An increase in pore pressure, such as might be induced by seepage or the appearance of bacterially formed gas, lowers the shear strength and promotes failure.

### 9.4.2  Loss of strength

When a mass of sediment that has been static for a period becomes involved in a slide, the motion normally causes a degree of internal dislocation of the material. The dislocation involves modification of the particle fabric that existed prior to failure, in some instance to the point of its total disruption, and may express itself macroscopically in the form of internal faults, folds and pull-aparts. The overall effect is that the disturbed sediment quickly loses strength, so that new and unexpected modes of behaviour become permissible, as in the metamorphosis of a slide into a debris flow.

The effect of disturbing a sediment, that is, remoulding it, can be assessed by measuring the sensitivity of the material, defined as the ratio of the strength of the undisturbed sample to that of the remoulded sample at the same water content (Skempton & Northey 1952, Skempton 1953). Sands and many muddy sediments afford sensitivities of 1–5 and are said to possess a low to medium sensitivity. Some muddy sediments, however, possess much larger sensitivities, up to the order of 100, and are called highly sensitive or extra-sensitive. These materials become viscous mobile slurries on being remoulded (Crawford 1963).

The loss of strength that accompanies the remoulding of muddy sediments is either reversible or irreversible. A reversible strength loss, normally associated with comparatively low sensitivity values, implies that a particle fabric of the same character as was present in the undisturbed deposits will eventually reform in the remoulded material (Fig. 9.7a). Materials behaving in this way – cream, emulsion paint, and many aqueous clay mineral dispersions – are said to be thixotropic. Highly sensitive materials are typified by an irreversible loss of strength (Fig. 9.7b). Remoulding in their case must permanently destroy the initial particle fabric.

The most sensitive deposits known are the so-called quickclays of Alaska, the Canadian Lowlands and Scandinavia (Mitchell & Markell 1974, Mitchell & Klugman 1979). These muddy sediments settled from

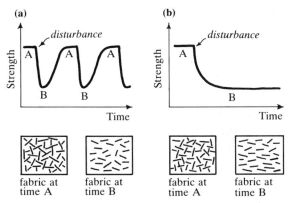

**Figure 9.7** Strength-time graphs for disturbed (a) thixotropic and (b) highly sensitive muddy sediments.

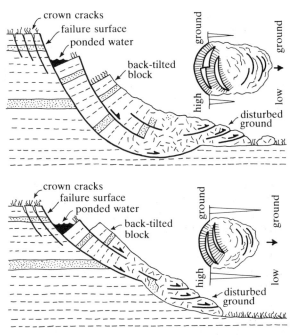

**Figure 9.8** Schematic representation in vertical section and plan of varieties of rotational slides.

glacial melt waters released into brackish seas. Texturally, quickclays are silts, consisting mainly of the common rock-forming minerals in finely crushed form and with little in the way of clay minerals (Bentley & Smalley 1978). Indeed, their sensitivity declines steeply with increasing clay mineral content.

What causes extreme sensitivity? One idea is that the particle fabric becomes unstable as the salt water originally present in the pores is replaced in the post-depositional period by fresh water. This idea is appealing for sediments dominated by clay minerals, but cannot easily be applied to quickclays, with their low content of such minerals. The fact that Canadian quickclays contain substantial amounts of amorphous matter in the form of grain coatings has prompted the other main idea on the origin of extreme sensitivity (Bentley & Smalley 1978). It is thought that these coatings allowed the mineral particles shortly after deposition to become cemented into a strong but unusually open packing. The later partial dissolution of this cement, however, led to a substantial weakening of the original bonds as well as to the realization of some of the extra porosity allowed by the open packing. The new packing would not only be metastable but also irrecoverable.

## 9.5 Sub-aerial and sub-aquatic slides

### 9.5.1 *Examples of rotational slides*
Rotational slides are restricted to the steeper sub-aerial and sub-aquatic slopes. On land, they are failures that occur on strongly curved, partly bowl-shaped glide surfaces, the upper parts of which create amphitheatre-like embayments in the affected cliffs or slopes. Below these embayments are aprons of disturbed ground, commonly

revealing pressure ridges or a chaos of blocks. Two kinds of rotational slide are generally recognized (Fig. 9.8) depending on whether the failure surface intersects the slope or emerges on the level ground. Slopes acquire a scalloped upper margin under repeated rotational sliding.

Figure 9.9 shows a crescentic rotational slide on the side of a large tidal creek emptying into the macrotidal Severn Estuary. The bank sediments forming the slide

**Figure 9.9** Multiple rotational slide affecting the saltings bordering a tidal creek (to right), Severn Estuary, England. Undisturbed cliff about 1.25 m tall.

Figure 9.10  Vertically striated glide surface (facing to right) intersected by a perpendicular dislocation (facing observer, also striated). Tidal creek, Severn Estuary, England. Blade of spade gives scale.

Figure 9.11  Rotated bedding in an old rotational slide, as exposed on the margins of a younger one. Tidal creek, Severn Estuary, England. Maximum height of cliff facing viewer about 1.5 m. Bedding visible on it dips 5–10° to the left at point 1, 25° to the right at point 2, 0–5° to the right at point 3, and 35° to the right at point 4.

are fissured strong silts and sandy silts which pass downwards into weak plastic muds. The slide involves certainly two and possibly three distinct sediment slices, each lying on a glide surface inclined at grass level at 70–80° towards the channel axis. Just above the water's edge can be seen tension cracks and incipient slip surfaces parallel to the main dislocations. A few tensional cracks (crown cracks) lie concealed in the meadow just beyond the lowermost glide surface. Large vertical faults arranged perpendicular to the head of the slide locally cut the slices. Figure 9.10 shows the striated surfaces exposed at the intersection of one such dislocation with a glide surface. The general effect of the slide is to rotate the bedding within the slices, as may be seen in Figure 9.11 illustrating the core of an old slide where it now forms the spur at the lateral margin of a younger one.

Deep rotational slides similar to this example but on a larger scale abound on coastal cliffs of clay. Figure 9.12 shows an impressive example developed in the London Clay of the Thames Estuary (Bromhead 1979), from which many other large slides are reported (Hutchinson 1965, Bromhead 1978). The till cliffs of East Anglia, and the Jurassic and Cretaceous clay cliffs of southern England, are likewise affected by many large rotational slips. Sliding is a recurrent process in such settings, for marine erosion, biting at the foot of

Figure 9.12  View from near cliff-top of the Miramar rotational slide affecting London Clay, Herne Bay, Thames Estuary. From Bromhead (1978) Q. J. Engng Geol. Lond. 11, 291–304, with permission of the Geological Society of London. Photograph courtesy of Dr E. N. Bromhead, Kingston Polytechnic.

the cliffs, is continually recreating the conditions for instability.

A more complicated type of cliff retreat due to rotational sliding is described by Barton (1973) from the English Channel coast west of the Isle of Wight. Three resistant horizons, each defining a bench, occur within the exposed Barton Clay. Each bench is affected by rotational slips, the glide surfaces flattening out downwards to run with the bedding.

Rotational slides like these are typified by a thickness of order 0.2 times the downslope length, and by a failure surface of circular to near-circular profile. There seem to be few sub-aquatic slides with these properties, but some possible examples are reported near the edge of the continental shelf off the Mississippi Delta (Prior & Coleman 1979). Most submarine slides must be classified as translational or, if the tilting at the head is regarded as sufficiently significant, as perhaps shallow rotational.

Rotational slides are occasionally preserved in the rock record. There is evidence for them in the Carboniferous deltaic deposits of western Ireland (Gill 1979) and Scotland (Kirk 1982).

### 9.5.2 A model for rotational slides

What factors promote rotational slides? Consider in Figure 9.13 a vertical downslope slice of unit width through a rotational slide assumed to be moving over a failure surface which, in the plane of the slice, is a circular arc. Hence the mass turns about the centre O of this arc. At the instant of failure, static equilibrium demands that

$$\frac{\text{moment of disturbing}}{\text{force about O}} = \frac{\text{moment of resisting}}{\text{force about O}} \quad (9.17)$$

Ignoring the possibility of an earthquake-induced failure, the disturbing moment must be related to the weight of the slice. This weight is equal to $A(\gamma - \varrho)g \, \mathrm{N\,m^{-1}}$, where $A$ is the cross-sectional area of the slice, $\gamma$ the total bulk density of the affected sedi-

ment, $\varrho$ the density of the fluid beneath which slip is taking place (negligibly small in the case of sub-aerial slides), and $g$ the acceleration due to gravity. Now the weight can be considered to act vertically downwards through the centre of gravity of the slice, at a normal distance $x$ from the centre of rotation. Hence

$$\frac{\text{moment of disturbing}}{\text{force about O}} = xA(\gamma - \varrho)g \quad (9.18)$$

Resistance to failure comes from the shear strength $\tau_s$ of the sediment, as given by the Coulomb criterion (Eqn 9.16). Now $\tau_s$ is a force per unit area, so the total force opposing failure is $\tau_s L \, \mathrm{N\,m^{-1}}$, where $L$ is the arc length of the glide surface. It follows that

$$\frac{\text{moment of resisting}}{\text{force about O}} = \tau_s rL \quad (9.19)$$

where $r$ is the radius of curvature of the failure surface. Combining Equations 9.18 and 9.19

$$xA(\gamma - \varrho)g = \tau_s rL \quad (9.20)$$

at the instant of failure.

Equation 9.20 contains too many unknown quantities for it to be easily used to predict failure, especially when it is recalled that $\tau_s$ comprises cohesion plus effective stress. The formula can, however, be used for back-calculations concerning actual slips. These calculations in many cases suggest that failure occurred under conditions of significantly raised pore-water pressure.

### 9.5.3 Examples of translational slides

Translational slides in all environments are typified by a very small thickness compared to downslope length. The other key feature is that the glide surface beneath the slide is essentially parallel to the original ground surface. However, this surface can be planar only in the case of a planar slope.

Some of the most striking mass movements to be found in the British Isles, including translational slides, occur at Bredon Hill in the Severn Valley and have been the subject of several accounts (Grove 1953, Watson 1971, Whittaker & Ivemy-Cook 1972). Bredon Hill, an outlier of the Cotswold Hills, consists of Liassic silts and clays capped by the resistant limestones of the Inferior Oolite (Fig. 9.14). The rocks dip at 3–5° towards the south and south-southwest. A bench on the precipitous western, northern and eastern slopes is underlain by the Marlstone Rock Bed (6 m) and largely smothered by slipped material from higher up. These

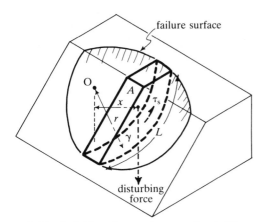

**Figure 9.13** Definition diagram for a rotational slide.

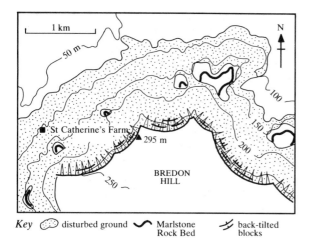

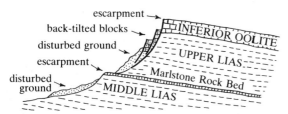

**Figure 9.14** Topographic and geological sketch of northern slopes of Bredon Hill affected by mass movements. Partly after Whittaker and Ivemy-Cook (1972).

**Figure 9.15** Air photograph of translational slide west of St Catherine's Farm, Bredon Hill. See Figure 9.16 for scale.

mass movements extend down to the low ground formed by the Lower Lias.

Near St Catherine's Farm on the north-west slopes is a spectacular tripartite translational slide (Figs 9.15 & 9.16). The bench underlain by the Marlstone Rock Bed supports an arcuate gathering region, where clay and limestone debris from an older and larger mass movement is remobilized as a complex of rotational slides within an amphitheatre-like crater. On the ground, fresh scars and rotated blocks similar to those in Figure 9.9 can be seen together with, on occasions, shallow-dipping striated glide planes and small temporary accretion lobes.

The mobilized debris travels down slope towards a steep wooded valley cut back into the Marlstone Rock Bed escarpment. Here the Rock Bed is probably faulted (non-diastrophically), a splinter giving rise to an island-like prominence at the head of the valley. This prow-shaped hillock divides the moving debris into two streams, each occupying a chute several metres wide of U-shaped cross-sectional profile and axially striated floor and sides (Grove 1953). Typically, the chute axes slope at 10–20° but locally are as steep as 50°. Debris moves down them in a sequence of spaced-out surges,

mainly during the winter and spring after prolonged rain or heavy snow-melt. The debris comprises an ill-sorted mixture of fragments up to cobble size of Upper Lias clay and some limestone in a viscous clay slurry.

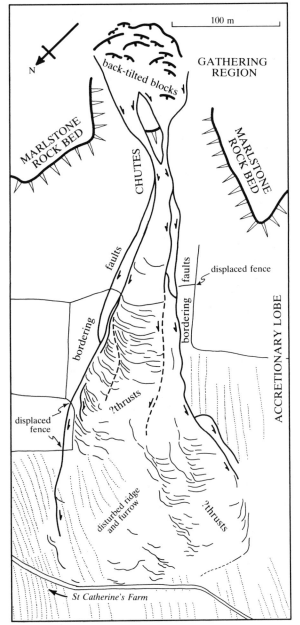

**Figure 9.16** Analysis of the main features of the translational slide shown in Figure 9.15.

The debris unites below the chutes to form the head of a long accretionary lobe that expands down the valley to terminate on gently sloping ground. The lobe moves spasmodically down slope on the average at a few decimetres per year, travelling faster where constrained by the valley walls than near its broad snout. It appears to overlie a principal glide surface in the form of a broad, shallow trough with sides inclined steeply inwards. Other less-active glide surfaces are detectable within the bounds of the lobe. Glide surfaces are exposed on scarps up to 1 m high and commonly show fresh slope-parallel striae. Movement over them is expressed by such additional features as (1) displaced fences, (2) displaced sheep runs, (3) bisected and displaced grass tussocks, and especially (4) plant roots, grass stems and bramble branches stretched in tension along the glide plane from the stationary to the moving side. In addition to exposed glide planes, the other features evident on the surface of the lobe are convex-downslope compressional ridges. These in many cases are associated with low-angle thrusts, expressed as overhangs of bare soil with stretched rootlets and occasionally striae. Where such features abound, the lobe as a whole is commonly upthrust on the principal glide surface.

This translational slide at Bredon Hill has been active over many hundreds of years. Fields in its vicinity are marked by ridge-and-furrow (Figs 9.15 & 9.16), a cultivation pattern dating from mediaeval and possibly mid-Saxon times. Some ridge-and-furrow were established on already disturbed ground within the confines of the slide, whereas others have been cut and overridden by the disturbance. The sluggishness and persistence of this slide may be contrasted with the abruptness and rapidity of movement of some translational slides in Northern Ireland (Prior & Douglas 1971, Prior & Graham 1974).

The violent and unpredictable quickclay translational slides affecting Alaska, the Canadian Lowlands and Scandinavia seriously hamper communications and planning and construction activities (Mitchell & Markell 1974, Mitchell & Klugman 1979). Well documented historical examples are those at Saint-Jean Vianney (Tavenas *et al.* 1971) and on the South Nation River (Eden *et al.* 1971).

Quickclay slides, involving silts of extreme sensitivity, occur along steep river bluffs and valley sides (Fig. 9.17). At the conclusion of motion one sees a large flask- or bottle-shaped crater extending back into the slope and connected through a narrow neck to a thin apron of debris spread out either in the river channel or over the lower ground beyond the slope. The first failure is a shallow rotational slide at the edge of the bluff or slope. Remoulding of the sensitive material quickly softens the lower part of the slipped mass, converting it into a slurry which rapidly flows away, leaving the slope unsupported. A new but larger shallow slide ensues and the process is repeated further back in the bluff or slope.

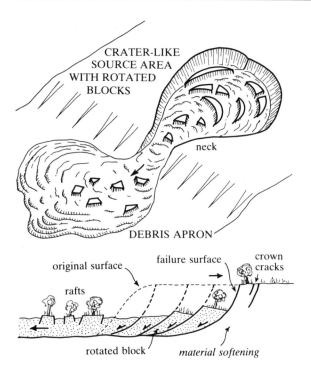

**Figure 9.17** Principal features of a quickclay slide.

In this way the failure spreads retrogressively back and sideways into the slope. Movement stops when the failure has penetrated so far back that the disturbed sediment can no longer escape from the crater. The crater, rimmed by a steep back-scarp representing the last glide plane, may be either virtually empty (Saint-Jean Vianney) or contain a series of partly back-tilted blocks, as at the South Nation River (Fig. 9.18), depending on the shape of the slope affected and the strength of the silt beneath. The viscous nature of the remoulded debris is dramatically expressed by the 'river' of slurry and tree-covered rafted blocks released by the South Nation River slide (Fig. 9.19). The largest recorded Canadian slide, in 1894, at Saint Albons on the Sainte Anne River, affected 6.4 km$^2$ of ground and had a volume estimated at 0.46 km$^3$. The raising of pore-water pressures following heavy rain or snow-melt seems mainly to trigger these failures.

Disintegrating retrogressive translational slides similar to quickclay slides occur in water depths up to 50 m in accreting parts of the Mississippi Delta, where bottom slopes are generally less than about 1° (Prior & Coleman 1978a, 1980, Prior & Suhayda 1979a, Prior *et al.* 1979). These slides have been investigated chiefly by side-scan sonar, combined with sediment sampling and some *in situ* studies of geotechnical properties and pore-water pressure. Their general character is summarized in Figure 9.20 and an example of a slide complex, in the form of a sonar mosaic, appears in Figure 9.21. As with quickclay slides, the head of the failure is a circular to irregular bowl-shaped source area composed of multiple rotated and partly disintegrated blocks on glide surfaces that level out downwards into a single slope-parallel dislocation. At the downslope end is an accretionary lobe, similar in character to the Bredon lobe with its pressure ridges. Source area and lobe adjoin in many cases but in others (Fig. 9.21) are connected by a long and relatively narrow channel or chute. Commonly, each lobe is fed from a number of tributary source areas. The connecting channels carry a mass of slurried sediment and still-undisrupted blocks, their banks showing multiple scarps indicative of enlargement by rotational sliding. Gas vents, 'mud' volcanoes and pressure ridges ornament the slopes adjoining many lobes.

These slides affect substantial volumes of sediment in the growing delta, the largest measuring 20–25 m in thickness and 10 km in length from source to lobe. Despite the low gradient, the slides move quite rapidly. A lobe surveyed by Prior and Suhayda (1979a) advanced approximately 2 km in one year. As with quickclay slides, failure is promoted by the large pore-water pressures known to exist in the rapidly accumulating, organic-rich deltaic sediments (Prior & Suhayda 1979b). Either depositional oversteepening or storm waves may trigger the slides. Whatever their cause, the Mississippi slides are a serious hazard, almost certainly responsible for the disappearance without warning of oil drilling platforms and the displacement of navigation marks.

Translational slides occur off other modern deltas and the effects of sliding should therefore be expected in the rock record. Gill (1979) identified several shallow slides in the well exposed Carboniferous deltaic rocks of the west of Ireland. The extensional disturbances described by Shirley (1955) may also represent translational sliding.

Substantial as may be these Mississippi sub-aquatic failures, they pale into insignificance beside the huge translational slides that occur beneath the oceans along the continental slope and rise, where gradients locally attain 5° (Dingle 1977, Prior & Coleman 1979, Woodcock 1979). Thicknesses of tens or hundreds of metres are normal, and volumes are measured in tens, hundreds and even thousands of cubic kilometres. The Agulhas slide – the largest so far described (Dingle 1977) – measures 168 km on average down slope, 700–800 km along the slope, and is about 300 m in

**Figure 9.18** Source area of the quickclay slide of 16–17 May 1971, South Nation River, Ontario, Canada. Movement towards the upper right, with South Nation River flowing away in background. From Eden *et al.* (1971) with permission. Copyright the National Research Council of Canada.

thickness. Its volume is estimated at $2.03 \times 10^4$ km$^3$. Although these slides involve sediments of little strength, their deep-seated glide surfaces probably reach down into deposits millions to tens of millions of years old. The much thinner slides of the Mississippi and other modern deltas are unlikely to involve sediments more than tens to hundreds of years in age.

Studies using echo sounding and shallow seismic profiling (Lewis 1971, Dingle 1977) reveal the distinctive morphology and internal structure of large submarine slides (Fig. 9.22). There are some similarities with the smaller sub-aquatic slides and with translational slides on land. The head is an area of tension, marked by tilted blocks on shallow partly rotational glide surfaces and by extensional conjugate faults bounding horst and graben. A few downslope-directed normal faults may occur up slope of the head-scarp. Down slope the glide surfaces flatten out and coalesce into a single slope-parallel failure surface. The slide toe comprises rounded compressional elevations above thrust surfaces.

### 9.5.4 A model for translational slides

How should we analyse translational slides? Assuming that the affected material moves as one, a simple model is suggested by the fact that such slides are very long compared to their thickness. Hence we may imagine that failure occurs on an infinite slope above a slope-parallel glide surface.

Consider in Figure 9.23 a vertical column within a uniformly thick translational slide, the column being of vertical height $y$ and unit basal area measured parallel to the slope. The affected sediment is of uniform total bulk density $\gamma$ and failure takes place on a uniform slope of angle $\beta$ beneath a fluid of density $\varrho$ (negligibly small in the case of sub-aerial slides). Static equilibrium requires, at the instant of failure, that the disturbing force acting on unit area of the failure surface just equals the resisting force. As with rotational slides, the disturbing force is the effective weight

$$\text{disturbing force} = (\gamma - \varrho)gy\cos\beta\sin\beta \quad \text{N m}^{-2} \quad (9.21)$$

**Figure 9.19** Slurry bearing tree-covered rafts of undisturbed sediment advancing towards viewer in the channel and valley of the South Nation River after slide of 16–17 May 1971. From Eden *et al.* (1971) with permission. Copyright the National Research Council of Canada.

The resisting force on the glide surface is the shear strength of the sediment

$$\text{resisting force} = \tau_s = c + \sigma_e \tan\phi \quad \text{N m}^{-2} \quad (9.22)$$

referring to the Coulomb criterion (Eqn 9.16). Equating forces

$$(\gamma - \varrho)gy\cos\beta\sin\beta = c + \sigma_e \tan\phi \quad (9.23)$$

at the instant of failure.

In order for slides to occur beneath low slopes, it is obvious that the cohesion and effective stress on the right-hand side of Equation 9.23 must in total be quite small. That the pore-water pressure can indeed compare with the sediment weight, making the effective stress very small or even negative, is clear from Prior and Suhayda's (1979b) *in situ* measurements off the Mississippi Delta.

## 9.6 Debris flows

### 9.6.1 General character and examples

Briefly, just to remind ourselves, a debris flow is a mixture of roughly equal parts of solid detritus and water moving under gravity in a state of general flow.

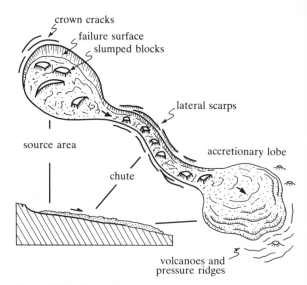

**Figure 9.20** Principal features of submerged translational slides in the shallows of the Mississippi Delta.

Sub-aerial debris flows are a major hazard and have consequently been widely and intensively studied. Three main varieties are recognized. Those found in temperate climates are generally triggered by prolonged as much as by heavy rainfall and may be formed of clay loosened from coastal cliffs (Hutchinson 1970), saturated clayey soils (Prior & Stephens 1972), or a wetted mixture of weathered rock and soil (Statham 1976). Movement is slow, commonly no more than a few metres per year, and also spasmodic and seasonal. Under semi-arid and arid conditions, debris flows are set off chiefly by rainstorms and by the rapid spring melting of snow (Fryxell & Horberg 1943, Bull 1964, Johnson & Rahn 1970, Morton & Campbell 1974, Campbell 1975). Movement is normally brisk and, in some instances, catastrophic. Lahars, the debris flows associated with volcanoes, were described by Scrivenor (1929) from Gunong Keleot Volcano in Java, and by Ulate and Cor-rales (1966) and Waldron (1967) from Irazu Volcano, Costa Rica. Lahars include the largest debris flows known and their motion is generally brief and violent. Scrivenor's Gunong Keleot flow travelled a distance of about 38 km at an average speed for much of the distance of approximately 18 m s$^{-1}$. Large debris flows have frequently arisen on the slopes of the volcanoes of Oregon and Washington in the northwestern USA (Crandell & Waldron 1956, Crandell 1971, 1980, Crandell *et al.* 1979). One of these mass movements, the Osceola Mudflow some 4800 years old, surged a distance of about 75 km from its source on Mount Rainer before stopping on the Puget Sound Lowlands as

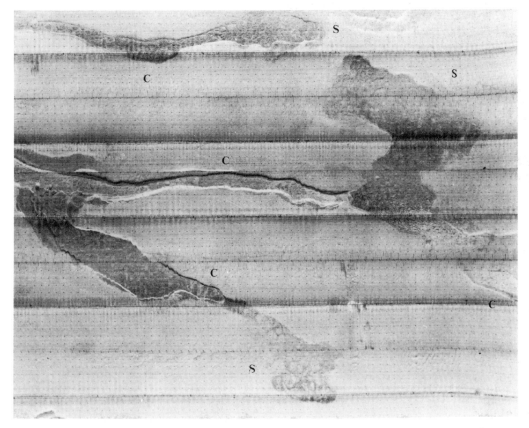

**Figure 9.21** Mosaic of side-scan sonar records showing slides composed of source areas (S) and chutes (C) on the shallow sea bed east of South Pass, Mississippi Delta. Bed slopes down to left and area measures 1.4 × 1.8 km. From Prior and Coleman (1980) with permission of Elsevier Science Publishers. Photograph courtesy of Professor D. B. Prior, Louisiana State University.

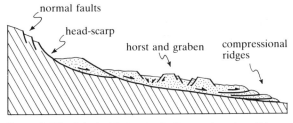

**Figure 9.22** Principal features of a large submarine slide, as seen in a vertical axial section.

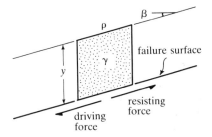

**Figure 9.23** Definition sketch for analysis of a translational slide.

a lobed sheet approximately 30 km long and 15 km wide. Lahars arise in three main ways. Some represent the effects of normal heavy rainfall on the loose ash mantling the slopes of the volcano. Others are triggered by eruption-induced rain. Many form because incandescent ejecta are erupted onto snow and ice.

It is generally possible to recognize three main parts to a debris flow (Fig. 9.24a), similar to the zones

distinguished from translational slides. Highest on the slope lies some kind of gathering or feeding area, from which debris is supplied to a channel, either a bedrock gully or valley or a chute formed in older debris flow deposits. A lobed sheet or a complex of narrow lobes forms the depositional zone on the more open or gently sloping ground down stream from the channel.

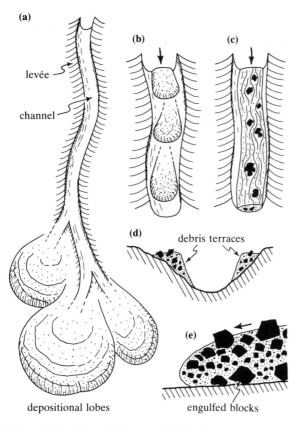

**Figure 9.24** Principal features of debris flows.

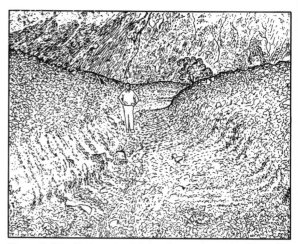

**Figure 9.25** U-shaped debris flow channel, Panamint Range, California. Drawn from a photograph by Johnson (1970).

**Figure 9.26** Slurry-like debris flow (right to left) about 7 m wide on Californian alluvial fan. After Morton and Campbell (1974) *Q. J. Engng Geol. Lond.* **7**, 377–84, by permission of the Geological Society of London.

These parts lie close together on the debris flows degrading the London Clay cliffs on the southern shores of the Thames Estuary (Hutchinson 1965, 1970). On the other hand, the U-shaped chute depicted in Figure 9.25 from the semi-arid Panamint Range, California, is part of a long channel that connects a depositional complex with a distant feeding ground. As Johnson (1970) emphasizes, such channels are bordered by tall levées (Fig. 9.24a), the equivalent of which in bedrock gullies and valleys is a perched terrace of debris flow deposit on each wall (Fig. 9.24d). Debris flows in gently graded channels tend to a fairly uniform thickness (Figs 9.24c & 9.26), but on slopes of 10° or so commonly form a series of large waves or discrete surges representing a mode of instability (Figs 9.24b & 9.27). Broadly, the motion of the viscous slurry appears to be laminar. The snouts of flows overhang, engulfing quantities of the ambient fluid (Figs 9.24e & 9.28). Cobbles and boulders supported on the tops of flows are carried forwards and downwards over the snouts, to disappear into the basal zone of the advancing masses.

Debris flow deposits are distinctive in several respects (Crandell & Waldron 1956, Bull 1964, Johnson 1970, Crandell 1971, 1980). They are ill-sorted and only weakly if at all bedded internally. A weak imbricate fabric may be evident amongst the larger clasts. Laboratory analyses confirm the poor sorting of debris flow deposits, showing that all grades of particle from clay normally up to cobbles and boulders are present. Sub-aerial debris flow deposits commonly include a variety of plant debris, amongst which are broken-off tree trunks and branches, and may bury the lower parts of the forest trees. Clay forms no more than a few per cent of some deposits, occurring chiefly as a slippery

**Figure 9.27** Steep stone front 2.4 m wide within a Californian debris flow. After Morton and Campbell (1974) *Q. J. Engng Geol. Lond.* **7**, 377–84, by permission of the Geological Society of London.

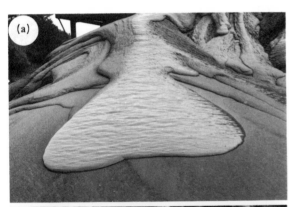

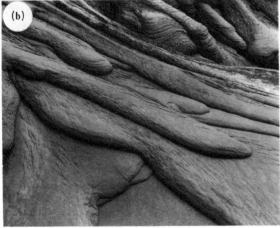

**Figure 9.28** Debris flows of muddy sand with overhanging snouts and flanks, as developed on the cone of sediment rejected from a gravel washing plant. Flow in (a) is active and up to about a metre wide. The thinner 'dead' flows in (b) are a few centimetres thick.

coating on the larger particles. In other flows it is just sufficient in amount to fill the voids between the larger particles, which largely support each other. Some flows contain as much as 80% clay, which then forms a matrix in which the coarser debris, including cobbles and boulders, may be described as floating. Johnson (1970) draws attention to the ability of debris flows to transport large boulders and blocks, which commonly appear to float on top of the moving mass.

Far less is known of sub-aquatic debris flows because of the difficulties of examining them directly. However, there is growing evidence from the deep oceans of their widespread role in the degradation of submarine slopes (Embley 1976, 1980, Damuth & Embley 1981, Damuth *et al.* 1983). Deep-ocean debris flows are recognizable on seismic profiles as lobe- or lens-shaped acoustically transparent deposits, and on side-scan sonar their signature is also distinctive. Core samples from within lobes show a chaotic unstratified mixture of unconsolidated sediments, in some instances accompanied by bedrock fragments.

### 9.6.2 Modelling debris flows

What is the rheological character of debris flows? This is a fundamental question to answer if we are to understand the motion and behaviour of these mass movements. A simple answer is unlikely, however, on account of the apparently contradictory properties possessed by debris flows.

It is significant that debris flows are capable of transporting cobbles, boulders and even huge blocks of rock. This observation becomes even more striking when it is appreciated that large clasts have remained dispersed throughout debris flow deposits, implying that the coarse debris did not sink to the base of the flow when forward motion ceased (the modest basal concentration is explicable by engulfment at the snout). Yet the cobbles and boulders in these cases were generally more dense than the viscous slurries enveloping them. Why did the debris not sink? A possible explanation is that the slurry surrounding each particle behaved like a solid and was sufficiently strong, relative to the distributed weight of the clast, to prevent it sinking.

An apparently contradictory property is illustrated in Figure 9.26, which shows a viscous slurry flowing in a laminar manner within a channel. Evidently debris flows can also behave like a liquid.

This paradox can be resolved if we think of debris flows as behaving as an ideal plastic substance. In Chapter 8 it was shown that some materials, when stressed above a certain stage, deform continuously for no increase of load. Above this stage, these materials behave plastically and cannot sustain an increased force. They therefore behave like a static fluid, which cannot sustain a tangential stress. Resistance to deformation in a fluid is expressed not by a yield strength but by the viscosity.

Hence it looks as though a rheological model for a debris flow should include expressions for both solid-like and fluid-like behaviour. The model for a Bingham plastic is the simplest of these, as follows

$$\tau = \tau_{yd} + \eta_a \frac{dU}{dy} \quad \text{N m}^{-2} \qquad (\tau \geqslant \tau_{yd}) \quad (9.24)$$

where $\tau$ is the tangential stress in the material, $\tau_{yd}$ the plastic yield strength, $\eta_a$ an apparent viscosity, sometimes called the Bingham viscosity, and $dU/dy$ the velocity gradient (or strain rate). Notice the important bracketed restriction, which states that the apparent viscosity contributes to the deformation only when the plastic yield strength is exceeded. We can use this rheological model to calculate the motion of a debris flow.

Most debris flows travel for part of the time within a channel, such as that in Figure 9.25. The simplest regular shape approximating the cross-sectional form of these chutes is a semi-circular arc, so let us analyse a uniform steady debris flow in a straight, sloping semi-cylindrical channel of radius $R$ (Fig. 9.29a). As with an ideal river (Sec. 5.2), consider the balance of forces acting on an element of the flow of downstream length $L$ and radius $r$. Two forces affect the element, one a driving force and the other a resisting force. The former is evidently the downslope component of the weight of the element:

$$\text{driving force} = (\tfrac{1}{2}\pi r^2 L)(\gamma - \varrho)g \sin\beta \quad \text{N} \qquad (9.25)$$

where the first bracketed term is the volume of the element, $\gamma$ the total bulk density of the debris flow, $\varrho$ the

density of the ambient medium (assumed stagnant), and $\beta$ the angle of the channel bed. The density of the ambient medium is included so that we can consider sub-aquatic as well as sub-aerial debris flows. Now the total resisting force is the product of the shear stress per unit area times the external area of the semi-cylindrical element of length $L$ and radius $r$, namely

$$\text{resisting force} = \tau(\pi r L) \quad \text{N} \qquad (9.26)$$

where $\tau$ is the unit shear stress and the bracketed term is the area. Since the motion is assumed uniform and steady, Newton's first law requires that no resultant force affects the element, whence

$$\tau(\pi r L) = (\tfrac{1}{2}\pi r^2 L)(\gamma - \varrho)g \sin\beta \qquad (9.27)$$

Cancelling out

$$\tau = \tfrac{1}{2}(\gamma - \varrho)g r \sin\beta \quad \text{N m}^{-2} \qquad (9.28)$$

This formula bears some relationship to Equation 5.6 for a river, and states that the shear stress increases with the excess density, the radial distance from the channel axis, and the channel slope.

Equation 9.24 for the Bingham plastic gives us another relationship for the shear stress. Elimination of the stress between Equations 9.24 and 9.28 yields

$$\tau_{yd} + \eta_a \frac{dU}{dr} = \tfrac{1}{2}(\gamma - \varrho)g r \sin\beta \quad \text{N} \qquad (9.29)$$

where $r$ is substituted for $y$ in the velocity gradient. On rearrangement

$$\frac{dU}{dr} = -\frac{1}{\eta_a}\left(\frac{(\gamma - \varrho)g \sin\beta}{2} r - \tau_{yd}\right) \quad \text{s}^{-1} \quad (9.30)$$

the negative sign appearing because the velocity decreases outwards from the flow axis. The velocity profile is obtained by integrating Equation 9.30

$$U = -\frac{1}{\eta_a}\left(\frac{(\gamma - \varrho)g \sin\beta}{4} r^2 - \tau_{yd}r\right) + C \quad \text{m s}^{-1}$$
$$(9.31)$$

where $C$ is an integration constant. This constant follows from the boundary conditions on the problem. If the debris flow behaves as a true fluid, then $U=0$ at the wall of the channel a distance $r=R$ from the flow axis. Substituting back into Equation 9.31

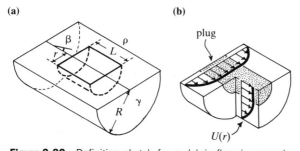

(a)

(b)

plug

$U(r)$

**Figure 9.29** Definition sketch for a debris flow in a semi-cylindrical channel.

$$C = \frac{1}{\eta_a}\left(\frac{(\gamma - \varrho)g\sin\beta}{4} R^2 - \tau_{yd}R\right) \quad \text{m s}^{-1} \quad (9.32)$$

It then follows that

$$U = \frac{1}{\eta_a}\left(\frac{(\gamma - \varrho)g\sin\beta}{4}(R^2 - r^2) - \tau_{yd}(R - r)\right) \quad (9.33)$$

showing that the velocity tends to increase outwards from the channel wall and has a parabolic profile (Fig. 9.29b). But we must now recall the restriction on Equation 9.24, which states that fluid-like behaviour is possible only where the yield strength is exceeded. Referring to Equation 9.28, there is a critical value for $r$ below which the stress due to the weight of the debris is less than the plastic yield strength. Hence only solid-like behaviour – plug flow – is possible in this near-axial part of the moving mass. To find the size of the plug, we make use of another boundary condition and put $dU/dr = 0$ into Equation 9.30, finding that

$$r_{pl} = \frac{2\tau_{yd}}{(\gamma - \varrho)g\sin\beta} \quad \text{m} \quad (9.34)$$

where $r_{pl}$ is the plug radius. The restriction on Equation 9.33 is therefore $r \geqslant r_{pl}$. Figure 9.29b shows the extent of plug flow within the channel. The plug moves as a single rigid body, at a velocity given by substituting $r_{pl}$ into Equation 9.33.

Debris flows on occasion spread to form extensive sheets, as in the case of the Osceola Mudflow (Crandell & Waldron 1956). For such flows, it is useful to have an analysis in terms of the uniform steady motion of debris in an infinitely extensive sheet of uniform slope. This can be done by considering the forces acting on a vertical column of debris of unit basal area. We obtain in the previous manner

$$U = \frac{1}{\eta_a}\left(\frac{(\gamma - \varrho)g\sin\beta}{2}(H^2 - y^2) - \tau_{yd}(H - y)\right) \quad \text{m s}^{-1} \quad (9.35)$$

and

$$y_{pl} = \frac{\tau_{yd}}{(\gamma - \varrho)g\sin\beta} \quad \text{m} \quad (9.36)$$

where $H$ is flow depth, $y$ depth below the flow surface, and $y_{pl}$ the thickness of the now slab-like plug. The restriction to be applied on Equation 9.35 is $y \geqslant y_{pl}$. Only beyond the plug is the velocity profile parabolic.

It is worth looking more closely at these relationships, particularly Equations 9.34 and 9.36. Essentially, they

state that a debris flow will come to rest if its thickness falls below that of the plug flow permitted by the ground slope and the plastic yield strength and excess density of the debris itself. Provided the flow was decelerated very gradually, the effective yield strength can be estimated by measuring the thickness and the slope of a debris flow deposit.

These models embrace both sub-aerial and sub-aquatic debris flows. In the case of sub-aerial flows, we can disregard the air as negligibly small in density compared to the debris itself, and for the same reason disregard the drag of the air on the upper surface of the flow. An aqueous mineral slurry and water are similar in density, however, so we cannot neglect the ambient medium in the case of sub-aquatic flows. Moreover, it is quite possible that the medium will exert a significant drag on a debris flow. If large enough, this drag may not only seriously retard the flow but as well induce fluid-like behaviour in its uppermost levels.

To judge the effect on the motion, we should compare the drag due to the medium on the upper surface of the flow to the drag exerted by the debris flow on its own base. Reminding ourselves of the quadratic stress law (Eqn 1.27), the drag of the medium equals $(f/8)\varrho U^2$, where $f$ is the Darcy–Weisbach friction coefficient for the top of the debris flow and $U$ the velocity of the plug. The drag on the flow bed (assumed channelized) equals $\frac{1}{2}(\gamma - \varrho)gR\sin\beta$, by Equation 9.28. For the representative quantities $f = 0.05$, $\varrho = 1025$ kg m$^{-3}$, $U = 1$ m s$^{-1}$, $\gamma = 1600$ kg m$^{-3}$, $R = 10$ m and $\beta = 1°$, the surface drag is only 1.3% of the bed drag. The retardation would therefore be insignificant and we are justified in neglecting the effect. Yield strengths for sub-aquatic debris flows are not known, but assuming their similarity to sub-aerial flows (Johnson 1970), values of the order of $10^2$–$10^3$ N m$^{-2}$ are plausible. As the drag of the medium is only 6.4 N m$^{-2}$ on the basis of the previous quantities, plug flow will persist to the upper surface of the debris flow. The only role of the aqueous medium is to lower the effective density of the debris flow.

## 9.7 Mass-movement associations

As more case studies are completed, it becomes increasingly clear that different types and scales of mass movement are closely associated spatially, but represent a range of ages. The temporal sequence is normally from large-scale to small-scale mass movements, the latter developing from the materials disturbed and transported by the former. Mass movements of the same type, however, may occur more than once in the temporal sequence. The various instances and kinds form an association of mass movements.

One important association, recognized by Watson (1971) and by Whittaker and Ivemy-Cook (1972), arose from the Jurassic clays and limestones of Bredon Hill. Bredon's northern slopes reveal at least six amphitheatre-like embayments (Fig. 9.14). At the head of each is a steep slope or cliff underlain by the Inferior Oolite, and to the sides, where adjoining embayments are cut, is a promontary ranging down the slope. Large rotated blocks and slices of Inferior Oolite are present below the main escarpment in several of the embayments. Hence each embayment would seem to be the site of a huge rotational slide, probably involving more than one glide plane, which in most cases ran out on the dip slope of the Marlstone Rock Bed. These slides perhaps date from early post-glacial times, when the climate was cooler and large snow patches persisted on the northern slopes of Bredon Hill, wetting the ground the year round. The younger failures, considerably smaller in scale, are continuing the task of degradation and range in age up to the present. Amongst them are debris flows as well as slides like those described above.

The continental slope off Nova Scotia revealed to Hill (1983) a similar association of mass movements. The evidence for it comes from a dense array of shallow seismic profiles, side-scan sonar sweeps and sediment cores. Figure 9.30 is a summary and interpretation. An embayment cut back into the continental shelf is bounded at the head by a scarp about 150 m high that cuts reflectors below the surface of the shelf. Beneath

the escarpment is a narrow belt of chaotic topography suggesting the presence of jumbled masses of sediment. The topography of the slope below them points to the presence of detached slide blocks up to 50 m thick. To judge from the many sharp regular reflectors they contain, these blocks are essentially undisturbed internally, although widely separated amongst ill-stratified material. An exceptionally large slide block measuring $3 \times 7$ km lies towards the bottom of the surveyed slope. Its mesa-like form may represent normal faulting as the block finally settled. The slope between the two distributions of slide blocks is dotted with scars suggesting the presence of small rotational or translational slides. A lobe-shaped acoustically transparent debris flow is also identifiable. Cores 56 and 60 from this intermediate portion of the slope reveal the presence of muddy sediments with balled-up and intricately folded sand layers. A wide channel transecting a depositional lobe cuts the escarpment and upper slope and a smaller channel occurs to the west. Cores 50 and 80, respectively from the large channel and depositional lobe, consist of undisturbed turbidite sands interbedded with thick muds. The slides are Late Pleistocene (Devensian) in age and the turbidites of late Devensian date; all arose when the continental slope was abundantly supplied with sediment, at a time of lowered sea level. Post-glacial deposition has not succeeded in concealing the slide topography.

## Readings

Barton, M. D. 1973. The degradation of the Barton Clay cliffs of Hampshire. *Q. J. Engng Geol. Lond.* **6**, 423–40.

Bentley, S. P. and I. J. Smalley 1978. Inter-particle cementation in Canadian post-glacial clays and the problem of high sensitivity. *Sedimentology* **25**, 297–302.

Bromhead, E. N. 1978. Large landslides in the London Clay at Herne Bay, Kent. *Q. J. Engng Geol. Lond.* **11**, 291–304.

Bromhead, E. N. 1979. Factors affecting the transition between the various types of mass movement in coastal cliffs consisting largely of overconsolidated clays with special reference to southern England. *Q. J. Engng Geol. Lond.* **12**, 291–300.

Bull, W. B. 1964. *Alluvial fans and near-surface subsidence in western Fresno County, California.* Prof. Pap. US Geol. Surv., no. 437-A.

Campbell, R. H. 1975. *Soil slips, debris flows, and rainstorms in the Santa Monica Mountains and vicinity, California.* Prof. Pap. US Geol. Surv., no. 851.

Carson, M. A. and M. J. Kirkby 1972. *Hillslope form and process.* Cambridge: Cambridge University Press.

Crandell, D. R. 1971. *Postglacial lahars from Mount Rainier, Washington.* Prof. Pap. US Geol. Surv., no. 677.

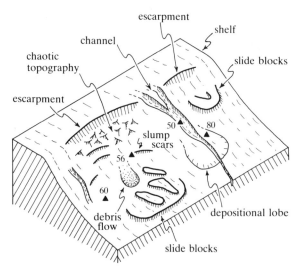

**Figure 9.30** Schematic representation of the mass-movement association on the continental slope off Nova Scotia (Hill 1983).

Crandell, D. R. 1980. *Recent eruptive history of Mount Hood, Oregon, and potential hazards from future eruptions.* Bull. US Geol. Surv., no. 1492.

Crandell, D. R. and H. H. Waldron 1956. A recent volcanic mudflow of exceptional dimensions from Mt Rainier, Washington. *Am. J. Sci.* **254**, 349–62.

Crandell, D. R., D. R. Mullineaux and C. D. Miller 1979. Volcanic-hazard studies in the Cascade Range of the western United States. In *Volcanic activity and human ecology*, P. D. Sheets and D. K. Grayson (eds), 194–219. New York: Academic Press.

Crawford, C. B. 1963. Cohesion in an undisturbed sensitive clay. *Geotechnique* **13**, 132–46.

Damuth, J. E. and R. W. Embley 1981. Mass-transport processes on Amazon cone: western equatorial Atlantic. *Bull. Am. Assoc. Petrol. Geol.* **65**, 629–43.

Damuth, J. E., R. D. Jacobi and D. E. Hayes 1983. Sedimentary processes in Northwest Pacific Basin revealed by echo-character mapping studies. *Bull. Geol. Soc. Am.* **94**, 381–95.

Dingle, R. V. 1977. The anatomy of a large submarine slump on a sheared continental margin (SE Africa). *J. Geol. Soc. Lond.* **134**, 293–310.

Dott, R. H. 1963. Dynamics of subaqueous gravity depositional processes. *Bull. Am. Assoc. Petrol. Geol.* **47**, 104–28.

Eden, W. J., E. B. Fletcher and R. J. Mitchell 1971. South Nation River landslide, May 16, 1971. *Can. Geotechn.* **8**, 446–51.

Embley, R. W. 1976. New evidence for occurrence of debris flow deposits in the deep sea. *Geology* **4**, 371–4.

Embley, R. W. 1980. The role of mass transport in the distribution and character of deep-ocean sediments with special reference to the North Atlantic. *Marine Geol.* **38**, 23–50.

Fleming, R. W. and A. M. Johnson 1975. Rates of seasonal creep of silty clay soil. *Q. J. Engng Geol. Lond.* **8**, 1–29.

Fryxell, F. M. and C. L. Horberg 1943. Alpine mud flows in Grand Teton National Park, Wyoming. *Bull. Geol. Soc. Am.* **54**, 457–72.

Gill, W. D. 1979. *Syndepositional sliding and slumping in the West Clare Namurian Basin, Ireland.* Spec. Pap. Geol. Surv. Ireland, no 4.

Grove, A. T. 1953. Account of a mudflow on Bredon Hill, Worcestershire, April 1951. *Proc. Geol. Assoc.* **64**, 10–3.

Hampton, M. A. 1972. The role of subaqueous debris flows in generating turbidity currents. *J. Sed. Petrol.* **42**, 775–93.

Hill, P. R. 1983. Detailed morphology of a small area on the Nova Scotia continental rise. *Marine Geol.* **53**, 55–76.

Hutchinson, J. N. 1965. *A survey of the coastal landslides of Kent.* Building Res. Station Note, no. EN/35/65.

Hutchinson, J. N. 1970. A coastal mudflow on the London Clay cliffs at Beltinge, north Kent. *Geotechnique* **20**, 412–38.

Johnson, A. M. 1970. *Physical processes in geology.* San Francisco: Freeman.

Johnson, A. M. and P. H. Rahn 1970. Mobilization of debris flows. *Z. Geomorph.* S **9**, 168–86.

Kirk, M. 1982. A gravity slide in the Lower Coal Measures at Swallow Craig near Kirkconnel, Dumfries and Galloway.

Scott. J. Geol. **18**, 317–22.

Lewis, K. B. 1971. Slumping on a continental slope inclined at $1°–4°$. *Sedimentology* **16**, 97–110.

Mitchell, R. J. and M. A. Klugman 1979. Mass instabilities in sensitive Canadian soils. *Engng Geol.* **14**, 109–34.

Mitchell, R. J. and A. R. Markell 1974. Flowsliding in sensitive soils. *Can. Geotechn.* **11**, 11–31.

Morton, D. M. and R. H. Campbell 1974. Spring mudflows at Wrightwood, southern California. *Q. J. Engng Geol. Lond.* **7**, 377–84.

Prior, D. B. and J. M. Coleman 1978a. Disintegrating retrogressive landslides on very-low-angle subaqueous slopes, Mississippi Delta. *Marine Geotechn.* **3**, 37–60.

Prior, D. B. and J. M. Coleman 1978b. Submarine landslides on the Mississippi delta-front slope. *Geosci. Man* **19**, 41–53.

Prior, D. B. and J. M. Coleman 1979. Submarine landslides – geometry and nomenclature. *Z. Geomorph.* **23**, 415–26.

Prior, D. B. and J. M. Coleman 1980. Sonograph mosaics of submarine slope instabilities, Mississippi River Delta. *Marine Geol.* **36**, 227–39.

Prior, D. B. and G. R. Douglas 1971. Landslides near Larne, Co. Antrim, 15–16th August. *Irish Geog.* **3**, 294–301.

Prior, D. B. and J. Graham 1974. Landslides in the Magho district of Fermanagh, northern Ireland. *Engng Geol.* **8**, 341–59.

Prior, D. B. and N. Stephens 1972. Some movement patterns of temperate mudflows: examples from northeastern Ireland. *Bull. Geol. Soc. Am.* **83**, 2533–44.

Prior, D. B. and J. N. Suhayda 1979a. Submarine landslide morphology and development mechanisms, Mississippi Delta. *Proc. 11th Conf. Ocean Technol.,* paper OTC 3482, 1055–8.

Prior, D. B. and J. N. Suhayda 1979b. Application of infinite slope analysis to subaqueous sediment instability, Mississippi Delta. *Engng Geol.* **14**, 1–10.

Prior, D. B., J. M. Coleman and L. E. Garrison 1979. Digitally acquired undisturbed side-scan sonar images of submarine landslides, Mississippi Delta. *Geology* **7**, 423–5.

Saunders, I. and A. Young 1983. Rates of surface processes on slopes, slope retreat and denudation. *Earth Surf. Processes Landforms* **8**, 473–501.

Scrivenor, J. B. 1929. The mudstreams ('Lahars') of Gunong Keleot in Java. *Geol. Mag.* **66**, 433–4.

Shirley, J. 1955. The disturbed strata in the Fox Earth Coal and its equivalents in the East Pennine Coalfield. *Q. J. Geol. Soc. Lond.* **111**, 265–79.

Skempton, A. W. 1953. Soil mechanics in relation to geology. *Proc. Yorks. Geol. Soc.* **29**, 33–62.

Skempton, A. W. and J. N. Hutchinson 1969. Stability of natural slopes and embankment foundations. *Proc. 7th Conf. Soil Mech. Foundn. Engng,* State-of-the-Art Volume, 291–340.

Skempton, A. W. and R. D. Northey 1952. The sensitivity of clays. *Geotechnique* **3**, 30–53.

Statham, I. 1976. Debris flows on vegetated screes in the Black Mountain, Carmarthenshire. *Earth Surf. Processes* **1**, 173–80.

Tavenas, F., J. Y. Chagnon and P. La Rochelle 1971. The Saint-Jean Vianney landslide: observations and eye-witness accounts. *Can. Geotechn.* **8**, 463–78.

Ulate, C. A. and M. F. Corrales 1966. Mud flows related to the Irazu Volcano eruptions. *J. Hydraul. Div. Am. Soc. Civ. Engrs* **92**, 117–29.

Varnes, D. J. 1958. *Landslide types and processes.* Spec. Rep. US Highway Res. Board, no. 29, 20–47.

Waldron, H. H. 1967. *Debris flow and erosion control problems caused by the ash eruptions of Irazu Volcano, Costa Rica.* Bull. US Geol. Surv., no. 1241-I.

Ward, W. H. 1945. The stability of natural slopes. *Geogr. J.* **105**, 170–91.

Watson, I. 1971. A preliminary report on new photogeological studies to detect unstable natural slopes. *Q. J. Engng Geol. Lond.* **4**, 133–7.

Whittaker, A. and H. C. Ivemy-Cook 1972. Geology of Bredon Hill, Worcestershire. *Bull. Geol. Surv. Gr. Br.,* **42**, 1–49.

Woodcock, N. H. 1979. Sizes of submarine slides and their significance. *J. Struct. Geol.* **1**, 137–42.

# 10 Changes of state

States of matter – fluidization – liquefaction – forces involved in deformation of liquid-ized sediments – load casts – convolute lamination – wrinkle marks – soft-sediment deformations in cross-bedded sands.

## 10.1 Introduction

In a perfect world, we should be surrounded by ideal solids, liquids and gases, but in the world we actually occupy we find that materials are not always what they seem.

Unconsolidated sediments, especially when water-saturated, have a knack of changing more or less abruptly from solid-like to fluid-like and back again. Consequently the sediments while fluid-like can be deformed by the action of quite small forces, to give us a class of features called soft-sediment deformation structures. These structures occur in a wide range of modern sedimentary environments and abound in the rock record. It is clear from bedding relationships, as well as from their presence in modern environments, that soft-sediment deformation structures are pene-contemporaneous features, arising during the build-up of sedimentary sequences and, in the case of some varieties, during the act of deposition itself.

Already we have met examples of sediments capable of abruptly changing their state. It was pointed out in Chapter 8 that certain muddy deposits are highly sensitive, being strong enough to support buildings and other engineered structures when solid-like, but chang-ing into mobile slurries when remoulded. Likewise sands are not always what they seem. Everyday experi-ence leads us to expect that a surface underlain by sand will generally be strong enough to support a person's weight, yet the dried grains can be poured like a liquid from one container to another, tending in the process to take the shape of the receiver. There may therefore be natural circumstances when even sand can temporarily lose its strength. Fiction at least abounds in dramatic rescues from (or deaths in) quicksand!

In this chapter we shall develop further the theme of changes of state in unconsolidated sediments, exploring some of the reasons for and the consequences of these changes, especially in the context of loose sands.

## 10.2 An experiment

Quicksand is easy to make. Take a well sealed, cylin-drical, screw-topped glass jar and half-fill it with clean very fine or fine sand. Cover the sand with water and agitate the mixture to remove air bubbles. Preferably under water, so that no more air becomes trapped, complete the filling of the jar with water and seal the top. Repeatedly invert the jar to obtain a uniform mix-ture and then carefully set it down on a table, which thereafter must not be disturbed. Using Plasticine, attach to each of two identical coins a piece of cotton thread so that the faces of the coins hang horizontally. Carefully measure the taut length of each attached thread, which should somewhat exceed the depth of the jar.

Holding the jar firmly, carefully unscrew the top and mark the glass with the level reached by the sand. Dip one of the coins by its thread into the water and, once all the air bubbles have been freed, gently lower it onto the sand, leaving the thread hanging over the side of the jar. You will find that the coin rests on the sand (Fig. 10.1a). Next jolt the table several times with a hammer, keeping the jar under observation. You will observe that (1) the sand becomes fluid-like, (2) the coin sinks and

**Figure 10.1** Changes of state in a metastable sand deposit. Coin supported by sand in (a) sinks in (b) after sediment was liquefied. (c) Resedimentation of the liquefied sand permits another coin to be supported.

disappears (Fig. 10.1b) and, more difficult to see, (3) an interface between the fluid-like dispersion above and the settled grains below rapidly ascends the jar. In the fluid-like portion, grains are moving slightly relative to each other and perhaps there is a certain amount of upward streaming of the water. Once settling is over, again mark the level of the sand on the glass, measure the length of thread remaining above the sand, and lower the second coin into the jar. You will find that the level of the sand has dropped slightly but that a coin can once more be supported on the surface (Fig. 10.1c). The second coin can usually be made to disappear, but more blows than before may be necessary.

What do these findings mean? As the coins are more dense than the sand in bulk, the fact that the sand can at times support them means that the aggregate, at these times, has a substantial yield strength. Almost as soon as the table is struck, however, the first coin sinks and disappears into a fluid-like grain dispersion which evidently has little or no strength. The coin is clearly responding to buoyancy forces in qualitatively the same way as any dense particle released into a lighter fluid. But this fluid-like state exists only over a limited period, for in its turn the second coin can be supported on the surface. Evidently, the duration of the fluid-like state at a given level in the jar is limited by the rise of the interface between the dispersion and the resettled particles to that level, the lower parts of the sand resolidifying earliest. Thus the first coin will be found by measurements of the thread not to have sunk to the bottom of the jar. Hence the interface must rise at a speed comparable to the sinking speed of the coin. No large-scale

stirring of the sand occurs while it is fluid-like, but there is some relative movement of particles, and the grains do repack to a slightly lower level in the jar. From this and earlier evidence we must conclude that (1) the initial particle packing and fabric are destroyed when the sand becomes fluid-like, (2) in the fluid-like state the grains do not support each other, (3) the resettled sand is different and closer in packing than the original, and (4) some pore fluid is lost from the sand layer.

What we did in the experiment was to change a sand from solid-like to fluid-like by the mechanism of liquefaction, a disturbing force having been repeatedly applied to a closed system. The system is described as closed because it was unnecessary to introduce fluid from outside in order to change the state of the aggregate. Below we shall meet another mechanism of liquefaction, now requiring a slightly open system. A third mechanism for changing the state of a loose sand – fluidization as discussed in Chapter 3 – necessitates a fully open system. Fluidization and the mechanisms of liquefaction can be lumped in the general process of liquidization.

## 10.3 What causes changes of state?

### 10.3.1 Fluidization

We learned from Chapter 3 of one way to change a loose grain aggregate from solid-like to fluid-like, namely, by fluidizing it with an upward stream of water or air. The mass becomes fluidized when the upward-acting fluid drag exerted on the grains exactly balances

**Figure 10.2** Sand fluidized by fresh water welling up through beach, Norfolk coast, England. Area measures approximately 0.4 × 0.5 m.

their downward-acting immersed weight. The superficial velocity required for fluidization can be calculated using Equation 3.13 and is, speaking generally, very small compared to the terminal fall velocity of a solitary grain in the unbounded fluid. Large upward discharges are not necessary to fluidize natural sediments, but we do require an open system, that is, a sustained external supply of the fluidizing medium. Fluidization is replaced by settling once this supply expires.

There are few sedimentary environments in which significant upward discharges of fluid are to be found. One is the beach and inshore zone, where ground water, recognizable mainly because it is fresh and cold, can be discharged under pressure from aquifers into the sea through localized springs (Fig. 10.2). Springs and seepages associated with fluidized sand can also be found along the toes of alluvial fans and river levées, where sloping layers of permeable sediment pinch out amongst impermeable muds, with the result that ground water under pressure is forced to the surface.

### 10.3.2 Liquefaction as a consequence of loading
Experience teaches that high porosity in loose sands goes with (1) a relatively small average number of contacts per grain, and (2) comparatively little grain interlocking. It is therefore small wonder that such poorly packed sands have little strength, for their grains can easily be loosened one from another and, once loosened, are free to collapse into tighter packings. All that is required to destroy the grain fabric is a sufficiently large impulsive or cyclically repeated force, as was the

case in the experiment just described. No pore fluid need be supplied from outside, and the system is closed.

The change of a sand from solid-like to fluid-like by the application of an impulse or repeated stress is called liquefaction as a consequence of loading. This mechanism can affect dry grains with air as the only intergranular fluid, but is common only in water-saturated sand.

Figure 10.3 summarizes what happens during the change of state. Imagine that sand grains of solids density $\sigma$ are loosely packed at a uniform fractional volume concentration $C$ in water of density $\varrho$, each to the same depth within a vessel fitted with a water-filled manometer tube. Prior to the application of any force, the grains must rest in some mutually supportive arrangement such as that sketched in Figure 10.3a, the entire grain weight being transferred through grain contacts to the base of the container. The fluid pressure at the manometer tapping must therefore be $p = \varrho g h_0 \, \mathrm{N\,m^{-2}}$, where $g$ is the acceleration due to gravity and $h_0$ the height of the water surface above this level. The manometer registers no difference of head. Once a disturbing force has been applied, however, the grains, singly and in groups, begin to loosen themselves from each other (Fig. 10.3b), until there comes a moment when none are directly supported from the base of the container (Fig. 10.3c). The pore fluid during this phase consists of water plus the loosened grains. Exploiting the definition of density, the pressure at the level of the manometer tapping now is

$$p = \varrho g h_0 + k(t)C(\sigma - \varrho)g h_0 \quad \mathrm{N\,m^{-2}} \qquad (10.1)$$

$$= g h_0 [\varrho + k(t)C(\sigma - \varrho)] \qquad (10.2)$$

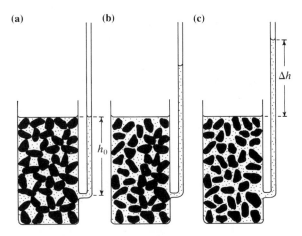

**Figure 10.3** A model for liquefaction.

where $0 \leqslant k(t) \leqslant 1$ is the fraction of grains that at any time $t$ are not directly supported through grain contacts from the base of the container. The second term in Equation 10.1 is the grain-related excess of pressure, $\Delta p = k(t)C(\sigma - \varrho)gh_0$, over the pressure due to the pore water alone. A head of water now appears in the manometer tube to balance the greater pressure of the mixture in the container.

As the grain fabric collapses progressively, $k$ will increase from 0 at the instant the force is applied to 1 at the moment when all the grains are disengaged, assuming no resettling meanwhile. The peak excess pressure is therefore substantial, as may be judged by comparing $C(\sigma - \varrho)gh_0$, the excess when all of the grains have become loosened $(k = 1)$, to $\varrho gh_0$, the pressure in the undisturbed state. For the representative values $C = 0.6$, $\sigma = 2650 \text{ kg m}^{-3}$ and $\varrho = 1000 \text{ kg m}^{-3}$, the excess is virtually equal to the undisturbed value. Such large excess pressures may lead to spectacular effects, such as the rapid flowage of liquidized sediment to regions of lower pressure.

Under what natural circumstances can liquefaction by loading take place? There are three principal origins for the necessary force: (1) shaking by either a tsunami or a landslide which strikes a surface underlain by waterlogged sediment (impulsive force), (2) ground shaking during an earthquake (cyclic stress), and (3) shaking of a lake or sea bed due to the passage of storm waves overhead (cyclic stress).

A tsunami is a large solitary wave created in a body of water as the result of either an earthquake shock, volcanic eruption, or the abrupt descent of a large avalanche from a bordering cliff or mountain. The arrival of one of these waves on the shores of a delta or similar body of sediment could liquefy loose sands, but just how effective the mechanism is we do not know. The fall of a landslide onto waterlogged sediments could also have a liquefying effect.

Ground shaking during earthquakes, representing a cyclic stress, is probably the most potent mechanism for the liquefaction of loose water-saturated sediment. Soft-sediment deformation structures in some modern and geologically young sediments can be directly linked to earthquakes (Sims 1975, Hempton & Dewey 1983).

Speaking generally, the shocks due to moderate to severe earthquakes have a frequency of 0.1–10 Hz, last for 10–100 s excluding aftershocks, and subject the ground to accelerations of the order of $0.2g \text{ m s}^{-2}$. The resulting liquefaction can be variously expressed, for example, by retrogressive slides, subsidence and the appearance of surface water, by spouts of water or sediment-laden water (volcanoes), by the deep sinking

of heavy engineered structures, and by the rise to the surface of such light objects as wooden piles and empty fuel storage tanks (Duke & Leeds 1963, Steinbrugge & Flores 1963, Seed 1968, Kuribayashi & Tatsuoka 1975, Youd & Hoose 1976, Okusa & Anma 1980, Youd & Wieczorek 1982). Analysing the effects of many earthquakes, mainly in Japan, India and the USA, Kuribayashi and Tatsuoka (1975) and Youd (1977) concluded empirically that the maximum distance $x$ from an earthquake epicentre to occurrences of liquefaction varied with the Richter earthquake magnitude $R_m$ as

$$\log x = 0.87R_m - 4.5 \tag{10.3}$$

as plotted in Figure 10.4. These data, of course, do not mean that all of the loose sediments lying within a radius of $x$ kilometres of an epicentre become liquefied, but they do indicate that significant liquefaction effects can be expected in responsive deposits located within that distance. Notice that earthquakes of magnitude 4–5 and less cause little or no liquefaction, the effects of which must therefore be confined to seismically active zones.

One way in which a cyclic stress can be applied to

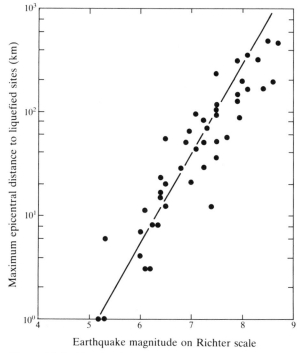

**Figure 10.4** Liquefaction distance as a function of earthquake magnitude (Richter). Data of Kuribayashi and Tatsuoka (1975) and Youd (1977).

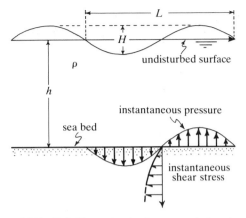

**Figure 10.5** Definition sketch for wave-induced cyclic stressing of sea bed.

sediments on a sea or lake bed is through the movement of waves over the water surface above (Fig. 10.5). At a place where the undisturbed water depth is $h$, the mean hydrostatic pressure at the bottom is $\varrho gh\,\mathrm{N\,m^{-2}}$. As each wave passes overhead, a pressure increase $\Delta p$ above the mean hydrostatic value is felt below the crest, while beneath the trough there is an equal and opposite pressure decrease $-\Delta p$. The places of maximum and minimum bottom pressure are therefore in phase with the profile of the surface wave. Because of the contrast in the bottom pressure between crest and trough, the sediment will experience a shear stress directed from beneath the crest towards the trough. But as the surface waves are progressive, the positions of the maxima and minima of pressure at the bottom must also be moving. The bottom at each place is therefore subject to a pressure force which varies on the same period as the waves, while the sediment is subject to a reversing shear stress varying on that period but with a phase lag.

It can be shown from the mathematical theory of waves (Wiegel 1964) that

$$\Delta p = \frac{\varrho gH}{2} \frac{1}{\cosh(2\pi h/L)} \cos(2\pi x/L - 2\pi t/T) \quad \mathrm{N\,m^{-2}} \tag{10.4}$$

where $H$ is the wave height, $h$ the undisturbed water depth, $L$ the wavelength of the waves, $x$ distance in the direction of wave propagation, $t$ time, and $T$ the wave period. Here 'cosh' represents one of the hyperbolic functions (Sec. 13.6.1) and the cosine term describes the wave profile. Equation 10.4 states that the pressure difference from hydrostatic increases linearly with wave height. Hence liquefaction due to wave-applied cyclic stresses is most likely to occur during storms. As $\cosh(2\pi h/L)$ increases as the water depth grows larger

relative to the wavelength, the excess pressure consequently declines with increasing depth, which would seem to limit wave liquefaction to comparatively shallow bottoms.

Is wave-related cyclic stressing a practicable mechanism for liquefaction? Equation 10.4 cannot be used straightforwardly to answer this question, because we must consider the rather complicated effects of changing water depth on the wavelength, height and stability of waves. For storm waves of $T = 10.9\,\mathrm{s}$, $L = 183\,\mathrm{m}$ and $H = 9.1\,\mathrm{m}$ in deep water, however, Henkel (1970) calculated that the pressure excess would reach a greatest magnitude of $9.6 \times 10^4\,\mathrm{N\,m^{-2}}$ when the wave had reached water of 14 m undisturbed depth. He further calculated that this excess would be enough to cause failure in bottom sediments of the strength found in the shallows of the Mississippi Delta (Ch. 9). The laboratory experiments of Mitchell *et al.* (1973) support Henkel's conclusion that waves are most effective when in depths comparable to their height.

What factors control the ability of a loose sediment to respond to a liquefying impulse or cyclic stress? Seed (1968) discusses this from the standpoint of sediments as materials, and Youd and Hoose (1977) examine the main geological controls. Liquefaction in water-saturated loose sediments is favoured by (1) a high magnitude for the disturbing force, (2) a large number of repetitions of the force, (3) a low particle concentration (open packing), found especially in cross-laminated or cross-bedded sediments, (4) a grain size for mineral sands towards the finer end of the sand class, and (5) little confining pressure (shallow burial). To these we may add with Youd and Hoose the recent deposition of the sediment, for it is observed that the potential for liquefaction decreases with increasing sediment age. Youd and Hoose also consider that fluvial sediments are more readily and frequently liquefied than any other.

### 10.3.3 Liquefaction by increase in pore fluid pressure

In liquefaction under impulsive or cyclic loading, the pressure increases because the grain fabric progressively collapses, that is, as a direct consequence of the changing state. However, it is possible to imagine circumstances in which an increase in pore fluid pressure could cause a change of state.

Consider conditions at the base of a laterally confined sand layer at the bottom of a water body to which it is freely connected. Imagine that the grains are elastic prolate spheroids packed in the regular but isotropic manner of Figure 2.20. Since the layer and water body are freely interconnected, the pore fluid pressure at the base of the layer is simply the hydrostatic pressure, and the basal grains carry the entire immersed weight of the aggregate above. Now the packing must be compressed on account of the weight, and at the base more than at any higher level. But as the layer is laterally confined, the packing can be diminished only in its vertical dimension. Assuming the grains themselves to be incompressible, their concentration must therefore be a little greater, and the porosity of the packing a little less, at the base of the layer as compared to the top.

We can change this situation by confining the sand layer beneath an impermeable bed of mud, so that the packing is no longer connected to the water body. On account of this seal, we are now free to vary the pore fluid pressure in the sand layer. If we can introduce into the layer a little water from some external source, the pore fluid pressure can be increased such that some of the grain weight is transferred to the fluid. The inflow is necessary because, in reducing the stress carried by grain contacts, the packing necessarily expands upwards and so increases very slightly in porosity. We can in fact raise the pore fluid pressure to such a degree that the entire grain weight is transferred from particle contacts to the fluid. The collapse of the grain fabric, that is, the liquefaction of the sand, is then inevitable, for the interparticle contacts become reduced to mathematical points.

These considerations therefore suggest that liquefaction may frequently occur in sand layers buried amongst muds, wherever high pore fluid pressures can be developed in external strata to which the sands are hydraulically connected.

## 10.4   What forces cause deformation?

### 10.4.1   Body forces
Having discovered that there exist several distinct ways by which to change the state of an unconsolidated sediment, what forces are available to deform that sediment in its liquidized condition?

Imagine the effect of a severe earthquake on dunes decorating the submerged sandy bed of a river, and of storm waves on sand waves covering the floor of a tideswept sea. These sand mounds would tend, under the action of gravity, to collapse and flatten once they had become fluid-like. The force driving their collapse is related to the pressure gradient denoted by the slope of their sides, as may be seen in Figure 10.6 showing an instantaneous vertical slice through a typical bedform. The slice is of unit width and length $\delta x$ measured

parallel with the net force. Let $h$ be the height of the slice so defined, that is, the local thickness of the liquidized bedform. Assuming hydrostatic conditions, the total force $F_1$ acting to the right on the left-hand face of the slice is the hydrostatic pressure summed over this face

$$F_1 = \gamma g \int_0^h (h - y)\, \mathrm{d}y \quad \mathrm{N\ m^{-1}} \qquad (10.5)$$

$$= \tfrac{1}{2}\gamma g h^2 \qquad (10.6)$$

in which $\gamma$ is the effective bulk density of the sand. On the right-hand face of the slice, of height $(h - \delta y)$, the horizontal force is similarly

$$F_2 = \gamma g \int_0^{h-\delta y} (h - \delta y - y)\, \mathrm{d}y \qquad (10.7)$$

$$= \tfrac{1}{2}\gamma g [h^2 - 2\,\delta y\, h + (\delta y)^2] \qquad (10.8)$$

The net force $\delta F$ is therefore $F_1$ less $F_2$, that is

$$\delta F = \tfrac{1}{2}\gamma g [2\,\delta y\, h - (\delta y)^2] \qquad (10.9)$$

tending to push the fluid-like sand to the right. But by Newton's first law, another force must act to balance the pressure force $\delta F$. This is the shear stress $\tau$ exerted over the base of the column, whence

$$\tau\, \delta x = \tfrac{1}{2}\,\gamma g [2\,\delta y\, h - (\delta y)^2] \qquad (10.10)$$

which in the limit becomes

$$\tau = h\left(\gamma g\, \frac{\mathrm{d}y}{\mathrm{d}x}\right) = h(\gamma g \tan \beta) \quad \mathrm{N\ m^{-2}} \qquad (10.11)$$

where $\beta$ is the slope angle. The term outside the bracket on the right-hand side of Equation 10.11 is simply the

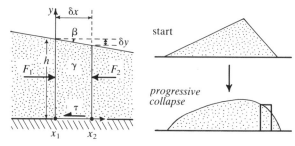

**Figure 10.6**   Definition sketch for deformation of a liquidized bedform.

local thickness of the liquidized bedform, while the bracketed term itself is the pressure gradient operating in the layer. Initially $\beta$ is large, up to 30–35° on the leeward side of a dune, but must diminish with increasing time as the bedform progressively flattens. Deformation under the action of pressure forces should therefore begin rapidly but afterwards continuously decline.

The preceding analysis is oversimplified in that deformation is assumed to be sufficiently slow that accelerations are negligible. Although strong accelerations are likely in the real case, the analysis does afford an idea of the magnitude of the body forces that can deform a liquidized bedform. For a typical dune of crest height equal to 1 m, leeward slope of 35° and effective bulk density of 1000 kg m$^{-3}$, the greatest value of the basal shear stress is $6.9 \times 10^3$ N m$^{-2}$.

### 10.4.2 Buoyancy forces

Under what circumstances can these forces cause the deformation of sediments rendered fluid-like?

One of these circumstances, explained more fully in connection with the origin of desert seif dunes (Sec. 11.4.3), arises where a change of state occurs in a multi-layered system composed of a vertical stack of water-saturated beds of alternately high and low bulk density. In most environments, sand makes up the high-density beds and mud the low-density ones, but either peat or drifted plant remains may in some instances substitute for mud. The water content of fresh water-saturated sands is 30–40%, whereas that of muds and organic beds can be twice as large. From the explanation given in Chapter 11 (see also Anketell *et al.* 1970), a heavy sand layer will tend to collapse down into a lighter mud below once the strength of each has been destroyed. The mud will similarly tend to rise, so that we end up with intrusions of the one bed in the other.

Deformation in response to buoyancy forces can also occur where there is a continuous upward increase in bulk density within a single layer of sediment, rather than discontinuously amongst a number of superimposed discrete layers. Such a bulk-density profile may either (1) be inherited from the time of deposition of the sediment (which may in other respects be vertically uniform), or (2) arise as a direct consequence of the change of state. The second possibility is of especial interest in the case of normally graded sand beds, which reveal a continuous upward decrease in grain size (Allen 1977).

The possibility that a gravitationally unstable profile of bulk density can arise as a direct consequence of liquidization has a simple basis. Once the liquidizing

mechanism is withdrawn from a water-saturated sand bed composed of particles all of the same density, the grains begin to sink at a rate increasing steeply with their diameter, assuming a reasonably uniform initial particle concentration. Hence large grains near the base of a normally graded bed will settle faster than smaller ones near the top. Consequently, the large grains will drift apart faster than the small particles. But the local bulk density of the sand–water mixture increases with the local grain concentration. A steeper rate of decline in concentration will occur where grains are separating rapidly than where they are drifting apart more slowly. Hence the local bulk density will decline more steeply with time in the lower than the upper part of the bed, and a profile of upward-increasing bulk density will tend to develop. The lower parts of the bed may consequently rise into the upper.

How can we shape these ideas into a mathematical model? Consider a short vertical column of height $h$ moving downwards with grains resettling from a mass of liquidized sand (Fig. 10.7). The grain size, and consequently the grain settling velocity $V(y, t)$, increase downwards in this dispersion, where $V$ is the velocity, $y$ vertical distance measured positively downwards and $t$ time. The particles moving with the upper face of the column are sinking at the velocity $V$, but those travelling with the lower face have the larger velocity $(V + h\, \partial V/\partial y)$, where $\partial V/\partial y$ is the local downward rate of increase of settling velocity. Let the local fractional particle volume concentration be $C$ at time $t$. From the definition of density

$$\gamma = \varrho + C(\sigma - \varrho) \quad \text{kg m}^{-3} \qquad (10.12)$$

where $\gamma$ is the overall bulk density of the dispersion, decreasing as the concentration declines. Now consider

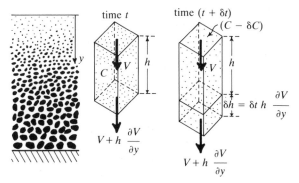

**Figure 10.7** Definition sketch for the development of instability in a liquidized sediment by differential particle settling.

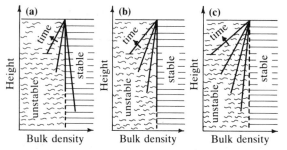

**Figure 10.8** Temporal evolution of profile of bulk density in a resedimenting liquidized bed in the case of initially (a) stable, (b) neutrally stable and (c) unstable profiles.

the column after a small time increment $\delta t$. The length at time $(t + \delta t)$ has increased to $(h + \delta h)$, where $\delta h = \delta t \, h \, \partial V / \partial y$ is also a small increment. This is on account of the greater velocity of the grains at the base of the column as compared to the top. But the number of grains present in the column is unchanged, whence their new concentration is $(C - \delta C)$, where $\delta C$ is another small increment. Making use of the definition of the fractional grain volume concentration (ratio of occupied to total space), it easily follows in the limit that

$$\frac{\partial C}{\partial t} = -C \frac{\partial V}{\partial y} \quad s^{-1} \qquad (10.13)$$

where $\partial C / \partial t$ is the time rate of change of concentration at the level of the column descending with the grains.

Equation 10.13, which in effect is a statement of continuity, shows that the time rate of change of grain concentration varies with the concentration itself and with the settling velocity gradient. In a normally graded bed, $\partial V / \partial y$ is positive, whence $\partial C / \partial t$ is negative and the concentration and hence the bulk density will tend to decrease downwards. The final effect will of course depend on the initial profile of bulk density (Fig. 10.8). However, a gravitationally unstable density profile could eventually arise if the initial profile was only weakly stable, and would certainly be formed from an initially uniform profile.

### 10.4.3 Current drag

A sand bed liquidized by one of the triggers described and lying beneath a current may become deformed wholly through the action of current drag, provided that the top of the bed is of negligible gradient. This proviso is important, because the drag force is small compared to the body forces promoted by even a gentle slope. In illustration, consider the drag due to a river with a mean velocity of $1 \, \mathrm{m \, s^{-1}}$ and a bed roughness equivalent to a

Darcy–Weisbach friction coefficient of 0.04. The quadratic law (Eqn 1.27) affords a mean bed shear stress of just $5 \, \mathrm{N \, m^{-2}}$. If the liquidized bed were 1 m thick, and of effective bulk density equal to $1000 \, \mathrm{kg \, m^{-3}}$, a surface slope of $1°$ would give rise in Equation 10.11 to a body force of no less than $172 \, \mathrm{N \, m^{-2}}$, almost 30 times more than the drag force.

## 10.5 For how long can deformation proceed?

This question is most easily answered in the case of a sand bed that became fluid-like in response to a mechanism of liquefaction. In this case, resettlement of grains, as noted in our introductory experiment, begins as soon as the destruction of the primary fabric is completed.

Events during resettlement can best be illustrated by allowing a not too concentrated dispersion of uniform glass beads or clean very well sorted sand in water to settle in a tall glass jar (Fig. 10.9). Two moving interfaces will be seen. The higher lies between the clear water A and the remaining dispersion B, while the lower separates the dispersion B from the resettled grains D (Fig. 10.10a). The interfaces move each at a constant rate, as sketched in Figure 10.10b, until eventually all of the dispersion is consumed and resettled grains only remain. Muddy sediments, however, can behave somewhat differently, and may while settling give rise to an intermediate layer C between B and D. Layer C is not in every case distinguished by sharp interfaces.

**Figure 10.9** From left to right, initial, intermediate and final stages in the laboratory sedimentation of a uniform particle dispersion. Interfaces are arrowed (see also Fig. 10.10).

**(a)**  **(b)**

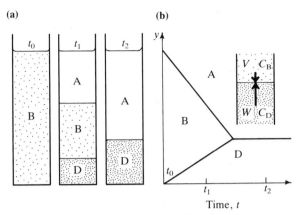

**Figure 10.10** Settling of a particle dispersion.

This experiment makes it clear that the extent of deformation of a liquefied bed will depend on the length of time over which state B can persist. How can we deduce this time?

Consider in Figure 10.10 the interface between states B and D. Taking the interface itself as the reference frame, continuity demands that the flux of grains approaching the interface from layer B should equal the flux passing downwards from the interface into layer D. Let the grains in layer B have a fractional concentration $C_B$ and a settling speed in the dispersion of $V$ relative to the ground. Let the speed at which the interface is travelling upwards relative to the ground be $W$, and let the concentration of the resettled grains be $C_D$. Putting $\sigma$ as the solids density, we have

grain flux towards interface from B =
$$\sigma C_B(V + W) \quad \text{kg m}^{-2}\text{s}^{-1} \quad (10.14)$$

where $(V + W)$ is the grain speed relative to the interface, and

grain flux downwards from interface into D = $\sigma C_D W$
$$(10.15)$$

On equating the fluxes and rearranging, we obtain

$$W = \frac{C_B V}{\Delta C} \qquad (10.16)$$

where $\Delta C = (C_D - C_B)$, the concentration difference between the dispersion and the resettled grains. Equation 10.16 reveals that the speed of ascent of the interface increases directly as the grain concentration and speed in the dispersion, and inversely as the concentration difference between the layers.

If the bed is uniform in grain concentration and den-

sity, and is liquefied throughout, the time for resettlement is simply $T = h/W$ s, where $h$ is the bed thickness. Substituting for $W$ from Equation 10.16, and then for $V$ from Equation 3.10, for particles in a dispersion, we can write

$$T = \frac{\Delta C\, h}{C_B V_0(1 - C_B)^n} \quad \text{s} \qquad (10.17)$$

where $V_0$ is the terminal velocity in an unbounded fluid and $n$ an exponent varying with the Reynolds number. For a bed of fine-grained quartz sand 1 m thick, it is plausible that $\Delta C = 0.025$, $C_B = 0.6$, $V_0 = 0.03$ m s$^{-1}$ and $n = 3$, affording a resettlement time of roughly 22 s. The resettlement time for sand beds generally is of the order of seconds to tens of seconds. Deformation is therefore unlikely to be of such a degree that primary lamination will have been destroyed before complete resettlement.

Whereas the time available for the deformation of a liquefied sand is comparatively short, that for a fluidized sediment can be very large. Fluidization occurs in an open system and the fluid-like state persists for so long as there is an adequate discharge of the fluidizing agent. The fluidized stage may be so prolonged that complete mixing of the sediment occurs, with the consequence that no primary lamination remains.

## 10.6 Complex deformations in cross-bedded sandstones

It is not uncommon in fluvial, tidal and aeolian sandstone formations to find cross-bedding sets in which the laminae are more or less strongly folded and even locally faulted. The open packing of avalanched sands (Ch. 2) makes them particularly susceptible to liquidization.

Figure 10.11 shows a relatively simple example, in which the laminae are deformed into open folds whose axial surfaces dip moderately to steeply against the direction of the inferred depositing current. A closely similar structure is described by Dalrymple (1979) from modern intertidal dunes that had been liquefied. A pattern of tighter and more complex folds is shown in Figure 10.12. Deformed cross-bedding is widely recorded from fluvial and tidal sandstones (Jones 1961, Gradzinski 1970, Young & Long 1977, Hobday & Von Brunn 1979, Miall 1979).

Deformations of great complexity occur on a large scale in many thickly cross-bedded aeolian sandstones (Rice 1939, Peacock 1966, Sanderson 1974, Steidtmann

189

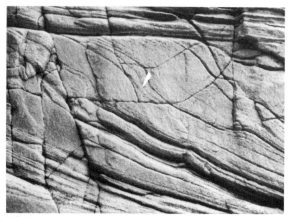

**Figure 10.11** Complex low-amplitude folds in upper part of cross-bedded sandstone, Upper Old Red Sandstone, southeast Ireland. About 0.5 m of sandstone is shown.

**Figure 10.13** Thrusts in a deformed cross-bedded sandstone, Hopeman Sandstone, Scotland. Scale is 0.1 m long. Photograph courtesy of Dr H. G. Owen, University College of Swansea.

**Figure 10.12** Complex deformed cross-bedding in Torridonian sandstone, western Scotland. Face is approximately 2 m tall. Note upward truncation of deformations. Photograph courtesy of Dr H. G. Owen, University College of Swansea.

1974, Doe & Dott 1980, Horowitz 1982). Not only are tight to open folds of widely varying orientations present, but it is common to discover folds with faulted limbs, particularly nappe-like features (Fig. 10.13). The large scale of these deformations is emphasized by the fact that locally more than one cross-bedding set is involved in the folding and faulting.

It is unlikely that these complex structures record the action of just one sort of deforming agent, but the evidence does point mainly to body forces. The simpler structures related to Figures 10.11 and 10.12 seem from their geometry to represent the compression and buckling of foresets during the partial collapse under their own weight of bedforms liquefied at the sedimentary surface. This is certainly the explanation of the structures that Dalrymple (1979) described from a modern tidal sand shoal, where the impact of breaking waves caused the liquefaction of emerging bedforms. The structures in aeolian sandstones require a more complex explanation (see Horowitz 1982). Not only were the water-saturated basal parts of some dunes deformed, but so also were buried cross-bedded sands ahead, beneath and behind the surface bedforms. The faulting accompanying the folds records the locally cohesive (brittle) response of these older and more thickly buried deposits. It should not occasion surprise to find soft-sediment deformation structures in aeolian sandstones. Youd (1977) points out that modern aeolian sands when water-saturated commonly become liquefied in response to earthquake shocks.

**Figure 10.14** Vertical section through load casts at the base of two thin sandstone beds, Bude Formation (Carboniferous), southwest England. Coin is 0.03 m across. Photograph kindly supplied by Professor J. D. Collinson, University of Bergen.

## 10.7 Load casts

These are mainly recorded from the sharp undersides of sandstone beds that overlie mudstones, but are also known from layered igneous rocks (Thy & Wilson 1980). As seen in a vertical section (Fig. 10.14), load casts appear as a row of flattened lobe-shaped masses of sandstone of a similar size, shape and spacing. Upwards between the sandstone lobes penetrate broad to narrow and flame-like fingers of mudstone. Load casts range in horizontal spacing or wavelength between a few millimetres and many decimetres, tending to increase in size with the thickness and vertical spacing of the affected sandstone beds. It is often possible in a large exposure to trace a sandstone with load casts into an unaffected portion of the same bed. The bedding within load casts and in the adjacent mudstones is deformed in comparison to undisturbed parts of the sequence. Lamination within the sandstone lobes tends to follow the shape of the lobes, except near their margins where stretched and severed laminae may be found. Any lamination visible within the mudstones invariably is sharply infolded within the upward-penetrating fingers. Hence load casts arise after the deposition of the affected beds.

Commonly, the mudstone weathers away to reveal the load casts in three-dimensional detail. The sandstone lobes are then seen as equant to elongated pillow-shaped masses, divided from each other by narrow clefts corresponding to the mudstone fingers. Animal tracks and trails, and other sedimentary markings, are commonly preserved in a distorted form on the surfaces of the pillows, another proof of the origin of load casts by soft-sediment deformation.

It has been emphasized by Anketell *et al.* (1970) that load casts are an example of the instability of an interface in a gravitationally unstable arrangement of layered sediments. The particular instability involved is called the Rayleigh–Taylor instability, and the forces driving the motion are those due to buoyancy (Sec. 11.4.3).

It is worth doing some simple experiments to test the correctness of this interpretation and to explore the system further. Place a large flat-bottomed plastic bowl on the bench and fill it almost to the top with a thin suspension of china clay sufficient after settling for a day to give a bed a few centimetres thick. Now gently sift dry plaster of Paris as quickly as possible into the bowl, until a layer similar in thickness to the mud has formed. When all has been added, strike the bench a series of vigorous blows, in order to liquefy the plaster and the mud. The plaster will collapse down into the mud, on account of its greater bulk density. After an hour or so, the resulting load casts can be freed from the bowl, washed and dried. The experiment should be repeated using sand instead of plaster of Paris; if sands of different colours are available, they can be alternated to simulate lamination. After the load casts have been formed, the bowl should be drained of water and partly dried out before the bed is carefully sectioned with a knife.

Figure 10.15 illustrates structures from representative

**Figure 10.15** View from below of experimental load casts of plaster of Paris into kaolinite mud. Scale bar is 0.05 m.

experiments. Notice the tendency for the pillow-like masses preserved on the interface to develop an equant, almost polygonal form. Notice too the narrow clefts between the pillows, corresponding to finger-like upward projections of the underlying mud. The internal lamination simulated by the layers of coloured sand tends to follow the shape of the pillows, and there is a certain amount of stretching of laminae, particularly towards the clefts. Thus these experimental structures closely resemble naturally occurring load casts.

It is worth asking why load casts remain attached to the underside of a sandstone. In some instances the sinking of the pillows was limited by a sandstone bed below, but in many cases there was no such constraint. The two possibilities suggesting themselves are that either the sand bed resolidified before deformation could proceed to the stage of detachment, or collapse stopped when the pillows had descended to the level of a strong mud. Whichever control applied, it is not difficult to calculate the terminal driving force. Consider the slice of unit width through the load casts of Figure 10.16. Following the analysis affording Equation 10.11, the total horizontal force driving the sand towards the mud is

$$F_s = \gamma_s g \int_0^h (h - y) \, dy = \tfrac{1}{2} \gamma_s g h^2 \quad \text{N m}^{-1} \quad (10.18)$$

while that acting from the mud towards the sand is

$$F_m = \gamma_m g \int_0^h (h - y) \, dy = \tfrac{1}{2} \gamma_m g h^2 \quad (10.19)$$

where $\gamma_s$ and $\gamma_m$ are respectively the bulk densities of the sand and mud layers, and $h$ the vertical distance between the base of a sand pillow and the top of a mud

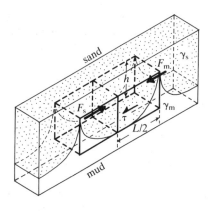

**Figure 10.16** Definition sketch for analysis of load casts.

finger. Putting $L$ as the spacing or wavelength of the pillows, the shear stress $\tau$ neutralizing the net pressure force is

$$\tau = \frac{g h^2}{L} (\gamma_s - \gamma_m) \quad \text{N m}^{-2} \quad (10.20)$$

Values representative of load casts are $h = 0.02$ m, $L = 0.1$ m, $\gamma_s = 2000$ kg m$^{-3}$ and $\gamma_m = 1700$ kg m$^{-3}$, affording a shear stress of approximately 12 N m$^{-2}$. The mud need not therefore have been particularly strong to have limited downward collapse. As a corollary, detachment of the sand can only occur when the mud below is particularly liquid.

## 10.8 Convolute lamination

Oceanic turbidity currents (Ch. 12) and flooding rivers (Ch. 5) are alike in tending to deposit thin sharp-based sand beds typified by an upward decline in grain size (normal grading) generally ending in mud. They further resemble each other in depositing the sediment rapidly, at rates of the general order of centimetres to decimetres per hour. Moreover, at certain stages of flow, deposition tends to occur on a current-rippled bed, grains avalanching in the lee of the bedforms. This distinctive combination of circumstances suggests that both turbidity current and river flood deposits should be prone to soft-sediment deformation.

This expectation is well justified. The distinctive soft-sediment deformation structure called convolute lamination (Fig. 10.17) abounds in the normally graded deposits of both turbidity currents (Kuenen 1953, Dzulynski & Smith 1963, McBride & Kimberley 1963, Mutti & Ricci Lucchi 1972) and river floods (McKee et al. 1967, Karcz 1972, Chakrabarti 1977). Typically, convolute lamination affects beds of very fine-grained sand grading up into silt. The structure consists of a regular series of generally upright folds which grow in wavelength with increasing thickness of the affected zone (Kühn-Velten 1955). The folds range in vertical cross-sectional shape from broad and open, through box-like, to that of an elastica in which axial traces are virtually parallel with the boundaries of the bed. Individual laminae thicken and thin markedly from one part of a fold to another, especially in the tighter and more complex structures. Folds seen in plan range from equant, like an assemblage of diapirs, to parallel and long-crested (Dzulynski & Smith 1963).

Three varieties of convolute lamination can be distinguished, depending on the inferred relative timing

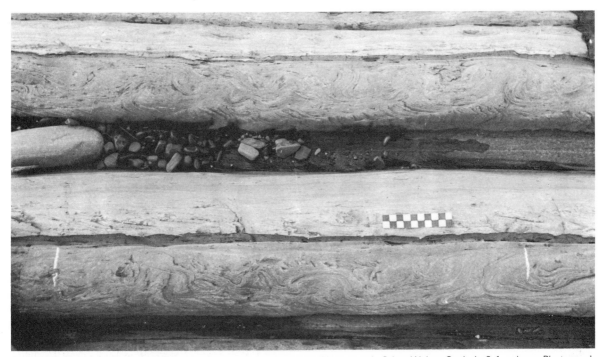

**Figure 10.17** Convolute lamination in two turbidite sandstones, Aberystwyth Grits, Wales. Scale is 0.1 m long. Photograph courtesy of Dr H. G. Owen, University College of Swansea.

of deposition and deformation (Fig. 10.18). In post-depositional convolute lamination, the fold amplitude decreases upwards and downwards from the centre of the affected bed or zone. The folds lack internal discordances, other than those related to, say, the normal erosion between current ripples. Given this combination

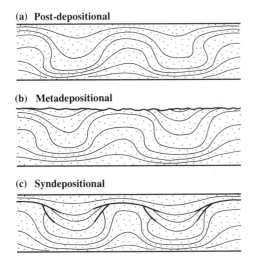

(a) **Post-depositional**

(b) **Metadepositional**

(c) **Syndepositional**

**Figure 10.18** Schematic types of convolute lamination.

of features, deformation could only have occurred after burial of the bed. A bed with metadepositional convolute lamination presents similar internal features, but has an undeformed erosional top. The structure may therefore have arisen either during or some time after deposition of the deformed bed. Syndepositional convolute lamination, the third variety of the structure, shows clear internal evidence of its origin during deposition. Anticlines rising spasmodically through the accumulating sediment were repeatedly beheaded, while the synclines between became infilled. It is important to note, however, that episodic truncations are generally restricted to the topmost one-quarter to one-fifth of the folded zone, pointing to the comparative lateness of deformation.

Geometrically, there are significant resemblances between the folds of convolute laminations and the folded interface in load casts. The generally upright nature of the folds suggests that convolute lamination arose under the same influence of buoyancy forces, but we are no longer able as with load casts to appeal to the gravitational instability of a two-layer system. Noting the association of convolute lamination with normally graded beds, however, it can be suggested that differential settling may have provided an unstable but continuously varying profile of bulk density, as described

**Figure 10.19** Vertical section in structure resembling convolute lamination, made by the laboratory deformation of a gravitationally unstable arrangement of layers of silicone putty. Specimen about 0.015 m thick. From Ramberg (1981) with permission. Photograph courtesy of Professor H. Ramberg, University of Uppsala.

by Equation 10.13 above. Unfortunately, there is so far no experimental proof that a normally graded sand bed can deform in response to this mechanism after being liquidized, but an indication that folding is to be expected comes from an experiment by Ramberg (1981). To augment the buoyancy forces, he centrifuged an arrangement of six parallel layers of silicone putty of inward-increasing density (1240–1620 kg m³). There resulted a series of folds of equant plan which in vertical section had the form of elasticas (Fig. 10.19), an extreme style of deformation but with representatives in some examples of convolute lamination (Fig. 10.17, lower bed). Although discrete layers were involved in Ramberg's experiment, the six of them were sufficient to have behaved collectively like a bed with a continuous upward increase of density (infinite number of layers).

What caused graded sand beds to become sufficiently fluid-like that buoyancy forces could deform them? The existence of varieties of convolute lamination suggests a multiplicity of triggers. Liquidization in the case of the syndepositional form could have occurred in response to push–pull pressure fluctuations related to the passage of large turbulence structures (Ch. 6) in the depositing

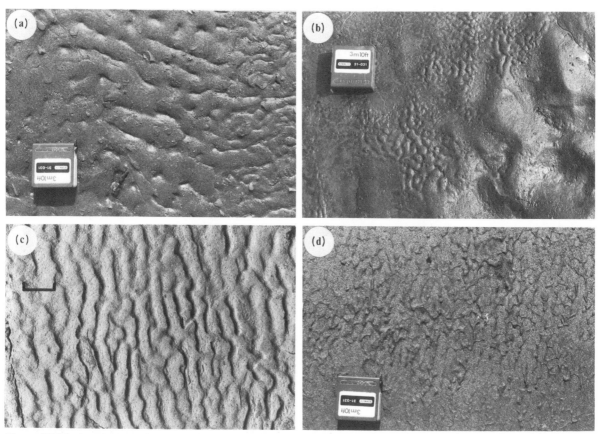

**Figure 10.20** Intertidal wrinkle marks, Severn Estuary, England. Surface appearance (a) on a planar sand lamina overlying mud, and (b) in association with a rippled sand layer overlying mud (a thin mud film coats the sand). Upper surface of deformed mud layer (c) brushed free from sand in a dried specimen, and (d) as partly exhumed in the field. Scale box in (a), (b) and (d) is 0.05 m square. Scale bar in (c) measures 0.01 m.

current. If this is correct, we have another possible mechanism of liquefaction due to a cyclic stress. Post-depositional convolute lamination suggests a response to either earthquake shock or an externally motivated increase in pore-water pressure.

## 10.9 Wrinkle marks

These soft-sediment deformation structures were first described from the mud-flats of the German North Sea coast, and have since been widely recognized in the Severn Estuary and inner Bristol Channel (Allen 1984). Wrinkle marks are distinctive in both origin and preservation.

Wrinkle marks, restricted to the intertidal zone, are visible at low tide in places where mud and sand become interbedded on a fine scale. As seen on the sedimentary surface, wrinkle marks are series of long equant flat-topped ridges of sand separated by narrow depressions, at the bottoms of which mud may be evident (Figs 10.20a & b). The longer ridges bear no particular orientation relative to the slope of the surface, but are more commonly contour-parallel than oblique or perpendicular. The structures are mainly developed in a sediment doublet composed of a lamina of mud, representing deposition from the high-water slack tide, overlain by a lamina of sand, left by the ebb stream and exposed on the sedimentary surface at low tide (Fig. 10.20a). In places the marks are associated with a sediment triplet, the sand lamina being sandwiched between two mud layers, the exposed upper layer being much the thinner (Fig. 10.20b). According to circumstances, the second mud lamina represents either slackening of the ebb stream or ponding of tidal water behind sand or mud shoals or rocks.

Samples returned to the laboratory and sectioned when moist show that the surface ridges and depressions are matched on the mud-sand interface below by antithetic hollows and ridges (Fig. 10.21). The sand therefore forms a series of bolster-shaped masses divided by fingers of mud, not unlike load casts. Furthermore, the lamination within the sand follows the shape of these bolsters. Frequently, current or wave action during subsequent tides removes the surface sand lamina to expose and scour the wrinkled mud-sand interface below (Fig. 10.20c). Part of an extensive mud-sand interface from which sand has been incompletely removed appears in Figure 10.20d. Hence only wrinkle-marked sediment triplets can be fully preserved, and even these are not immune to subsequent partial destruction.

When do wrinkle marks form? Wrinkle marks seem to form after the ebbing tide has exposed the sediment-ary surface but while the sediment is still water-saturated. First, the marks do not become visible at a site until it has been exposed for some tens of minutes. Secondly, the ridges and depressions on the exposed surface of a doublet are much too delicate to have survived passage through the zone of wave action moving with the descending tide. Thirdly, even more delicate organism trails and feeding traces are commonly preserved on the ridged surfaces, but are unrelated in character to the presence of the ridges and depressions and, in some cases, are actually severed by the depres-sions. The tracks therefore antedate the wrinkle marks but, like them, could not have survived passage through the wave zone.

How do wrinkle marks form? These structures resem-ble load casts in two ways over and above general shape. First, they are linked to a gravitationally unstable system of sand over mud. Secondly, lamination within the sand bolsters generally parallels the deformed inter-face. Hence wrinkle marks are soft-sediment defor-mation structure. They are almost certainly triggered either by high pore fluid pressures created as water drains out of and off the intertidal flats, or in conse-quence of liquefaction of the mud by waves during the

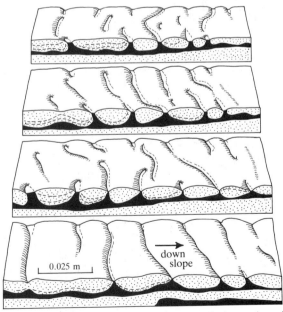

**Figure 10.21** Dissection of wrinkle-marked mud-sand couplet from Severn Estuary, England.

withdrawal of the tide. We cannot appeal to earth-quakes, for neither the German nor British tidal flats lie in a seismically active area.

## 10.10 Overturned cross-bedding

Figure 10.22 shows examples of overturned cross-bedding, a common soft-sediment deformation structure restricted to single sets and observed chiefly from fluvial sandstones (Robson 1956, Jones 1962, Rust 1968, Hendry & Stauffer 1975, 1977). This structure is particularly simple in geometry, consisting of a single, similar, flat-lying fold with a weakly to moderately curved but never cusped hinge. Typically, the fold limbs are about equal in length, the hinge appearing roughly half-way up the affected bed, but either limb can be considerably the longer. In a long upper limb the laminae may be stretched out and difficult to separate. Normally, the upper limb is truncated by an undeformed

**Figure 10.22** Overturned cross-bedding in four units (arrowed) seen in a vertical section, Kap Holbaek Formation, north Greenland. Geological hammer for scale. Photograph courtesy of Professor J. D. Collinson, University of Bergen.

erosion surface, commonly underlying another cross-bedding set.

It is difficult to avoid the conclusion that overturned cross-bedding is a soft-sediment deformation structure (Allen & Banks 1972). That the folds are of penecontemporaneous origin is clear from their upward truncation by younger beds; their smoothly flowing shape suggests that the sand was fluid-like when deformed. A horizontal force was involved, to judge from the flat-lying nature of the folds. This force, many workers believe, was the drag of sand-laden currents building cross-bedded dunes.

Simply by comparing in Figure 10.22 a folded with an unaffected cross-stratum, an important conclusion can be drawn regarding the deformation involved in overturned cross-bedding. As sketched in Figure 10.23, the amount of horizontal movement of an element in the cross-stratum is a steeply increasing function of its height above the base of the set. Therefore the problem of overturned cross-bedding is that of predicting the function that describes this displacement.

On what basis should we make this prediction? The horizontal displacement of an element in a liquidized cross-bedded sand acted on by a shear stress applied to the top of the bed was considered by Allen and Banks (1972) to depend on the interaction between (1) the velocity profile induced within the fluid-like portion of the bed, and (2) the duration of the fluid-like state at the level of the element, as limited by the upward movement of the surface of resettlement (see Figs 10.9 & 10.10). We therefore consider in Figure 10.24 the horizontal movement of an element P lying at a constant depth $y$ relative to the top of a liquidized bed of water-saturated sand of thickness $h_0$ at time $t = 0$ when resettlement began. The bed is considered to be formed of sand of uniform solids density, particle diameter and fractional grain volume concentration $C_B$. Hence the grains in the fluid-like portion of the bed sink at a uniform steady

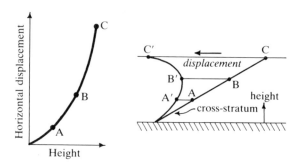

**Figure 10.23** General implications of overturned cross-bedding.

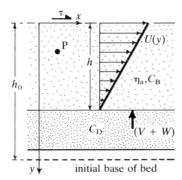

**Figure 10.24** Definition diagram for displacement of an element in overturned cross-bedding.

speed $V$ relative to the ground. The resettled sand takes the concentration $C_D$ and the surface of resettlement moves upwards through the bed at a uniform steady speed $W$ relative to the ground. Finally the bed is acted on by a steady shear stress $\tau$ in the direction of the cross-bedding dip azimuth. As argued above, this stress could be applied by a river or similiar current.

It is reasonable to assume a linear velocity profile in the fluid-like sand, provided that the motion is laminar. In terms of Figure 10.24, the equation for a linear velocity profile at any instant is

$$U = U_{max}\left(1 - \frac{y}{h}\right) \quad \text{m s}^{-1} \tag{10.21}$$

and we have from Section 1.8 the definition

$$\tau = -\eta_a \frac{dU}{dy} \quad \text{N m}^{-2} \tag{10.22}$$

where $U$ is the instantaneous velocity at depth $y$ within the fluid-like sand, $U_{max}$ the steady surface velocity, $h$ the instantaneous thickness of the fluid-like layer, and $\eta_a$ the apparent viscosity of the sand–water mixture. Integrating Equation 10.22 under the boundary conditions $U = 0$ at $y = h$ and $U = U_{max}$ at $y = 0$, we obtain $U_{max} = (\tau/\eta_a)h$. Substituting this value for $U_{max}$ in Equation 10.21 gives

$$U = \frac{\tau}{\eta_a}(h - y) \quad \text{m s}^{-1} \tag{10.23}$$

for the instantaneous velocity profile within the fluid-like portion of the bed. To establish whether the use of this velocity profile is justified, we need to calculate a representative Reynolds number for the bed. The apparent viscosity of the sand will be of order $10^4$ times

the viscosity of water (Allen 1982), whence $U_{max}$ for $\tau = 2.5 \text{ N m}^{-2}$ (a reasonable value for a river current) will be of order $0.25 \text{ m s}^{-1}$. For a sand bed of thickness 1 m and total bulk density $2000 \text{ kg m}^{-3}$, the Reynolds number is therefore of order 50, making the linear velocity profile plausible.

But what is happening meanwhile to the interface between the settled grains and the remaining liquid portion of the bed? The movement of the interface is continually reducing the value of $h$ and clearly limits the displacement of P under the influence of the velocity profile (Eqn 10.23). As the liquidized bed is uniform in all respects, this interface ascends relative to the top of the fluid-like layer at the uniform speed $(V + W)$ and

$$h = h_0 - t(V + W) \quad \text{m} \tag{10.24}$$

Substituting for $W$ from Equation 10.16 above

$$h = h_0 - tV\left(1 + \frac{C_B}{\Delta C}\right) \tag{10.25}$$

which may be rearranged to give

$$t = \frac{(h_0 - y)}{V[1 + (C_B/\Delta C)]} \quad \text{s} \tag{10.26}$$

In Equations 10.25 and 10.26, $t$ is the time required for the surface of resettlement to ascend to P at the level $y$, and $\Delta C$ is the concentration difference between the two layers. The displacement of P will accordingly cease after the elapse of time $t$.

We now have all the information necessary to calculate the horizontal displacement $x$ of the element P. First substitute from Equation 10.25 for $h$ in Equation 10.23, to obtain the velocity profile in terms of time. As $U = dx/dt$, the integral of this new equation under the boundary conditions $x = 0$ at $t = 0$ is the displacement of P in terms of time, that is

$$x = \frac{\tau}{\eta_a}(h_0 - y)t - \frac{1}{2}\frac{\tau}{\eta_a}V\left(1 + \frac{C_B}{\Delta C}\right)t^2 \quad \text{m} \tag{10.27}$$

To obtain the displacement of P in terms of vertical position, we need only substitute for $t$ from Equation 10.26, obtaining

$$x = \frac{1}{2}\frac{\tau}{\eta_a}\frac{(h_0 - y)^2}{V[1 + (C_B/\Delta C)]} \tag{10.28}$$

Thus the displacement varies directly as the applied

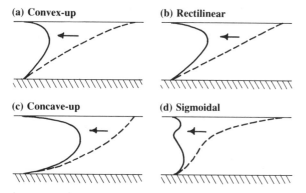

**(a) Convex-up**  **(b) Rectilinear**

**(c) Concave-up**  **(d) Sigmoidal**

**Figure 10.25** Variation in the final shape (vertical streamwise profile) of overturned cross-bedding as a function of initial foreset shape, given a constant displacement pattern.

shear stress, inversely as the apparent viscosity of the sand, and inversely as its settling velocity. The displacement increases as the concentration difference grows larger. The most interesting result, however, is that the displacement increases as the square of the initial height $(h_0 - y)$ above the base of the bed, much as observed in the real case (Figs 10.22 & 10.23).

But Equation 10.28 only tells us the displacement of P from its starting position. The final shape of the overturned cross-stratum can be obtained only by adding to this expression an equation for the initial shape. The scope for choice is fairly wide. Figure 10.25 illustrates the kinds of results obtainable for convex-up, straight, concave-up and sigmoidal starting shapes. Note that a sigmoidal cross-stratum may yield a second flat-lying fold.

## Readings

Allen, J. R. L. 1977. The possible mechanics of convolute lamination in graded sand beds. *J. Geol. Soc. Lond.* **134**, 19–31.

Allen, J. R. L. 1982. *Sedimentary structures,* Vol. II. Amsterdam: Elsevier.

Allen, J. R. L. 1984. Wrinkle marks: an intertidal sedimentary structure due to aseismic soft-sediment loading. *Sed. Geol.* **41**, 75–95.

Allen, J. R. L. and N. L. Banks 1972. An interpretation and analysis of recumbent-folded deformed cross-bedding. *Sedimentology* **19**, 257–83.

Anketell, J. M., J. Cegla and S. Dzulynski 1970. On the deformational structures in systems with reversed density gradients. *Roczn. Pol. Tow. Geol.* **40**, 3–30.

Chakrabarti, A. 1977. Upward flow and convolute lamination. *Senckenbergiana Maritima* **9**, 285–305.

Dalrymple, R. W. 1979. Wave-induced liquefaction: a modern example from the Bay of Fundy. *Sedimentology* **26**, 835–44.

Doe, T. W. and R. H. Dott 1980. Genetic significance of deformed cross-bedding – with examples from the Navajo and Weber Sandstones of Utah. *J. Sed. Petrol.* **50**, 793–812.

Duke, C. M. and D. J. Leeds 1963. Responses of soils, foundations and earth structures to the Chilean earthquake of 1960. *Bull. Seism. Soc. Am.* **53**, 309–57.

Dzulynski, S. and A. J. Smith 1963. Convolute lamination, its origin, preservation and directional significance. *J. Sed. Petrol.* **33**, 616–27.

Gradzinski, R. 1970. Sedimentation of dinosaur-bearing Upper Cretaceous deposits of the Nemegt Basin, Gobi Desert. *Palaeont. Pol.* **20**, 147–227.

Hempton, M. R. and J. F. Dewey 1983. Earthquake-induced deformational structures in young lacustrine sediments, East Anatolian Fault, southeast Turkey. *Tectonophysics* **98**, T7–14.

Hendry, H. E. and M. R. Stauffer 1975. Penecontemporaneous recumbent folds in trough cross-bedding of Pleistocene sands in Saskatchewan, Canada. *J. Sed. Petrol.* **45**, 932–43.

Hendry, H. E. and M. R. Stauffer 1977. Penecontemporaneous folds in cross-bedding: inversion of facing criteria and mimicry of tectonic folds. *Bull. Geol. Soc. Am.* **88**, 809–12.

Henkel, D. J. 1970. The role of waves in causing submarine landslides. *Geotechnique* **20**, 75–80.

Hobday, D. K. and V. Von Brunn 1979. Fluvial sedimentation and palaeogeography of an early Palaeozoic failed rift, southeastern margin of Africa. *Palaeogeog. Palaeoclimatol. Palaeoecol.* **28**, 169–84.

Horowitz, D. H. 1982. Geometry and origin of large-scale deformation structures in some ancient windblown sand deposits. *Sedimentology* **29**, 155–80.

Jones, G. P. 1961. Sedimentary structures in the Bima Sandstone. *Records Geol. Surv. Nigeria* 1959, 5–14.

Jones, G. P. 1962. Deformed cross-stratification in Cretaceous Bima Sandstone, Nigeria. *J. Sed. Petrol.* **32**, 231–9.

Karcz, I. 1972. Sedimentary structures formed by flash floods in southern Israel. *Sed. Geol.* **7**, 161–82.

Kuenen, P. H. 1953. Graded bedding with observations on the Lower Palaeozoic rocks of Britain. *Verh. K. Akad. Wet.* (I) **20** (3), 1–47.

Kühn-Velten, H. 1955. Subaquatische Rutschungen im höheren Oberdevon des Saurlandes. *Geol. Rndsch.* **44**, 1–25.

Kuribayashi, E. and F. Tatsuoka 1975. Brief review of liquefaction during earthquakes in Japan. *Soils Foundns* **15** (4), 81–92.

McBride, E. F. and J. E. Kimberley 1963. Sedimentology of Smithwick Shale (Pennsylvanian), eastern Llano region, Texas. *Bull. Am. Assoc. Petrol. Geol.* **47**, 1840–54.

McKee, E. D., E. J. Crosby and H. L. Berryhill 1967. Flood deposits, Bijou Creek, Colorado, June 1965. *J. Sed. Petrol.* **37**, 829–51.

Miall, A. D. 1979. *Mesozoic and Tertiary geology of Banks Island, Arctic Canada.* Mem. Can. Geol. Surv., no. 387.

Mitchell, R. J., K. K. Tsui and D. A. Sangrey 1973. Failure of submarine slopes under wave action. *Proc. 13th Conf. Coastal Engng* Vol. 2, 1515–41.

Mutti, E. and F. Ricci Lucchi 1972. Le torbiditi dell'Appennino settentrionale: introduzione all'analisi di facies. *Mem. Soc. Geol. Ital.* **11**, 161–99.

Okusa, S. and S. Anma 1980. Slope failures and tailings dam damage in the 1978 Izu–Ohshima–Kinkai earthquake. *Engng Geol.* **16**, 195–224.

Peacock, J. D. 1966. Contorted beds in the Permo-Triassic aeolian sandstones of Morayshire. *Bull. Geol. Surv. Gr. Br.* **24**, 157–62.

Ramberg, H. 1981. *Gravity, deformation and the Earth's crust*, 2nd edn. London: Academic Press.

Rice, R. C. 1939. Contorted bedding in the Trias of the NW Wirral. *Proc. Liverpool Geol. Soc.* **17**, 361–70.

Robson, D. A. 1956. A sedimentary study of the Fell Sandstone of the Coquet Valley, Northumberland. *Q. J. Geol. Soc. Lond.* **116**, 241–58.

Rust, B. R. 1968. Deformed cross-bedding in Tertiary–Cretaceous sandstones, Arctic Canada. *J. Sed. Petrol.* **38**, 87–91.

Sanderson, I. D. 1974. Sedimentary structures and their environmental significance in the Navajo Sandstone, San Rafael Swell, Utah. *Geol. Stud. Brigham Young Univ.* **21**, 215–46.

Seed, H. B. 1968. Landslides during earthquakes due to soil liquefaction. *J. Soil Mech. Founds Div. Am. Soc. Civ. Engrs* **94**, 1055–122.

Sims, J. D. 1975. Determining earthquake recurrence interval from deformational structures in young lacustrine sediments. *Tectonophysics* **29**, 141–52.

Steidtmann, J. R. 1974. Evidence for eolian origin of cross-stratification in sandstone of the Casper Formation, southernmost Laramie Basin, Wyoming. *Bull. Geol. Soc. Am.* **85**, 1835–42.

Steinbrugge, K. V. and R. Flores 1963. The Chilean earthquake of May, 1960: a structural engineering viewpoint. *Bull. Seism. Soc. Am.* **53**, 225–307.

Thy, P. and J. R. Wilson 1980. Primary igneous load-cast deformation structures in the Fongen–Hyllingen layered basic intrusion, Trondheim region, Norway. *Geol. Mag.* **117**, 363–71.

Wiegel, R. L. 1964. *Oceanographical engineering*. Englewood Cliffs, NJ: Prentice-Hall.

Youd, T. L. 1977. Discussion of 'Brief review of liquefaction during earthquakes in Japan' by E. Kuribayashi and F. Tatsuoka. *Soils Foundns* **17** (1), 82–5.

Youd, T. L. and S. M. Hoose 1976. Liquefaction during 1906 San Francisco earthquake. *J. Geotech. Engng Div. Am. Soc. Civ. Engrs* **102**, 425–39.

Youd, T. L. and S. N. Hoose 1977. Liquefaction susceptibility and geologic setting. *Proc. 6th World Conf. Earthquake Engng* Vol. 6, 37–42.

Youd, T. L. and G. F. Wieczorek 1982. *Liquefaction and secondary ground failure*. Prof. Pap. US Geol. Surv. no. 1254, 223–46.

Young, G. M. and D. G. F. Long 1977. A tide-influenced delta complex in the Upper Proterozoic Shaler Group, Victoria Island, Canada. *Can. J. Earth Sci.* **14**, 2246–61.

# 11 Twisting and turning

Mixing layers – jets – sand volcanoes – corkscrew vortices (secondary flows) – seif dunes – sand ribbons – horseshoe vortices at bluff bodies – current crescents – drifts – horseshoe vortices in hollows – flute marks – flow patterns on current ripples and dunes.

## 11.1 Introduction

The fluid streams operating in many natural environments do not follow simple paths but behave in a complicated manner, forming either transverse or paired streamwise vortices, or some combination of the two. Streamwise vortices, possessing transverse as well as longitudinal velocity components, are frequently called secondary flows.

One of these situations – flow through a curved channel – is so important sedimentologically that it was discussed at length in Chapter 5. We saw that during flow through a curved channel, a secondary current arises because the pressure and centrifugal forces acting on the viscous fluid are locally unbalanced. The imbalance is ideally expressed as a single corkscrew vortex curving round the channel, but in most actual bends two or more vortices are present for some of the time.

Paired corkscrew vortices arise naturally for a wide variety of reasons. As in rivers, flow curvature in some instances determines the motion, but in other cases these secondary currents form in rectilinear flows. Paired corkscrew vortices will shape longitudinal bedforms where bed materials can be eroded and transported.

An important situation in which transverse vortices can be formed is where two fluids are in relative motion, as in stratified water bodies and in the jets formed by erupting igneous and sand volcanoes. These vortices are important for sediment dispersal and, when accompanying separated flows on ripples and dunes, significantly influence bedform character.

Active current ripples and dunes, flute marks scoured in mud beds, and bluff bodies (e.g. pebbles, shells, grass tufts) standing on sediment surfaces present us with innumerable examples of flow separation (Sec 1.10). The vortices induced typically include both transverse and streamwise types, and so exhibit a complex three-dimensional internal motion. They take their form and internal flow pattern largely from the shape of the bed feature to which they are coupled and, at the same time, contribute to that shape by helping to control the spatial variation of sediment erosion and transport.

## 11.2 Mixing layers

### 11.2.1 Two experiments

The following experiment (Thorpe 1968, 1973) illustrates the general character of the mixing layers that can form between miscible fluid streams in relative motion.

Take a straight tube of clear plastic approximately 1.5 m long and about 0.05 m in diameter, and close one end firmly with a rubber bung. Half-fill the tube with a saturated solution of common salt diluted a few times with plain water and coloured with potassium permanganate. Holding the tube vertical, carefully fill the remaining half with water and tightly seal with a second bung. The tube can be conveniently filled from a tap using a long rubber tube to the end of which either a piece of sponge or loose ball of thin cloth has been wired. Now slowly and carefully tilt the tube about one end, so that it comes to rest across a wood block placed towards one end (Fig. 11.1a). Holding the tube in one hand and the wood block in the other, slide the block a distance along the tube, so that the tube assumes naturally an opposite tilt. A surge of coloured brine will

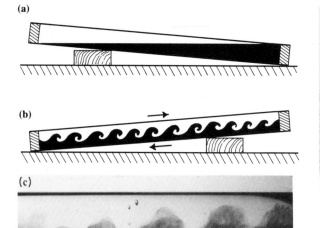

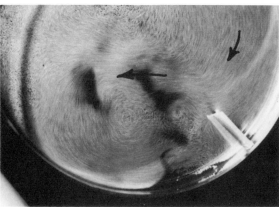

**Figure 11.2** Experiment showing vortices in the mixing layer produced as a consequence of flow separation at a vertical barrier in water rotating (clockwise) in a flat dish.

**Figure 11.1** Interfacial mixing (Kelvin–Helmholtz instability) between water and salt solution (black) in a tilted tube: (a) start of experiment; (b) idealized vortices developed at interface; (c) typical experimental vortices.

flow down the tube, while up the tube will move a comparable surge of water, at an equal and opposite velocity. It will be noticed that the sharp interface between the counter-flowing miscible streams suddenly breaks into transverse waves, which quickly grow in amplitude to assume an intricate spiral form before dissolving into turbulence (Figs 11.1b & c). These transverse vortices remain approximately stationary, because they form at the interface between two opposite streams of roughly equal speed. The growth of the vortices to fill the tube, as if two carpets originally laid edge-to-edge were together rolled up from the join, clearly effects an almost perfect blending of the fluids. The vortical zone is justifiably called a mixing layer.

The next experiment illustrates a rather different but common situation in which a mixing layer will form. Fill with water to a depth of a few centimetres a large crystallizing basin or pneumatic trough and sprinkle over the surface a little fine aluminium powder. Gently stir the water with a smooth circular motion. When the flow has steadied, place upright against the inside of the vessel a strip of wood or stiff metal a few centimetres wide so as to project into the current like a groyne (Fig. 11.2). The flow separates at the edge of this projecting body, in the lee of which a sluggish recirculating current will be seen. An expanding row of near-vertical vortices will be found along the boundary between the recirculating fluid and the faster external stream. These

vortices (1) rotate in harmony with the outer stream, like roller bearings, (2) grow in size with increasing distance from the point of separation and (3) have an internal spiral motion. Vortices of this kind are not restricted to liquids. Figure 11.3 depicts a similar row of vortices in the expanding mixing layer produced between co-flowing streams of gas juxtaposed at a sharp edge (Brown & Roshko 1974).

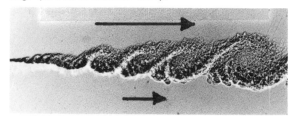

**Figure 11.3** Mixing layer developed between two parallel gas streams flowing from left to right at unequal speeds. Photography courtesy of Professor A. Roshko, California Institute of Technology.

### 11.2.2 A model for mixing layers

What causes mixing layers? Figure 11.4 illustrates the kinds of velocity profiles that can be measured from these layers. Figure 11.4a represents Brown and Roshko's (1974) co-flowing gas streams, Figure 11.4b the separated flow of our last experiment, and Figure 11.4c flow in the tilted tube. Their common feature is an inflection point. The third profile (c) corresponds physically to flow in a frame of reference moving with the vortices, a convenient frame in which to analyse mixing.

Consider in Figure 11.5 two horizontal streams of a miscible inviscid fluid flowing at equal and opposite

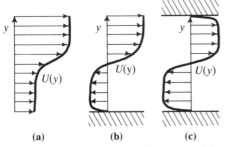

**Figure 11.4** Forms of velocity profile across mixing layers.

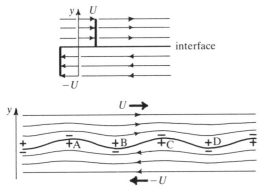

**Figure 11.5** Instability at the interface between two fluid streams of contrasting speed.

uniform steady velocities $U$ and $-U$. As the fluids are inviscid, we can imagine the streams to be separated by a sharp interface. What would be the effect of imposing a small sinusoidal disturbance, that is, a row of incipient vortices, on this interface? The answer comes from Bernoulli's equation (Sec. 1.5.3), stating that the total energy on a streamline is a constant. For the horizontal flow in Figure 11.5, we can write

$$\tfrac{1}{2}\varrho U^2 + p = \text{constant} \quad \text{J m}^{-3} \qquad (11.1)$$

where $\varrho$ is the fluid density, $U$ its velocity and $p$ the pressure. At A and C the upper stream is compressed above the crests of the interfacial disturbances, with the result that its velocity is locally increased but pressure decreased, whereas the lower stream is expanded, the local magnitude of its velocity being reduced but its pressure increased. Hence by Equation 11.1 there exists a resultant pressure tending to push the crests of the disturbances still further upwards. Similarly, at B and D the streamline expansions and contractions imply a resultant pressure tending to push the troughs still lower. The combined effect of the resultant pressures is therefore increasingly to distort the interface which, in

this inviscid case, is clearly unstable to all disturbances. This tendency for the interface between streams in relative motion to roll up is called Kelvin–Helmholtz instability, after its early investigators.

But this model supposes a single inviscid fluid. What happens when, as in the tilted tube (Fig. 11.1), the fluids are viscous and differ in density? It turns out that whereas viscosity has little effect on interfacial stability, a gravitationally stable density gradient is a powerful stabilizing factor. In the general case of fluid streams in which the velocity and density vary continuously upwards, the stability of the system is described by the ratio

$$Ri = -\frac{(g/\varrho)(d\varrho/dy)}{(dU/dy)^2} \qquad (11.2)$$

where $Ri$ is the Richardson number, and $g$ the acceleration due to gravity. In this non-dimensional grouping, the numerator represents the stabilizing effect of the density gradient, and the denominator the destabilizing effect of the interfacial shear. As velocity and density appear as gradients in the above expression, the Richardson number as defined by Equation 11.2 strictly describes only a point in the flow. However, experiments show that interfacial instability occurs provided that $Ri < 0.25$ throughout a substantial region.

### 11.2.3 Implications for sandy bedforms

Flow separation is inevitable to leeward of current ripples, dunes in water and beneath the wind, sand and gravel bars in river and tidal currents, and steep-fronted deltas such as build into glacial lakes, due to the sharp change in slope at the crest or brink. The resultant mixing layers (see Fig. 11.2) are important in dispersing to leeward the bedload sediment carried to the crests of these structures.

The role of the mixing layer is best understood by examining the leeward flow at two levels of detail, for example, as in Raudkivi's (1966) laboratory experiments. Figure 11.6a shows how the time-averaged streamwise velocity varies spatially over current ripples. The flow separates at each crest, creating a large zone of sluggishly recirculating fluid in the trough to leeward, where reattachment occurs. The instantaneous flow structure, however, is like that sketched in Figure 11.6b, based on various experiments (see Fig. 11.3). A mixing layer expanding at about 15° curves down stream from the ripple crest. The configurations within it are growing transverse vortices which, as they are carried down stream, may pair and coalesce (Meynart 1983) and can even become stretched longitudinally into corkscrew

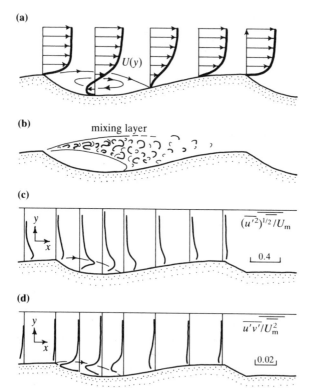

**Figure 11.6** Flow separation and its consequences at sandy bedforms. (a) Schematic variation in the form of streamwise velocity profile. (b) Shape and vortical character of the mixing layer. (c) Intensity of streamwise turbulence component over laboratory current ripples (data of Raudkivi 1966). (d) Reynolds stress over laboratory current ripples (data of Raudkivi 1966).

patterns (Jimenez 1983). According to the experiments of Kiya and Sasaki (1983), the vortex period averages $1.67x/U_{\mathrm{m}}$, where $x$ is the downstream length of the separation bubble and $U_{\mathrm{m}}$ the mean velocity of the external stream. The vortices as developed on a dune bed would not therefore be distinguishable in period from bursting streaks (Sec. 6.5), as the separation zone on dunes is just a few times longer than the dune height, and the height is a small fraction of the flow depth. The boils that Jackson (1976) related to streak bursting could have recorded mixing-layer vortices. These vortices are somewhat disorderly by the time they reach the reattachment region. Their presence and growth means that the fluctuating components of velocity (Sec. 6.2.1) are several times larger within the mixing layer than in the turbulent flow outside. As illustrated by Raudkivi's (1966) experiments (Fig. 11.6c,d), the Reynolds stress $\overline{u'v'}/U_{\mathrm{m}}^2$ and turbulence

intensity $(\overline{u'^2})^{1/2}/U_{\mathrm{m}}$ are significantly more than in the external stream, $u'$ and $v'$ being the fluctuating components of velocity measured parallel to respectively $x$ and $y$.

Because of the disturbed nature of mixing layers, as shown by Raudkivi's (1966) and Brown and Roshko's (1974) experiments, we should expect bedload grains carried to the crest of a bedform and injected there into a mixing layer to become widely dispersed while settling to leeward. Allen (1968a,b) demonstrated the correctness of this inference in various laboratory experiments, some involving a bedload layer ejected over the front of a dune-like structure divided transversely into narrow sampling compartments, and other grains injected into a mixing layer from a point source. Deposition on the leeward slope of the dune-like feature fell off rapidly (see Fig. 4.22) according to

$$R = mx^{-n} \quad \mathrm{kg\ m^{-2}\,s^{-1}} \tag{11.3}$$

where $R$ is the dry-mass deposition rate, $m$ a dimensional coefficient, $x$ the horizontal distance down stream from the crest and $n > 1$ an exponent. Very close to the crest, however, the formula is invalid, because here the deposition rate reached a measurable maximum. The exponent increased moderately rapidly with the ratio $V_0/U_{\mathrm{m}}$, where $V_0$ is the terminal fall velocity characteristic of the bedload grains at zero concentration, and $U_{\mathrm{m}}$ the mean flow velocity above the crest. Hence an increase in flow velocity for the same bedload material spreads the grains over a greater distance in the lee of a bedform, whereas an increase in the coarseness of the bedload for a fixed flow velocity causes a contraction towards the crest of the zone over which the grains settle. These trends affect the shape and texture of ripple, dune, bar and delta foresets. For example, with fine-grained sediments at large flow velocities, substantial deposition may take place in the reattachment region, resulting in thick bottomset beds. At low velocities and with coarse sediments, deposition is restricted to the vicinity of the bedform crest, so that grains reach the bottom of the leeward slope only after avalanching (Sec. 4.7.2).

### 11.2.4 Making flute marks

Flute marks are a common and widely developed structure produced by the erosion of mud beds, particularly, it is thought, where currents were powerful enough to suspend a good deal of sand. Figure 11.7 is representative of flute marks as preserved in the rock record. Moulded beneath an overturned sandstone bed, we see a series of similarly oriented, smoothly rounded, heel-shaped hollows that open out down stream.

How do these structures arise? With their sharp upstream rims, and asymmetrical streamwise profiles, the hollows which must originally have been present on the mud bed clearly resemble in shape the more strongly three-dimensional types of current ripple and dune (Sec. 4.6). Hence the flutes when being actively eroded should have contained a separated flow associated with a mixing layer. Another clue comes from the observation that immature flute marks commonly reveal traces of irregularities, such as animal burrows, that existed on the mud prior to erosion, and which served as the points of origin of the flutes (Colton 1967). The erosion process itself involved wholly or mainly cutting wear (Sec. 8.6), to judge from the delicate criss-crossing striae carried by some flute moulds.

The character of the mixing layer associated with a separated flow affords an appealing explanation for

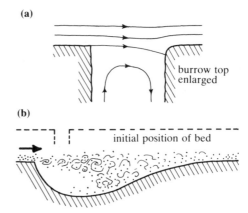

**Figure 11.8** Origin of flute marks.

flute marks. Consider the response to a powerful sand-laden current of a smooth flat mud bed carrying a small depression on its surface, say, an animal burrow (Fig. 11.8a). The current will separate along the sharp upstream edge of the burrow and impinge at a steep angle on the downstream wall. Because of the steeper attack, the rate of cutting wear by the sediment load on the downstream margin will exceed that on the bed surrounding the burrow, and the burrow will therefore expand down stream. At this stage, mixing-layer vortices will have little influence on the erosion, and may not be present at all (see Fig. 11.3), because the layer contained within the burrow is only short. An erosion proceeds, however, permitting the separated flow to grow and the mixing layer to lengthen and become more turbulent, the differential erosion due to the steepened attack of the grains will be enhanced by the increased grain momentum resulting from the action of the growing transverse vortices (Fig. 11.8b). The flute will go on expanding.

A simple and instructive analogy is provided by the way a plaster of Paris bed responds to a turbulent water stream (Allen 1971). Plaster of Paris is hydrated calcium sulphate and dissolves more quickly into the stream as the intensity of turbulence is raised (consider why sugar is *stirred* into a drink of tea or coffee). From three long pieces of wood make a straight channel about 0.1 m wide and deep and approximately 0.5 m long. Close off one end but at the other leave a sill about 0.015 m below the channel rim. Cast into the channel a plaster of Paris bed about 0.05 m thick, avoiding a thick mixture. On the axis of the channel, and 0.15–0.20 m from the sill, drill into the surface of the hardened bed a shallow hole several millimetres across. Now run water into the channel so that a vigorous stream a few centimetres deep is

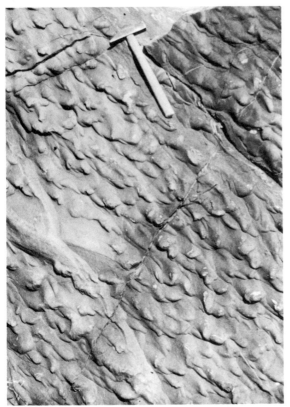

**Figure 11.7** Moulds of flute marks on the underside of a turbidite sandstone, Vlambourous Formation, Cyprus. Current from lower right. Photography courtesy of Dr N. H. Woodcock, University of Cambridge.

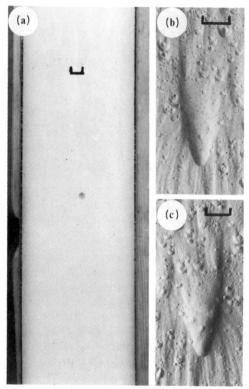

Figure 11.9 Experiments on flute marks (scale bars 0.01 m long). (a) Small pit on smooth plaster bed contained in channel. (b) Pit enlarged into a flute-like hollow (current from bottom to top). (c) Plasticine mould of flute-like mark.

formed, inspecting the bed every few hours. The drill mark becomes even larger (Fig. 11.9), until eventually the structure resembles the flute moulds of Figure 11.7. Differential solution resulting from the enhanced turbulence of the mixing layer contained in the mark is the explanation for the differential growth of the hollow.

## 11.3 Jets

### 11.3.1 An experiment

Fill a wide glass tank at least 0.25 m deep with cold water, letting the apparatus stand to dampen the turbulence. Meanwhile clamp a narrow-bore funnel so that the tip of its vertical spout just dips below the water surface. Thoroughly disperse about 0.025 kg of china clay in approximately 0.5 litre of water using a crystal of sodium hexametaphosphate. Steadily pour the dispersion into the funnel and watch the flow from its spout (Fig. 11.10). Concerning this milky stream, notice (1) the smooth-sided and slightly convergent early part,

Figure 11.10 Jet of water made milky with kaolinite spreading down into a tank of still water. Jet has maximum width of about 0.18 m.

(2) the uniformly expanding conical later section, (3) the presence on the surface of the cone of vortices increasing in size with the width of the cone, (4) the shape and sense of rotation of the vortices, which point to larger velocities along the axis of the cone than elsewhere, and (5) if the tank is deep enough, the gradual dilution of the clay dispersion. What you have produced is a jet, namely, the stream formed by ejecting a fluid from an orifice. Our jet was driven partly by the excess pressure of the fluid in the funnel, and partly by its density excess. A jet driven upwards, however, would be similar in appearance, as in the case of the hot ash-laden gases forced under pressure from a volcano (Fig. 11.11).

The jet of Figure 11.10 may be classified as turbulent, for obvious reasons, and also as axisymmetric, as it flows from a circular orifice and is conical in form. A two-dimensional jet results when fluid is ejected uniformly from a parallel-sided slot, but such flows are of little sedimentological interest.

### 11.3.2 Modelling a steady axisymmetric turbulent jet

Figure 11.12 represents an axial section in a steady axisymmetric turbulent jet of incompressible fluid discharging uniformly into an infinitely extensive stagnant region of the same fluid. How should we model the flow?

It is useful to begin with some definitions and general laboratory results (Rajaratnam 1976). The jet at its origin is of radius $r_0$ and uniform velocity $U_0$, measured parallel to the axis, the velocity profile having a 'top-hat' shape. Over an axial distance of about $5r_0$ – the so-called development region – the velocity measurable at the axis remains equal to $U_0$. At greater distances, the largest value $U_{max}$ of the time-averaged velocity $U$ measured parallel to the $x$-direction is found at the axis itself, but is less than $U_0$ and declines with increasing $x$. At each cross section $U$ declines with increasing radial

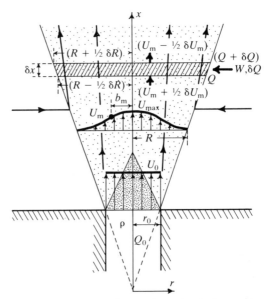

**Figure 11.12** Definition diagram for a circular jet entering a calm fluid.

distance $r$ approximately according to a cosine function. A bell-shaped velocity profile therefore characterizes sections beyond the development region. It is convenient to identify a radial distance $b_m$, where the velocity measured parallel to the $x$-direction attains its mean value $U_m$. Assuming that $U(r)$ follows a cosine function strictly, and putting $R$ as the jet radius, it follows straightforwardly that $b_m = 0.363R$.

One way of modelling the jet is to suppose that the volumetric discharge is constant from the one cross section to another, and therefore equal to the orifice discharge $Q_0 = (\pi r_0^2)U_0$ m$^3$ s$^{-1}$, where the bracketed term is the cross-sectional area. Now, sufficiently far from the origin, the jet is no longer affected by the orifice shape and the initial velocity profile; we can also suppose, because of the mixing vortices, that viscosity is of negligible influence. The quantities defining the motion are therefore the discharge $Q_0$, the axial distance $x$ from the virtual origin, and a representative velocity and transverse dimension, say, $U_m$ and $b_m$. This set contains only two quantities with the dimensions of length, so the length scale $b_m$ describing the jet must depend only on $x$, allowing us to write

$$b_m = kx \quad \text{m} \tag{11.4}$$

where $k$ is a proportionality constant to be determined empirically. The experimental jet (Fig. 11.10) expands at constant rate, so Equation 11.4 may be accepted as

**Figure 11.11** Aerial view of lower part of eruption column (jet) ejected from Mount St Helens, western USA, 18 May 1980. Photograph courtesy of Professor L. F. Radke, University of Washington.

a valid statement. If the volumetric discharge is conserved, then the discharge $Q = (\pi R^2)U_m$ at any $x$ is given by

$$Q = Q_0 = \text{constant} \quad \text{m}^3\,\text{s}^{-1} \tag{11.5}$$

Substituting into this equation from relations stated above, replacing $R$ by $2.752b_m$ according to the previous definition, and rearranging yields

$$U_m = (r_0^2 U_0)\frac{1}{(2.752kx)^2} \quad \text{m s}^{-1} \tag{11.6}$$

whence the velocity of the jet declines as $x^{-2}$ and the term $(r_0^2 U_0)$ emerges as a characteristic of the jet.

The jet velocity does indeed decline with increasing distance, but not according to Equation 11.6, which, although of interest, incompletely describes the flow. By conserving the effluent discharge, we have in effect said that the boundary of the jet is impervious. Two experimental observations (Fig. 11.10) stand in the way of our supposition, for (1) the vortices clearly draw in the ambient medium, and (2) the jet in the process becomes visibly diluted, again pointing to the absorption into it of the surrounding fluid.

The jet can be completely modelled only from dynamical considerations. As the ambient medium is of infinite extent, there is no external body on which the jet can exert any force, whence by the principle of momentum conservation the flow or flux of momentum along it must be the same for all cross sections, including that at the origin. Introducing $\varrho$ as the fluid density, the momentum flux here is $M_0 = (\pi r_0^2)(\varrho U_0)U_0$ N, where $(\varrho U_0)$ is the momentum per unit volume, $U_0$ the rate of its transport, and $(\pi r_0^2)$ the cross-sectional area. Conserving the momentum flux, the value $M = (\pi R^2)\varrho U_m^2$ through a cross section at any distance $x$ is

$$M = M_0 = \text{constant} \quad \text{N} \tag{11.7}$$

Rearrangement after substitutions for $M_0$, $M$ and $R$ yields

$$U_m = (r_0 U_0)\frac{1}{2.752kx} \quad \text{m s}^{-1} \tag{11.8}$$

as the jet velocity. The velocity accordingly declines as $x^{-1}$, instead of the steeper rate given by Equation 11.6.

We now have two important results concerning turbulent jets; namely, their width increases linearly with distance from the virtual origin, while the velocity decreases inversely as the distance. One more important result is to come.

What is the physical meaning of the difference between Equations 11.6 and 11.8? Equation 11.6 does not correctly describe the motion, but it is nonetheless a valid statement about the fate of the particular fluid which came from the orifice. Because the $U_m$ of this equation declines more rapidly than the $U_m$ of Equation 11.8, we can infer that ambient fluid is being drawn into the expanding jet, as is suggested by our observations on dilution and the role of the vortices. The difference between Equation 11.8 and Equation 11.6 can therefore be interpreted as describing for any cross section the fraction of the total discharge which represents entrained fluid. The orifice discharge in fact rapidly declines in importance as the jet spreads, for when $x$ is large $x^{-2}$ compared to $x^{-1}$ is seen to be negligibly small.

The mean horizontal velocity $W$ of the entrained fluid where it passes through the margin of the jet can easily be calculated by applying the principle of continuity. In the jet of Figure 11.12, the small increase in discharge $\delta Q$ between the ends of a thin cross-sectional slice $\delta x$ in thickness must exactly equal the small increment of discharge inward through the margin of the same slice. The increment through the margin is

$$\delta Q = W(2\pi R)\,\delta x \quad \text{m}^3\,\text{s}^{-1} \tag{11.9}$$

where $(2\pi R)\,\delta x$ is the area of the margin. Between the ends of the slice

$$\delta Q = \pi(R + \tfrac{1}{2}\delta R)^2(U_m - \tfrac{1}{2}\delta U_m) - \\ \pi(R - \tfrac{1}{2}\delta R)^2(U_m + \tfrac{1}{2}\delta U_m) \tag{11.10}$$

where $R$ is the average radius, and $\delta R$ and $\delta U_m$ are small increments respectively of the radius and mean velocity. Multiplying out the brackets in Equation 11.10, and equating this expression for $\delta Q$ to Equation 11.9, we find in the limit that

$$W = U_m\frac{dR}{dx} - \tfrac{1}{2}R\frac{dU_m}{dx} \quad \text{m s}^{-1} \tag{11.11}$$

As we already have relationships for $R$ in terms of $b_m$, and for $b_m$ and $U_m$ each in terms of $x$, we can substitute from these and from their differentiated forms into Equation 11.11 to write finally

$$W = -\tfrac{3}{2}(r_0 U_0)\frac{1}{x} \tag{11.12}$$

the negative sign indicating that the ambient medium loses fluid. Thus like $U_m$ the entrainment velocity is directly proportional to the strength of the jet, as measured by $(r_0 U_0)$, and inversely proportional to axial distance. Dividing Equation 11.12 by Equation 11.6 allows $W$ to be compared with $U_m$, that is, $W/U_m \simeq 4.12k$. As $k$ has an experimental value of about 0.075, $W$ is approximately one-third of $U_m$ and very roughly one-sixth of $U_{max}$. In the ambient medium away from the jet, it is easy to show, again from continuity, that the velocity of fluid moving towards the jet declines as $W(R/r)$. Hence inflow velocities may be ignored in sedimentological problems involving jets, except where grains of small terminal fall velocity are concerned.

Contours of equal $U$ within the jet are approximately pear-shaped (Fig. 11.13), and can be calculated by combining Equation 11.8 for $U_m$ at a radial distance $b_m$ with the cosine function for $U(r)$. Many workers have measured the distribution of the fluctuating components of velocity within turbulent jets (Rajaratnam 1976). As may be surmised from the large mixing vortices visible on their exterior (Figs 11.10 & 11.11), levels of turbulence are high, reaching a maximum on the axis and declining radially outwards in a similar manner to the time-averaged velocity.

### 11.3.3 Application to sand volcanoes

The typical sand volcano (Fig. 11.14) is a flat conical mound of interbedded sand and silt centred on a near-vertical vent from which water and sediment were spouted upwards into a large body of water, either stagnant or gently moving. Sand volcanoes are largest and most numerous on the tops of large masses of slipped sediment which, as they resettled, provided both the ejected water and the sediment (Gill & Kuenen, 1958, Gill 1979). Modern parallels are afforded by the sand boils and mud vents found on top of submarine slides off the deltas of rivers, for example, the Mississippi (Prior & Suhayda 1979) and the Klamath (Field et al. 1982). Small volcanoes can be found in some numbers on the tops of sand beds that de-watered after being rapidly deposited with a metastable grain fabric (Rust 1965, Burne 1970). Axial slices through sand volcanoes generally show that (1) the sides gradually steepened as the volcano grew, and (2) the coarsest sediment accumulated nearest to the vent. To what extent do the properties of an axisymmetric jet explain these various characteristics?

The grains flushed through the vent can be thought of as a pollutant which marks the fluid ejected from the

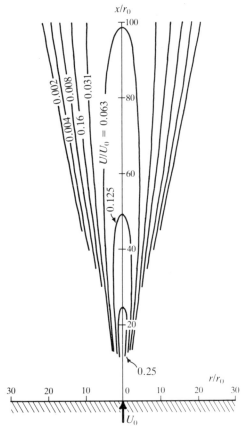

**Figure 11.13** Calculated lines of equal relative velocity in an axial section of a circular jet.

**Figure 11.14** Large sand volcano (note secondary cone on outer flank) lying on thick sediment slide, Carboniferous, Co. Clare, Ireland. Photograph courtesy of Professor T. Elliott, University of Liverpool.

orifice, where their fractional volume concentration is $C_0$. Taking the particles to be weightless, their flux must be the same from one cross section to another, whence in the same manner as before

$$(\pi R^2)C_m U_m = (\pi r_0^2)C_0 U_0 = \text{constant} \quad (11.13)$$

where $C_m$ is the mean concentration at a cross section. It would therefore seem that the local time-averaged concentration $C$ of the grains should vary in the same way as the local time-averaged velocity $U$, decreasing axially as $x^{-1}$ and declining according to a cosine function radially away from the axis. This being the case, the velocity contours of Figure 11.13 become equivalent to lines of equal sediment concentration.

The assumption that the particles are weightless gives us some idea of their distribution within the jet but is clearly unsound beyond this stage; each grain in fact has its particular terminal fall velocity $V$. Under the dispersing action of the vortices, a grain can neither reach upwards against gravity higher than the axial location where $U_{max} = V$, nor spread outwards further than the greatest radial distance corresponding to $U = V$. As this radial distance increases with declining $U$ (Fig. 11.13), particles of a given density settling to the bed should decrease in size with increasing radial distance from the vent. Lines of equal grain size in the volcano should therefore plot as circles centred on the vent, provided that a stagnant water body received the jet. A cross flow, however, would bend the jet over, creating an elliptical to parabolic pattern of fall-out. Similarly, the general radially outward decline in grain concentration within the jet suggests that the downward flux of grains will also decline radially, so permitting the slopes of the volcano to steepen as building proceeds. A cross flow should introduce an asymmetry into the grain flux similar to that suggested for grain size. It is from the air-fall deposits of igneous volcanoes that the most detailed measurements of grain size and thickness of erupted material have come (Walker & Croasdale 1971). These measurements support the above suggestions.

## 11.4 Corkscrew vortices

### 11.4.1 Corkscrews and their effects
Corkscrewing fluid motions abound in natural environments and give rise to a wide variety of sedimentary structures. In the general case (Fig. 11.15a), the motion consists of streamwise, oppositely rotating, spiral vortices lying side by side in pairs on the bed. Hence there are alternate zones of convergence and divergence

amongst the bottom currents, matched by corresponding divergent and convergent flows at some height above the boundary. Typically, the transverse wavelength of the vortices (e.g. convergence to convergence) is 2–4 times the height. Streamlines generally diverge by no more than a few degrees from the mean flow direction, the corkscrews being long-pitched. The transverse components of the velocity and mean bed shear stress are therefore small compared to the streamwise values. It is in zones of divergence at the bed that the velocity profile is steepest and the mean bed shear stress greatest. The absolute scale of corkscrew motion varies enormously, from the order of 0.01 m in the shallowest aqueous flows up to several kilometres in the atmosphere.

What are the likely effects of corkscrew vortices on sedimentary surfaces? This may be expected to vary with the composition of the surface and the character of the transported sediment. A surface underlain by deep loose sand (Fig. 11.15b) could become shaped into streamwise ridges and hollows, as grains are moved from zones of divergent currents towards zones of convergence. In this case Equation 5.40 may be applied in reverse, to yield the angle by which the bottom currents diverge from the mean flow direction. A bed of deep mud over which nothing but suspended silt and clay are transported can respond to corkscrew vortices only through either fluid stripping (Ch. 8) or differential deposition (Ch. 7). In these cases also streamwise hollows will arise beneath the zones of divergent current, because of the greater shear stress there (Fig. 11.15c). If a small amount of coarse sediment is present, however, the opposite relief could arise (Fig. 11.15d) particularly where the mud is strong enough to respond only by corrasion. In this type of wear (Sec. 8.6), the erosion rate depends not only on the speed and angle of attack of the tools, but also on their areal density, which will be greatest at zones of convergence.

### 11.4.2 Structures attributed to corkscrew vortices
Longitudinal or seif dunes (Fig. 11.16) are an important type of aeolian bedform in many deserts (Glennie 1970, Cooke & Warren 1973, McKee 1979). Their crestal length is extreme compared to other dimensions, the height varying between a few tens and few hundreds of metres, and the spanwise wavelength from hundreds of metres to several kilometres. Dunes join down wind at Y-shaped junctions, which in shape form an arrow pointing along the wind direction. The crests of seif dunes are slightly sinuous in plan, with avalanche faces appearing alternately to one side and then the other.

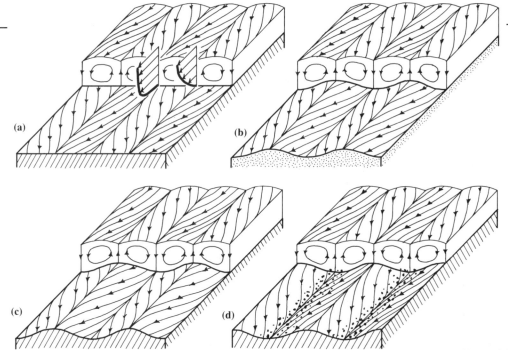

**Figure 11.15** Corkscrew vortices (secondary flows) and their effects. (a) General character of motion. (b) Shape of deformable granular surface adjusted to the motion. (c) Shape of a mud bed adjusted to the motion in the absence of bedload. (d) Shape of a mud bed adjusted to the motion in the presence of bedload particles.

**Figure 11.16** Air photograph of longitudinal dunes in the Simpson Desert, Australia. Wind towards top. Scale bar is 1 km.

The cross-strata underlying these faces therefore have azimuths that are steeply inclined to the trend of the crest (McKee & Tibbitts 1964). Longitudinal dunes are found in a variety of wind regimes (Lancaster 1982). The larger complexes of such dunes occur where the sand-driving winds come from a broad but single band of directions, roughly paralleled by the crests of the structures (Brookfield 1970). In certain special cases, however, the crestal trend bisects the angle between almost diametrically opposed, either diurnal or seasonal sand-transporting winds (Tsoar 1983).

The very much smaller structures shown in Figure 11.17 are sand ribbons. A strong wind blowing over ill-graded sand will sort the particles into linear streamwise bands of alternately well graded finer particles and well graded coarser grains (Fig. 11.17a). Notice the longer wavelength of the ballistic ripples developed on the coarse-grained bands. The ribbons of Figure 11.17b arose as an ebbing tide transported a small quantity of sand across stable gravel. On parts of the continental shelf where sand is in short supply, tidal currents shape similar longitudinal sand ribbons measuring many tens of metres in spanwise wavelength and some kilometres in length (Kenyon 1970). These ribbons are so large that in many cases they carry flat dunes.

Figure 11.18 shows linear streamwise structures shaped on mud beds in response to the bedload transport of coarser sediments. The widely spaced erosional furrows and ridges shown in Figure 11.18a arose parallel to tidal currents as small amounts of gravel and sand were driven across a surface of soft mud. During high spring tides, when the currents are strongest, the gravel becomes heaped into small barkhan dunes spaced a few metres apart along the furrows. Similar combinations of grooves and dunes, but orders of magnitude larger in scale, occur where the muddy floor of the deep

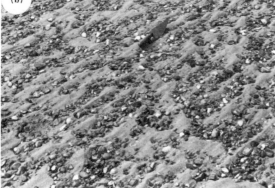

**Figure 11.17** Sand ribbons on Norfolk coast, England, shaped by (a) the wind and (b) a tidal current. Trowel 0.28 m long points in flow direction.

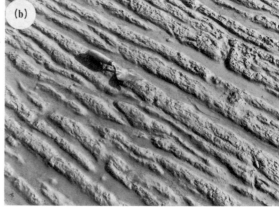

**Figure 11.18** Corrasional grooves in mud formed parallel to tidal flow, Severn Estuary, England. (a) Note patches of gravel in grooves. Current from upper right. Spade for scale. (b) Sand present in grooves. Current from upper left. Trowel 0.28 m long.

ocean experiences erosion (Lonsdale & Malfait 1974, Lonsdale & Spiess 1977). Corresponding structures from the desert, occurring on a scale comparable to seif dunes, are described by Hagerdorn (1968) from around the Tibesti Mountains in the Sahara. They are giant streamwise furrows corraded by the wind into hard level-bedded rocks, and in many instances contain a row of barkhan dunes. The small ridges and furrows of Figure 11.18b were cut as the tide drifted sand, at times heaped into current ripples, across a level surface of stiff mud. The undersides of sandstone beds commonly preserve similar structures (Friend 1965).

### 11.4.3 Origins of corkscrew vortices

Naturally occurring corkscrew vortices originate in four main ways: (1) because of flow curvature, (2) in consequence of thermal instability, (3) because of transverse differences of boundary roughness induced during sediment transport, and (4) by a complex interaction between wave- and wind-related currents.

Curvature-related corkscrew vortices are mainly restricted to the atmospheric boundary layer and to the larger tidal streams in shelf seas, that is, to flows extensive enough to be significantly affected by the Earth's rotation. Corkscrews originating in this way have an uncertain sedimentological role, may account for seif dunes (Wipperman 1969), and could explain the large nearly flow-parallel tidal sandbanks grouped together in some shelf seas (Stride 1982).

When the wind blows at a modest speed over a water surface, the complicated interaction between the resulting wave- and wind-related currents gives rise in the subsurface to corkscrew motions which in shallow water may reach down to the bed (Leibovich 1983). These corkscrews, called Langmuir circulations, are expressed at the water surface as wind-parallel bands of flotsam or smooth water, formed where floating debris and surface contaminants are concentrated in zones of current convergence. Whether Langmuir circulations are sedimentologically important is also unclear.

Thermal instability in the atmosphere boundary layer is a major cause of corkscrew vortices and provides a satisfactory general explanation for the more extensive fields of seif dunes. Figure 11.15b is the appropriate flow model. The idea here is that air brought into contact with the bare heated surface of a desert, or the warmer waters of an ocean or lake, will itself heat up

and so expand and become of lower density than its surroundings. Consequently, parcels of warm air will tend to rise but, because they find themselves sheared by a horizontal wind, in the process will become stretched out into corkscrews somewhat like the blob of fluid discussed in connection with bursting streaks (Sec. 6.3).

An experimental analogy is easily made. Take a large rectangular glass tank and carefully fill it to the brim with cold water, so that a convex-up meniscus is formed. On one side of a glass sheet large enough to span the tank paint a uniform layer of a thin mixture of finely powdered potassium permanganate and water. Let the paint dry thoroughly. Now carefully lay the glass painted-side down across the water surface and at once observe the fluid beneath the glass. For a few seconds the potassium permanganate merely dissolves into the water, creating a uniformly thin dense layer of coloured liquid immediately beneath the glass. Suddenly, the interface between the coloured layer and the less dense plain water below begins to warp, followed by the descent of mushroom-shaped pendants of coloured fluid into the tank from scattered points on the underside of the glass (Fig. 11.19). There is a definite circulatory motion within these pendants, downwards near the axis and upwards near each outer margin. The coloured layer behaves in a similar manner to the heated layer of air, and it is only for experimental convenience that the density stratification was inverted. In essence, the mushroom is a ring-shaped vortex. But why should the heated layer in the atmosphere, and the interface between the potassium permanganate solution and the plain water below, have proved unstable and become involved in a convective motion? There are two routes to an answer.

One approach is to examine the energy of a density-stratified system, which can only be potential, since initially there is no motion. Consider in Figure 11.20a two horizontal layers of stagnant fluid of density respectively $\varrho_1$ and $\varrho_2$ and thickness respectively $h_1$ and $h_2$,

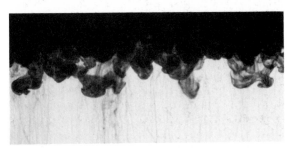

**Figure 11.19** Inverted mushroom-like bodies of heavy potassium permanganate solution descending from a horizontal layer into plain water.

213

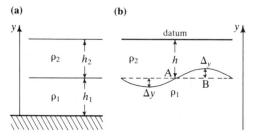

**(a)**　　　　　**(b)**

**Figure 11.20** Definition sketch for gravitational instability.

and for convenience put $h_2 = h_1$. The potential energy per unit volume of a fluid element is therefore $\varrho g y$, where $y$ is its vertical height above a datum, say, the bottom of the lower fluid layer. Hence the total potential energy of a fluid column of unit basal area is

$$\text{potential energy} = \varrho_1 g \int_0^{h_1} y \, dy + \varrho_2 g \int_{h_1}^{(h_1 + h_2)} y \, dy \quad \text{J m}^{-2}$$

$$(11.14)$$

whence, remembering that we put $h_2 = h_1$,

$$\text{potential energy} = \frac{g h_1^2}{2} (\varrho_1 + 3\varrho_2) \quad (11.15)$$

the terms of Equation 11.14 appearing in the upward order of the layers. Now the stability of the layered arrangement depends on whether or not its total potential energy is a minimum. If the total is not a minimum, then some potential energy is available for conversion into fluid motion, which leads to a new and more stable layering. Inspection of Equation 11.15 shows that the energy is a minimum for $\varrho_1 \geqslant \varrho_2$, which means that the arrangement is most stable when the heavy fluid layer is below. In our experiment, the creation of a heavy layer above a lighter one resulted in instability, as Equation 11.15 requires. The effect of heating the bottom of the atmosphere is similar.

We can otherwise look at the distribution of hydrostatic pressure (Fig. 11.20b). There are two stagnant fluid layers as before, but the horizontal interface between them is disturbed from its rest position by small vertical increments $\Delta y$. Let the undisturbed interface lie at a depth $h$ below an arbitrary horizontal datum over which the hydrostatic pressure is uniform, and let us calculate the pressure at this depth. The additional pressure at A is therefore $p = \varrho_2 g h$. At the upwarp B the excess pressure is the sum of $\varrho_2 g(h - \Delta y)$, representing the upper fluid, and $\varrho_1 g \, \Delta y$, due to the increment filled with the lower fluid. The pressure difference $\Delta p$ between

A and B, representing a horizontal pressure gradient, is therefore

$$\Delta p = \varrho_2 g h - [\varrho_2 g(h - \Delta y) + \varrho_1 g \, \Delta y] \quad \text{N m}^{-2} \quad (11.16)$$

$$= (\varrho_2 - \varrho_1) g \, \Delta y \quad (11.17)$$

The pressure difference is positive when $\varrho_1 < \varrho_2$. It corresponds to a turning force acting from A towards B and tending to steepen still further the disturbed interface. Thus the initial disturbance is amplified and the layers eventually overturn. The force acts from B towards A, however, when $\varrho_1 > \varrho_2$ and so lowers the slope of the disturbed interface, until it is once again flat. Only this arrangement is stable. The same result emerges if we consider a downwarp rather than an upwarp on the interface.

The stretching out of the convective vortices of our experiment into corkscrews under the influence of horizontal shear is elegantly demonstrated in the laboratory work of Chandra (1938) and Krishnamurti (1975). Similar corkscrews, frequently called roll vortices, are particularly common in the atmosphere. Haines (1982) and Haines and Smith (1983) show that they can arise during forest fires, such instances affording a rather complete demonstration of the characteristics to be expected of the atmospheric vortices developed in more important ways. The burning of the crowns of forest trees is commonly restricted to broad bands parallel to the wind, whereas in the intervening belts of woodland the fire is restricted to ground vegetation and litter. Moreover, the orientation of scorch marks clearly shows that the ground wind had a transverse component of motion directed from the belts of unburned crowns to the bands of burnt ones. In this case, the fire provided heated air which, in rising into and structuring the wind, controlled the pattern of burning. Natural convection is widely prevalent where air becomes heated in blowing over a ground or water surface at a sufficiently higher temperature (Brown 1980). The corkscrews are made visible chiefly by the occurrence of parallel bands of clouds – so-called cloud streets (Fig. 11.21) – formed at the zones of divergent motion in the upper air (see Fig. 11.15a). The presence of corkscrews can also be inferred where balloons tracked by radar follow helical spiral paths, as above the plains of Idaho (Angell et al. 1968). The measured transverse spacing of cloud streets ranges between 2 and 8 km, and is not different from the transverse wavelength of the larger seif dunes (Hanna 1969). The Australian longitudinal dunes are smaller, however,

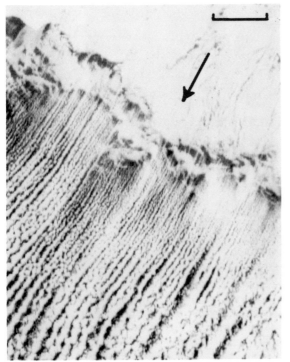

**Figure 11.21** Cloud streets in north-east wind flowing off Arctic pack ice onto Norwegian Sea. Scale bar is 50 km. Photograph courtesy Dr P. Baylis and copyright University of Dundee.

roughness are already fully established. How is the pattern initiated?

Where sand is moving in small quantities over a stable gravel surface, as in Figure 11.17b, any chance local accumulation of grains in the form of a smooth patch will at once initiate the gradients in the Reynolds stresses described above, and so attract more sand because of inflow towards the patch. The case of Figure 11.17a is a little different, for here the fine and coarse grains are simultaneously in motion. However, arguing as before, any chance concentration of either coarse or fine grains should spark off the same self-perpetuating process. The emergence of a dominant wavelength amongst the ribbons is more difficult to explain but probably represents the depth-limited geometry that maximizes the gradients of the controlling stresses.

The origin of the corkscrews on which mud furrows appear to depend is so far mysterious. If the furrows shown in Figure 11.18 originated by corrasion where bottom-travelling particles had been concentrated (Fig. 11.15d), then a roughness-related secondary flow can at once be ruled out, for these zones of concentrated particles would have proved the rougher surface and the tools would have been dispersed away from them. Thermal convection can also be ruled out in tidal environments, so it looks as though there are mechanisms of secondary flow awaiting discovery.

but it is not known if for some reason the convective motions above these deserts take a reduced scale.

The other main known mechanism for producing corkscrew vortices depends on spatial variations in the turbulence field in a fluid flow above a boundary of transversely varying roughness. This mechanism (Pantin *et al.* 1981) satisfactorily explains sand ribbons such as those in Figure 11.17, where smooth and rough surfaces alternate across the flow. To see its basis, let us describe the turbulence using the co-ordinate system and velocity notation of Section 6.2.1. The quantity $(w'^2 - v'^2)$ represents a pressure acting transversely across the flow. Laboratory experiments show that near the bed this pressure is greater above a rough than a smooth surface. Hence in the flow over beds such as those in Figure 11.17, there is a force thrusting near-bed fluid sideways from the rough to the smooth bands of sediment. Under steady conditions, this force is balanced by a transverse gradient in the turbulence shear stress $\varrho v'w'$, which represents the vertical flux of cross-stream momentum. This is the situation for an equilibrium bed, on which the transverse differences of

## 11.5 Horseshoe vortices due to bluff bodies

### 11.5.1 An experiment

Take a pneumatic trough or large crystallizing basin and glue upright to its floor a stout circular cylinder cut from dowelling, at a distance inwards of about one-third the diameter of the trough. At about 45° in the intended upstream direction from the cylinder, glue in the corner of the trough a long row of large potassium permanganate crystals spaced several millimetres apart. Also glue a few of the crystals around the base of the cylinder. Quickly fill the trough with cold water, wait for a few seconds for turbulence to dampen, and gently stir with a smooth circular motion. As is partly shown in Figure 11.22, you will see that (1) the coloured streaks from the corner of the tank are deflected around the cylinder and kept well away from its base, (2) streaks originating at the front and sides of the cylinder, and perhaps some from up stream, contribute to one or more powerful vortices which spiral around the base of the cylinder, and (3) coloured fluid wells up in the rear of the cylinder.

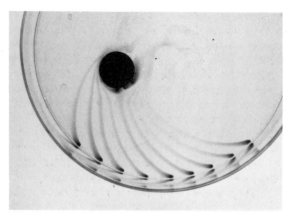

**Figure 11.22** Experimental streamline pattern around a cylinder mounted on a smooth boundary.

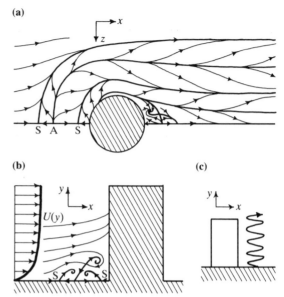

**Figure 11.23** Fluid motion associated with a cylinder mounted on a plane surface: (a) on the bed, (b) in the axial streamwise plane, and (c) off the axial streamwise plane. S, separation point; A, attachment point.

What you have produced is a rather complicated flow dominated by one or more horseshoe vortices which wrap around the upstream part of the cylinder and trail down stream from it. Such vortices are well known from experiments with wall-mounted bluff bodies in laminar flows (Hunt *et al.* 1978, Langston & Boyle 1982) and similar ones arise under turbulent conditions. The complete flow pattern is difficult to establish experimentally, but its more important aspects can be illustrated by reference to the plane of symmetry and the bed (Fig. 11.23). In the plane of symmetry, the flow separates from the bed and forms a number of unequal vortices. These spiral round the base of the cylinder and become stretched out down stream into corkscrews which give rise on the bed to streamwise zones of convergence and divergence. Rearward of the cylinder are two symmetrically disposed points from which the flow spirals vertically upwards. Scorch marks related to these upwashes were used by Haines (1982) to establish the pattern of ground flow during crown fires.

### 11.5.2 Origin of horseshoe vortices

Why should the flow as it nears the cylinder in Figure 11.23 give rise to vortices? The answer lies in the fact that the cylinder extends out into a boundary layer in which there is a vertical gradient of streamwise velocity. Assuming that no vortices had been formed, consider the behaviour of the fluid moving along a streamline in the plane of symmetry of the motion and at a height $y$ above the bed. Far up stream the velocity on the streamline is $U$. But as the streamline lies in the plane of symmetry, the fluid elements moving along it are decelerated as they approach the cylinder and finally caused to stagnate where the streamline joins the body.

**Figure 11.24** Effects of horseshoe vortex generated by current flowing towards top round a cockle shell (about 0.02 m across). Beach on Norfolk coast, England.

The motion-related pressure on the cylinder at this stagnation point is therefore $p = \frac{1}{2}\varrho U^2$, where $p$ is the pressure and $\varrho$ the fluid density. Now as $U$ increases upwards with increasing $y$, the pressure similarly increases up the leading face. Consequently, there is a pressure gradient $\mathrm{d}p/\mathrm{d}y$ along the front of the cylinder, to which the fluid responds by moving downwards and forwards to meet the oncoming stream. Flow separation inevitably occurs at the convergence of those streams, but it is due as much to the downwash on the front face of the cylinder as to the deceleration of the near-bed fluid as it approaches the body. In consequence of the shape of the cylinder, the resulting vortices are swept around the flanks and trailed for some distance down stream.

### 11.5.3 Current crescents and drifts

What are the likely effects of a motion such as that in Figure 11.22 on a bed of sediment? A lot would depend on whether sand or mud was involved, and on what grades of material were in transport, but one general effect should be the scouring of a U-shaped hollow beneath the vortices (repeat the last experiment putting sand in the trough). In the case of a sand bed, enhanced deposition might also be expected where the near-bed flow was slowed down in zones of convergence, particularly in the sheltered lee of the bluff body.

The features called current crescents are perhaps the commonest and most widespread of the sedimentary structures produced by horseshoe vortices at bluff bodies, for example, shells and stones on sandy surfaces. They can be useful indicators of local current direction. The crescent shown in Figure 11.24 arose at a cockle shell on a sandy beach and preserves evidence for the occurrence of multiple vortices. Structures similar to this can result from the differential corrasion of a mud bed in which there are hard zones, for exam-

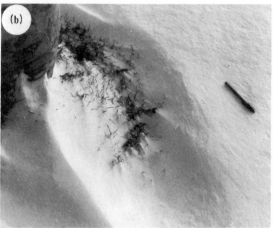

**Figure 11.25** Current crescent in snow accumulated round a tree trunk. (a) General view. (b) Detail of left-hand furrow to show divergent small-scale features. Pen 0.15 m long gives scale.

**Figure 11.26** Drifts in (a) sand blown from upper left round a tangle of vegetation and other debris (trowel 0.28 m long), Norfolk coast, and (b) in mud down stream of seaweed tussocks (spade for scale), Severn Estuary, England.

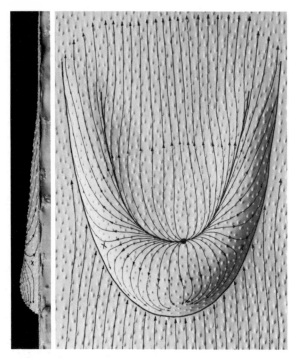

**Figure 11.27** Mould of an artificial flute mark 0.17 m wide and 0.025 m deep, to show the pattern of motion mapped over the bed in side elevation (left) and plan (right).

ple, small concretions or burrow fills, and are commonly preserved in rocks (Friend 1965). The crescent shown in Figure 11.25 was formed by a process largely of differential deposition, as dry snow accumulated around the base of a small tree. It stands little chance of preservation, but is instructive in showing a poniard-shaped drift to leeward as well as the expected U-shaped furrow. The orientation of the small-scale features produced at individual grass blades clearly proves a divergent bottom flow in the furrow (Fig. 11.25b), as we expect from the experimental results summarized in Figure 11.23.

Where there is no U-shaped furrow at a bluff body, but just a depositional tail to leeward, the term drift is appropriate. The drift shown in Figure 11.26a formed as a grass tussock trapped sand blown by the wind over a mud-hardened intertidal sand flat. Notice how the wind ripples on the drift tend to bend inwards towards its axis, indicating a slight convergence of the bottom currents (see Fig. 11.23). Mud drifts formed to leeward of seaweed tussocks attached to a stable gravel surface are shown in Figure 11.26b. These drifts, resembling sandy ones formed in rivers (Karcz 1968), are com-

**Figure 11.28** Pattern of motion on the surface of a bed of current ripples.
(a) Plaster of Paris replica of the rippled bed, showing artificial pits elongated in the direction of the local bottom current.
(b) Reconstruction of bottom streamlines on the bed shown in (a). Flow depth is 0.095 m and mean flow velocity 0.22 m s$^{-1}$

plicated internally and took a long period to form as deposition and erosion alternated under a highly variable tidal regime.

## 11.6 Horseshoe vortices at flute marks, current ripples and dunes

These bedforms are all associated with separated flows, the separation occurring because of the adverse pressure gradient generated by the sharp change in bed slope at the rim or crest. The separated regions are vortices of a complicated internal structure. As was pointed out, the vortex pattern and the bed shape are interdependent and change in harmony as the structure evolves, wholly erosively in the case of flute marks, but through local

218

Figure 11.28b

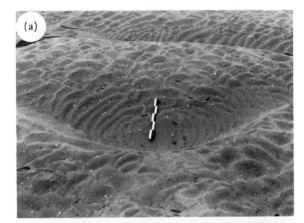

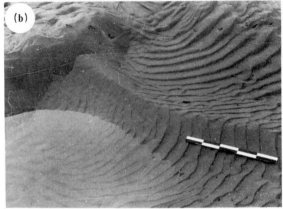

**Figure 11.29** Patterns of current ripples illustrating (a) divergent bottom flow in the trough of a strongly three-dimensional dune (external flow bottom to top), and (b) the convergent bottom flow at a streamwise spur (external flow from upper left). Scale bar is 0.5 m.

erosion and deposition in the case of the sandy bedforms.

The ideal flute mark (Fig. 11.7) on a mud bed is a parabolic hollow, deepest and steepest-sided at the upstream end. We can establish the flow pattern in such a hollow by reproducing the mark in plaster of Paris and exploiting the fact that the plaster is slightly soluble in water. Small pits made in a surface of plaster of Paris with a sharp object such as a carpenter's nail become elongated parallel to the local bottom current when exposed to a turbulent stream, in a manner already discussed (see Fig. 11.9). Allen (1971) used this technique to map the flow pattern on the floor of a model flute mark (Fig. 11.27), showing that a complex vortex was present. The current recirculates in the upstream part of the vortex but spirals along each limb. The general effect is of a divergent flow over the floor of the U-shaped hollow.

The same technique applied to current ripples demonstrates the complexity of their separation vortices (Allen 1969). Figure 11.28a shows a bed of ripples in fine-grained sand that was reproduced in plaster of Paris, the cast with pits cut in its surface being after-wards exposed to the same current as made the ripples. The map of bottom currents (Fig. 11.28b), compared to the original bed, shows that hollows on the surface coincide with divergent currents, and that the longitudinal ridges, as well as the ripple crests themselves, lie in zones of convergence. The divergence zones are where local grain transport paths begin, and it is here that one would expect the greatest erosion and deepening of the bed. Deceleration occurs towards convergences, however, and at these an element of deposition should prevail.

Although dunes (Sec. 4.6) are an order of magnitude or more larger than current ripples, the two kinds of bedforms give rise to similar patterns of separated flow (Allen 1968a). Just as we used flute-like features to prove the flow pattern on a rippled bed, so ripples can in turn be exploited to establish the local currents on dunes. Figure 11.29a is a view down stream into the deep trough of a crescentic tidal dune. The current ripples in the trough define a pattern of ovals with a common centre, their asymmetry indicating a divergent bottom flow, outwards in all directions from that centre. Hence the flow pattern in the closed trough of this strongly three-dimensional dune closely resembles that in the closed hollows locally developed on the rippled bed (Fig. 11.28). Another parallel in terms of bed shape and related flow pattern is illustrated in Figure 11.29b. To judge from the ripple pattern, the separated flows to leeward of this downstream-projecting dune crest converged over the bed to build a sharp-crested streamwise ridge.

# Readings

Allen, J. R. L. 1968a. *Current ripples*. Amsterdam: North-Holland.

Allen, J. R. L. 1968b. The diffusion of grains in the lee of ripples, dunes and sand deltas. *J. Sed. Petrol*. 38, 621–33.

Allen, J. R. L. 1969. On the geometry of current ripples in relation to stability of fluid flow. *Geogr. Annlr* A 51, 61–96.

Allen, J. R. L. 1971. Transverse erosional marks of mud and rock: their physical basis and geological significance. *Sed. Geol*. 5, 167–385.

Angell, J. K., D. H. Pack and C. R. Dickson 1968. A Lagrangian study of helical circulations in the planetary boundary layer. *J. Atmos. Sci*. 25, 707–17.

Brookfield, M. 1970. Dune trends and wind regime in central Australia. *Z. Geomorph*. S 10, 121–53.

Brown, G. L. and A. Roshko 1974. On density effects and large structure in turbulent mixing layers. *J. Fluid Mech*. 64, 775–816.

Brown, R. A. 1980. Longitudinal instabilities and secondary flows in the planetary boundary later. *Rev. Geophys. Space Phys*. 18, 683–97.

Burne, R. V. 1970. The origin and significance of sand volcanoes in the Bude Formation (Cornwall). *Sedimentology* 15, 211–28.

Chandra, K. 1938. Instability of fluids heated from below. *Proc. R. Soc. Lond*. A 164, 231–42.

Colton, G. W. 1967. Late Devonian current directions in western New York with special reference to *Fucoides graphicus*. *J. Geol*. 75, 11–22.

Cooke, R. U. and A. Warren 1973. *Geomorphology in deserts*. London: Batsford.

Field, M. E., J. V. Gardiner, A. E. Jennings and B. D. Edwards 1982. Earthquake-induced sediment failures on a 0.25° slope, Klamath River delta, California. *Geology*. 10, 542–6.

Friend, P. F. 1965. Fluviatile sedimentary structures in the Wood Bay Series (Devonian) of Spitsbergen. *Sedimentology* 5, 39–68.

Gill, W. D. 1979. *Syndepositional sliding and slumping in the west Clare Namurian basin, Ireland*. Spec. Pap. Geol. Surv. Ireland, no. 4.

Gill, W. D. and P. H. Kuenen 1958. Sand volcanoes on slumps in the Carboniferous of County Clare, Ireland. *Q. J. Geol. Soc. Lond*. 113, 441–57.

Glennie, K. W. 1970. *Desert sedimentary environments*. Amsterdam: Elsevier.

Hagerdorn, H. 1968. Über äolische Abtragung und Formung in der Südest-Sahara. *Erdkunde* 22, 257–69.

Haines, D. A. 1982. Horizontal roll vortices and crown fires. *J. Appl. Meteorol*. 21, 751–63.

Haines, D. A. and M. C. Smith 1983. Windtunnel generation of horizontal roll vortices over a differentially heated surface. *Nature* 306, 351–2.

Hanna, S. R. 1969. The formation of longitudinal sand dunes by large helical vortices in the atmosphere. *J. Appl. Meteorol*. 8, 874–83.

Hunt, J. C. R., C. J. Abell, J. A. Peterka and H. Woo 1978. Kinematical studies of the flow around free or surface-mounted obstacles; applying topology to flow visualization. *J. Fluid Mech*. 86, 179–200.

Jackson, R. G. 1976. Sedimentological and fluid-dynamic implications and turbulent bursting phenomena in geophysical flows. *J. Fluid Mech*. 77, 531–60.

Jimenez, J. 1983. A spanwise structure in the plane shear layer. *J. Fluid Mech*. 132, 319–36.

Karcz, A. 1968. Fluviatile obstacle marks from the wadis of the Negev (southern Israel). *J. Sed. Petrol*. 38, 1000–12.

Kenyon, N. H. 1970. Sand ribbons of European tidal seas. *Marine Geol*. 9, 25–39.

Kiya, M. and K. Sasaki, 1983. Structure of turbulent separation bubble, *J. Fluid Mech*. 137, 83–113.

Krishnamurti, R. 1975. On cellular cloud patterns. Part 2: laboratory model. *J. Atmos. Sci*. 32, 1364–72.

Lancaster, N. 1982. Linear dunes. *Prog. Phys. Geog*. 6, 475–504.

Langston, L. S. and M. T. Boyle 1982. A new surface-streamline flow-visualization technique. *J. Fluid Mech*. 125, 53–7.

Leibovich, S. 1983. The form and dynamics of Langmuir circulations. *Annu. Rev. Fluid Mech*. 15, 391–427.

Lonsdale, P. and B. Malfait 1974. Abyssal dunes of foraminiferal sand on the Carnegie Ridge. *Bull. Geol. Soc. Am*. 85, 1697–712.

Lonsdale, P. and F. N. Spiess 1977. Abyssal bedforms explored with a deeply towed instrument package. *Marine Geol*. 23, 57–75.

McKee, E. D. 1979. *A study of global sand seas*. Prof. Pap. US Geol. Surv., no. 1052.

McKee, E. D. and G. C. Tibbitts 1964. Primary structures of a seif dune and associated deposits in Libya. *J. Sed. Petrol*. 34, 5–17.

Meynart, R. 1983. Speckle velocimetry study of vortex-pairing in a low-*Re* unexcited jet. *Phys. Fluids* 26, 2074–9.

Pantin, H. M., D. Hamilton and C. D. R. Evans 1981. Secondary flow caused by differential roughness, Langmuir circulations, and their effect on the development of sand ribbons. *Geo-Marine Lett.* **1**, 255–60.

Prior, D. B. and J. N. Suhayda 1979. Application of infinite-slope analysis to subaqueous sediment instability, Mississippi delta. *Engng Geol.* **14**, 1–10.

Rajaratnam, N. 1976. *Turbulent jets*. Amsterdam: Elsevier.

Raudkivi, A. J. 1966. Bed forms in alluvial channels. *J. Fluid Mech.* **26**, 507–14.

Rust, B. R. 1965. The sedimentology and diagenesis of Silurian turbidites in south-west Wigtownshire, Scotland. *Scott. J. Geol.* **1**, 231–46.

Stride, A. H. 1982. *Offshore tidal sands*. London: Chapman and Hall.

Thorpe, S. A. 1968. A method of producing shear in a stratified flow. *J. Fluid Mech.* **32**, 693–704.

Thorpe, S. A. 1973. Cat in the lab. *Weather* **28**, 471–5.

Tsoar, H. 1983. Dynamic processes acting on a longitudinal (seif) dune. *Sedimentology* **30**, 567–78.

Walker, G. P. L. and R. Croasdale 1971. Two Plinian-type eruptions in the Azores. *J. Geol. Soc. Lond.* **127**, 17–55.

Wipperman, F. 1969. The orientation of vortices due to instability of the Ekman-boundary layer. *Beitr. Phys. Atmos.* **42**, 225–44.

# 12 Sudden, strong and deep

## 12.1 Some experiments

What will happen when a quantity of fluid made more dense by having sediment particles mixed with it is released into a larger stagnant volume of the same but plain fluid? On general grounds, some kind of relative motion is to be expected, but precisely what form of motion, and under the control of which particular forces, may depend on the boundary conditions. A few simple experiments in a domestic bath can help to answer this and related questions.

Fill the bath to a depth of about 0.15 m with the coldest water available from the tap and let the water grow still. Two-thirds fill a one-litre heat-resistant glass beaker with boiling water strongly coloured with methylene blue or potassium permanganate. Have to hand a second but empty beaker. Handling the container of hot water with strong tongs, and immersing it almost to the brim, slowly release its contents into the cold water at one end of the bath. The coloured hot water will spread rapidly down the bath, the front of the mass remaining sharp but developing lobes. Gradually, press the second beaker upright down into the water affected by the flow and look out for colour changes. The hot water is evidently floating as a thin layer on the cold.

It is only natural that the hot water should float, for the density near the boiling point is some 4% less than at 10°C. This density contrast also suggests that the motion may have been driven by a hydrostatic pressure difference acting between the hot current and the cold water at the same level ahead of it. What is surprising about so small a density difference, however, is that

it should permit (1) such a swift current to form, and (2) so little mixing between the current and its surroundings.

Now try in the bath the effect of introducing water in which mud and sand are dispersed. One can either use about 0.05 kg of garden soil mixed with roughly 0.75 litre of water, or an equivalent quantity of china clay and very fine sand (colour the mixture if the bath is white). Because of the relatively high density of the mineral matter (about 2650 kg m$^{-3}$), the new mixture will exceed in density the cold water into which it will be released, in contrast to the hot water of lower density.

First carefully spoon a little of the freshly stirred sediment–water mixture onto the surface of the water. What happens is not easily seen, but a few repetitions will show that the blob rapidly sinks and, passing through an intermediate shape like an inverted mushroom, becomes a ring-shaped vortex which may break up into smaller and themselves unstable drops. Like the heavy particles analysed in Chapter 3, the blob of mixture sinks because of its greater weight compared to the plain water displaced. But as the blob is deformable, it eventually loses (but only gradually) its discrete character and ultimately blends with the surroundings. We can infer that a continuous downward stream of the mixture would result if the supply at the water surface was maintained.

Finally, pour the well stirred mixture into the shallow end of the bath. This should be done slowly, with the beaker immersed almost to its brim, and at roughly the same rate as the current transports the heavy mixture over the sloping floor of the bath. Aside from being the other way up, the current is similar to that formed from

**Figure 12.1** Examples of gravity currents. (a) Head of turbidity current (0.5 m wide) made from dispersion of kaolinite advancing towards observer. (b) Aerial view of *nuée ardente* descending slopes of St Augustine Volcano, Alaska, February 1976. Photograph courtesy Professor L. F. Radke, University of Washington.

hot water, but with more features visible. A distinct head is recognizable (Fig. 12.1a) with a sharp top and overhanging front divided spanwise into lobes and clefts. A good deal of mixing, mainly in the form of large transverse billows, can be seen to occur along the top and to the rear of the head. With any luck, the current will reach the other end of the bath before sediment loss checks it. What forces drove this current? Clearly a pressure force could have been involved, as in the case of the hot water current, but to this we must add a body force due to the sloping floor of the bath. Lastly, very slowly drain off the water and let the layer of deposited sediment dry out. This layer will prove on dissection to be normally graded, the smaller particles of each mineral overlying the larger.

## 12.2 Kinds of gravity current

What we made in these experiments were kinds of grav-

ity or density currents, that is, the flow of one fluid within another caused by the density difference between the fluids (Simpson 1982). Strictly, rivers on the land (Ch. 5) are of this class, but it is conventional to reserve the term gravity current to flows that involve only a relatively small density difference, as in the experiments just described. Note that the density contrast has no particular cause. It may in practice be due to either a difference of temperature, or a difference of bulk composition, or to both.

Our experiments point to the existence of three classes of gravity current, to which a fourth may for completeness be added (Fig. 12.2). Overflows arise where the moving fluid is less dense than the ambient medium. Natural examples are virtually restricted to the seas and oceans, in the form of plumes of fresh but usually muddy water at river mouths (Garvine 1977). Underflows exceed the surroundings in density and occur in considerable variety. In the atmosphere they are found, for example, as sea-breeze fronts (Simpson 1967) and as sudden, radially spreading outwashes of cold air from thunderstorms (Lawson 1971, Idso 1976). Although it is the greater density of the cold air that drives such atmospheric flows, in arid lands a contribution can be made by entrained dust (Fig. 12.3). Notice how closely the sharply overhanging and lobed front of this dust storm resembles the experimental case (Fig. 12.1a). Other atmospheric underflows are powder-snow avalanches (Mellor 1978), *nuées ardentes* (Stith *et al.* 1977) and base surges (Moore 1967). *Nuées ardentes* are violent surges of hot gases and incandescent ash and rock that form when magma bursts from a shallow depth beneath the flanks of a volcano (Fig. 12.1b). Base surges are much cooler mixtures of gases, water droplets or steam, and ash that follow eruption into an aqueous environment. In all three cases the driving density contrast depends on the presence of dispersed particles. The turbidity current is the main kind of sediment-driven

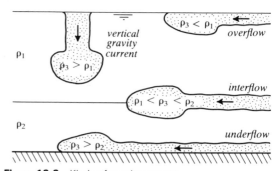

**Figure 12.2** Kinds of gravity current.

**Figure 12.3** Dust-laden atmospheric gravity current (cold front), New South Wales, Australia. From Garratt (1984) with permission of the Royal Meteorological Society. Photograph courtesy of Dr J. R. Garratt, CSIRO (Australia).

underflow encountered in oceans, seas and lakes. They are familiar from the laboratory (Fig. 12.1a), but no-one has ever observed a full-sized prototype and lived to tell the tale. The evidence for their occurrence in the oceans is purely circumstantial, taking the form chiefly of the sudden and unexpected but sequential breaking of submarine telegraph cables (Heezen *et al.* 1954, Krause *et al.* 1970). The scale of turbidity currents may well approach that of atmospheric gravity currents (Fig. 12.3), and their frontal appearance should certainly be similar. Sedimentologically, turbidity currents, *nuées ardentes* and base surges are the main agents for the transport of coarse debris within calm water bodies and the atmosphere. Irvine (1980) suggests that crystal-laden gravity currents may operate in magma chambers within the Earth's crust.

Interflows can occur between water layers of contrasted density in a stratified ocean or lake (Holyer &

Huppert 1980, Britter & Simpson 1981) and may in some cases may be due to a sediment-related density difference. The sediment concentration must be very low, however, and the particles themselves extremely small, to prevent dissipation of the current. Finally, vertical density currents are theoretically possible, but it is doubtful that this class is represented naturally. Man-made vertical currents may occur where sludge or tailings are discharged into the sea from barges and floating outfalls.

## 12.3 Difficulties with gravity currents

A fluid dynamicist attempting to describe gravity currents would categorize them as non-uniform, unsteady, non-linear, free-boundary flows, and would probably then throw up his hands in despair. He would evidence horror as well if told that the currents of greatest sedimentological interest flowed by virtue of dispersed sediment, which became deposited during the course of the motion, so hastening the dissipation of the flow. As if this were not enough, it would have to be admitted that turbidity currents, *nuées ardentes* and base surges did not depend on just one force. These important flows mostly travel over sloping ground, and so may be driven by a changing balance of body and pressure forces.

What all this means is that a comprehensive theoretical model of a gravity current is not yet achievable, and that the best we can at present do mathematically is to explore a few aspects of the motion under rather severe constraints. Much of our understanding has come to date from well designed laboratory experiments. In what follows we shall be concerned exclusively with gravity current underflows.

## 12.4 Drag force and mean velocity of a uniform steady gravity current

It is easy to imagine that heavy fluid could be continuously supplied to a uniform slope beneath a stagnant fluid in such a manner that a uniform steady gravity current was set up. The problem of this flow will be recognized as similar to that of the ideal river considered in Section 5.3, before the atmosphere was neglected in the analysis.

As the flow is down an open slope, the element we shall consider is a section of the flow of unit width and fixed streamwise length (Fig. 12.4). Arguing as before, the actuating force per unit width is the downslope component of the immersed weight of the element

$$\text{driving force} = x h_B (\varrho_2 - \varrho_1) g \sin \beta \quad \text{N m}^{-1} \quad (12.1)$$

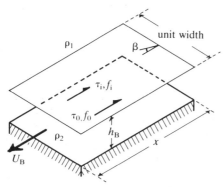

**Figure 12.4** Definition sketch for a uniform steady gravity current.

where $x$ is the length of the element, $h_B$ its thickness, $\varrho_1$ and $\varrho_2$ the density of respectively the ambient medium and the current, $g$ the acceleration due to gravity, and $\beta$ the slope angle. The driving force therefore increases as the thickness and excess density of the current, and as the sine of the angle of the slope. The resisting force comes from friction between (1) the current and its bed, and (2) the current and the medium (neglected in the case of the river on the land). For the element shown in Figure 12.4, the resisting force per unit width is

$$\text{resisting force} = x\tau_0 + x\tau_i = x(\tau_0 + \tau_i) \quad (12.2)$$

where $\tau_0$ and $\tau_i$ are the shear stresses per unit area respectively at the bed and the fluid interface. Using the quadratic stress law (Eqn. 1.27) we now introduce

$$\tau_0 = \frac{f_0}{8} \varrho_2 U_B^2 \qquad \tau_i = \frac{f_i}{8} \varrho_1 U_B^2 \qquad \text{N m}^{-2} \quad (12.3)$$

where $f_0$ and $f_i$ are the Darcy–Weisbach friction coefficients respectively for the bed and the interface, and $U_B$ is the mean flow velocity of the current. As the flow is steady and uniform, Newton's first law of motion requires that the driving and resisting forces should balance. Eliminating $x$ between Equations 12.1 and 12.2, and introducing the stresses from Equation 12.3, we have

$$U_B = \left(\frac{8 \sin \beta}{f_0 + f_i}\right)^{1/2} \left(\frac{\Delta \varrho}{\varrho_2} g h_B\right)^{1/2} \quad \text{m s}^{-1} \quad (12.4)$$

which may be rearranged to

$$\frac{U_B}{[(\Delta \varrho / \varrho_2) g h_B]^{1/2}} = \frac{8 \sin \beta}{(f_0 + f_i)} \quad (12.5)$$

where $\Delta \varrho = (\varrho_2 - \varrho_1)$. Equations 12.4 and 12.5 correspond to respectively Equations 5.8 and 5.9 for a river, and have the same implications. The main differences for a gravity current are that $g$ is in effect density-adjusted and that there is significant friction at a second boundary. Equation 12.5 therefore defines a densiometric Froude number.

Figure 12.5, based on Equation 12.4, shows that uniform steady gravity currents can attain high speeds on the slopes encountered in the oceans, provided that the currents are sufficiently thick and dense. That such sediment-driven river-like currents actually arise there is strongly suggested by the wide distribution and common occurrence of huge channels of various shapes on the

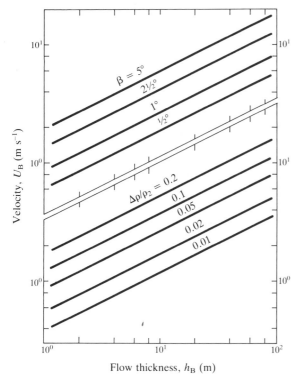

**Figure 12.5** Speed of a uniform steady gravity current for plausible conditions at (a) fixed density difference $\Delta\varrho = 0.05$, and (b) fixed slope $\beta = 1°$. In all calculations $f_i = f_0 = 0.05$.

depositional sites known as deep-sea fans and cones. Figure 12.6 shows part of a side-scan sonar record of a channel roughly 2 km wide from the Amazon deep-sea cone. This channel, as tortuous as the Mississippi River, has a cross-sectional area (and perhaps discharge) an order of magnitude larger. A gravity current flowing round one of its bends would have a considerably steeper transverse slope than in the case of a river. Arguing as in the case of a river (but without neglecting the

ambient medium), we would write Equation 5.27 for a gravity current as

$$\frac{dh_{\mathrm{B}}}{dr} = \frac{\varrho_2}{\Delta\varrho} \frac{kU_{\mathrm{B}}^2}{gr} \qquad (12.6)$$

the difference between the two equations being the large factor $\varrho_2/\Delta\varrho$. In order to retain the flow, the outer bank of a meander in a deep-sea channel would have to stand metres or tens of metres higher than the inner one (Komar 1969).

The main difficulty in using Equation 12.4 and its derivatives is the selection of values for the friction coefficients $f_0$ and $f_i$. At present this can only be done empirically. Middleton (1966b) shows experimentally that $f_1/f_0$ in the case of smooth lower flow boundaries increases gradually with the densiometric Froude number, from approximately 0.5 at a Froude number of 1 to roughly 1 when the Froude number is of order 3.

## 12.5 Shape and speed of a gravity-current head

Similarity of shape is a striking feature of the heads of gravity currents, whether in the laboratory (Fig. 12.1) or the field (Fig. 12.3). The top of the head slopes downwards increasingly steeply in the direction of travel, finally overturning at a level relatively near the ground. The overturning is a viscous effect, as explained below, but the shape of the upper part of the head is uninfluenced by the bed, and so must represent an equilibrium between the forces acting on the interface from the ambient medium and those exerted by the current itself. Benjamin (1968) neglected viscosity to apply Bernoulli's equation (Sec. 1.5.3) to the interface, and so calculated theoretically the shape of the head (Fig. 12.7), the horizontal and vertical distances being plotted in terms of the head height $h_{\mathrm{H}}$. Because viscosity is

**Figure 12.6** Part of side-scan sonar record (area measures 14 km × 62 km) of meandering channel on Amazon deep-sea cone in 3.5 km of water. Down slope towards the left. Compare with Figure 5.1. Photograph courtesy of Mr R. H. Belderson and reproduced by permission of the Institute of Oceanographic Sciences.

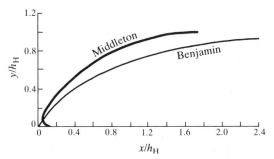

**Figure 12.7** Theoretical shape of a gravity-current head (Benjamin 1968) compared with the average shape of laboratory-scale turbidity currents (Middleton 1966a).

neglected, there is no overhang and the head joins the bed at the acute angle $\pi/3$. This theoretical shape may be compared with the average head determined by Middleton (1966a) from laboratory-scale turbidity currents. The agreement is quite good, considering the difficulty of positioning overhanging heads in Figure 12.7.

What controls the velocity of the head? We are most interested in what may be called the 'deep-water' case, when the thickness of the ambient medium (ocean, atmosphere) is so large compared to the current that it is not a relevant scale.

Benjamin (1968) explored this case by analysing the motion at a constant velocity $U_H$ of an empty cavity into a stagnant inviscid fluid of density $\varrho$ contained beneath a horizontal roof (Fig. 12.8). By superimposing an equal and opposite velocity $-U_H$ on the fluid, the cavity is brought to rest and the motion appears steady. The pressure exerted by the moving fluid at S, which is a stagnation point, is equal to $\frac{1}{2}\varrho U_H^2 \, \mathrm{N \, m^{-2}}$ by Bernoulli's equation. However, this pressure must be equalled by the pressure in the cavity, which is easily seen to be the hydrostatic pressure $\varrho g h_B$, where $h_B$ is the thickness of the cavity behind the head and the net fall in the level of the free surface limiting the cavity. Equating the pressures under Newton's first law, since the motion is steady, we find

$$U_H = 2^{1/2}(gh_B)^{1/2} \quad \mathrm{m \, s^{-1}} \qquad (12.7)$$

the density cancelling out. Substituting for the empty cavity a lighter fluid of density $\varrho_2$ moving into a heavy fluid of density $\varrho_1$, the same argument affords

$$U_H = 2^{1/2}\left(\frac{\Delta\varrho}{\varrho_1} gh_B\right)^{1/2} \qquad (12.8)$$

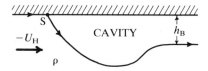

**Figure 12.8** Arrested cavity beneath a roof.

where $\Delta\varrho = (\varrho_1 - \varrho_2)$. This equation states that the head advances at a velocity increasing with the square root of the density difference and the square root of the thickness of the trailing flow. Notice that the density ratio in this equation contains the denominator $\varrho_1$ and not $\varrho_2$ as appears in Equation 12.4 for the uniform steady gravity current. Equation 12.8 yields velocities for the head of the order of $1{-}10 \, \mathrm{m \, s^{-1}}$ for density differences of a few per cent, provided that $h_B$ is of order $10{-}100 \, \mathrm{m}$.

Another approach is to arrest the head by superimposing on the ambient medium the equal and opposite velocity $-U_H$ and then to consider the flow forces acting on the now steady head assumed to have no internal circulation (Fig. 12.9). The medium then appears to exert on the head a driving force per unit width given by

$$\text{driving force} = C_D h_H \varrho_1 \frac{U_H^2}{2} \quad \mathrm{N \, m^{-1}} \qquad (12.9)$$

where $C_D$ is a drag coefficient and $h_H$ the head height. The net resisting force per unit width is evidently the excess hydrostatic pressure integrated from the bed to the top of the flow behind the head

$$\text{resisting force} = \tfrac{1}{2}(\varrho_2 - \varrho_1)gh_B^2 \qquad (12.10)$$

As the motion is steady, the two forces balance and

$$U_H = \left(\frac{1}{C_D}\frac{h_B}{h_H}\right)^{1/2}\left(\frac{\Delta\varrho}{\varrho_1}gh_B\right)^{1/2} \quad \mathrm{m \, s^{-1}} \qquad (12.11)$$

this new formula being identical in form to Equation 12.8 except for the coefficient of proportionality. At large Reynolds numbers, when viscosity can be ignored, this coefficient should be constant and of the same order as Benjamin's. Its constancy seems assured from the general observation that the drag coefficient of bluff

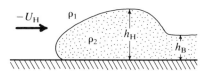

**Figure 12.9** Arrested gravity-current head.

228

bodies is invariant at large Reynolds numbers (e.g. Fig. 3.6). The magnitude of the coefficient is less certain but a value of order of unity seems possible. Experiments by Simpson and Britter (1979) suggest that $h_B/h_H$ is of order 0.25 in the 'deep-water' case. Comparison of a gravity-current head with an elliptical cylinder in cross flow suggests that $C_D$ could be similar or even somewhat smaller, particularly when allowance is made for the fact that the head in reality possesses an internal circulation.

## 12.6  Why does the nose overhang?

The heads of gravity currents carry an overhanging nose (Figs 12.1 & 12.3). What controls the nose height, and why does the overhang occur?

Simpson and Britter (1979, 1980) measured the nose height $h_N$ from many horizontal gravity currents in the laboratory and from numerous thunderstorm outwashes, comparing the relative height $h_N/h_H$ with the Reynolds number $U_H h_H \rho_2/\eta_2$, where $\eta_2$ is the viscosity of the current. They found that the relative height gradually declined with increasing Reynolds number, to become constant at approximately 0.12 at Reynolds numbers greater than about $10^4$ (Fig. 12.10). Hence this is the value of the nose height applicable to volcanogenic gravity currents and to turbidity currents.

Why does the overhang develop? Two facts are relevant: (1) the head represents a sudden intrusion of rapidly moving fluid along the bottom of an otherwise stagnant one, and (2) the ambient medium is viscous and therefore must oppose the intrusion. These suggest that the ambient medium in a thin zone at the bottom

cannot be deformed rapidly enough to be thrust upwards over the advancing head, but instead adheres to the bed to become squeezed beneath the head. That this in fact occurs is easily proved by running an underflow made of salt solution through water in a bath into which a few potassium permanganate crystals were sprinkled and left for a minute or two. Illuminating the bath with a bright overhead light, so that the current is visible as a shadowgraph, it will be seen that the coloured fluid is swept partly if not wholly under the head. Thinking of the head as stationary, the salt current is clearly receiving a discharge of ambient medium. Furthermore, the stagnation point on the head cannot be situated at the bed, as in Benjamin's (1968) inviscid model (Fig. 12.8), but must, as Benjamin (1968) and Allen (1971) supposed, lie either at or near the tip of the nose and therefore above the ground (Fig. 12.11a). Allen estimated the discharge of ambient medium into the head as approximately one-fifth of the discharge represented by the dense current. Simpson (1982) obtained experimentally the more accurate upper limit of 0.13 times the discharge of dense fluid.

## 12.7  Lobes, clefts and sole marks

Another singular feature about a gravity-current head is a spanwise series of regularly spaced lobes and clefts (Figs 12.1a & 12.3). Simpson (1969, 1972) investigated these experimentally and found from shadowgraphs (Fig. 12.12) that the clefts (1) extended a substantial relative distance backwards into the head, (2) are continually absorbed by neighbours, and (3) are initiated by the subdivision of comparatively large lobes. Cleft

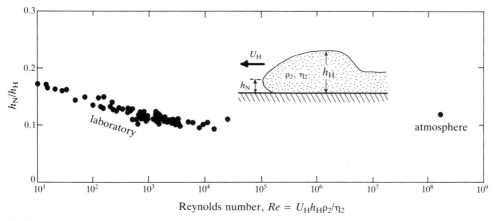

**Figure 12.10**  Nose height of a gravity-current head as a function of Reynolds number in the laboratory and for one field case. Data of Simpson and Britter (1979, 1980).

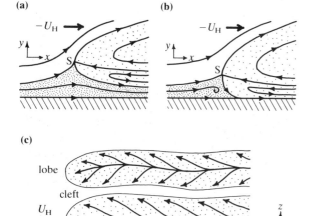

**(a)** $-U_H$

**(b)** $-U_H$

**(c)**

lobe

cleft

$U_H$

lobe

cleft

lobe

$z$

$x$

**Figure 12.11** Speculative patterns of motion associated with gravity-current heads. (a) Relative to the head in a vertical streamwise plane through a cleft. (b) Relative to the head in a vertical streamwise plane through a lobe. (c) Relative to the ground on the bed beneath lobes.

initiation and absorption are well displayed in Figure 12.13, showing one of Simpson's sequences of shadowgraphs of an advancing current. For Reynolds numbers smaller than about $10^4$, Simpson found

experimentally that

$$\frac{z}{h_N} = 12\, Re^{-0.16} \qquad (12.12)$$

where $z$ is the transverse spacing of the clefts. Extrapolated to prototype scale, this formula yields a relative cleft spacing of 0.63 for $Re = 10^8$. Such spacings seem rather on the low side in comparison with, say, atmospheric gravity currents such as that in Figure 12.3. It may be that, like the relative nose height, the relative cleft spacing is independent of Reynolds number at sufficiently large numbers. If the same threshold applied in both cases, the relative cleft spacing is 2.7 at large Reynolds numbers, in better agreement with Figure 12.3.

Why should spanwise lobes and clefts develop? The clue lies in the greater density of the current than the layer of ambient medium squeezed beneath it at the head. This layer therefore combines with the overriding current to form a gravitationally unstable system which attempts to invert (Allen 1971), as explained for sand emplaced on mud and then liquidized (Sec. 10.7) and for air heated at a desert surface (Sec. 11.4.3). The similarity between an overriding gravity-current head and the rise of heated air into a strong wind is particularly close, for in each case Rayleigh–Taylor instability is accompanied by unidirectional shear.

Are lobes and backward-reaching clefts therefore associated with corkscrew-like secondary currents (see Fig. 11.15)? This seems probable on two counts. First, the initially stagnant ambient medium trapped in the

**Figure 12.12** Shadowgraph (horizontal plane) of lobes and clefts developed at head of a laboratory gravity current. Scale bar is 0.01 m. From Simpson (1972) *J. Fluid Mech.* **53**, 759–68, by permission of the Cambridge University Press. Photograph courtesy of Dr J. E. Simpson, University of Cambridge.

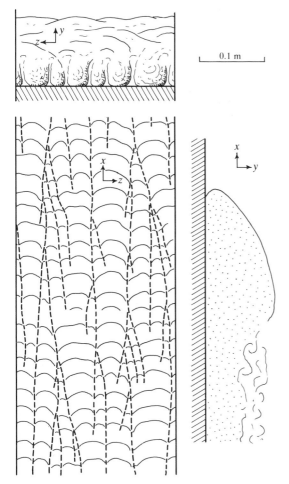

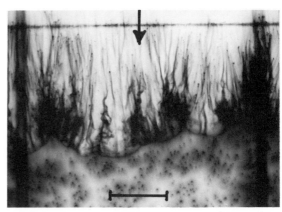

**Figure 12.14** View from below through transparent flume of gravity current spreading over a thin layer of potassium permanganate solution (specks are undissolved crystals). Lobes and clefts at nose pass up stream into convergent and divergent streaks of colour indicating large corkscrew vortices. Scale bar is 0.1 m long.

**Figure 12.13** Loci of clefts on advancing head of a laboratory gravity current shown at intervals of 0.5 s. Adapted from Simpson (1972).

clefts is likely to have a smaller velocity relative to the ground than the current in the advancing lobes. Secondly, when a salt current is run into stagnant water coloured over the lowest millimetre or so with potassium permanganate, the descent of the lobes through the ambient medium squeezed beneath the head drives the coloured fluid laterally into the clefts (Fig. 12.14). The steamline pattern shown in Figure 12.11a can therefore apply only to clefts, and the pattern at lobes must be different (Fig. 12.11b). Moreover, the fact that the lobes push the coloured fluid away suggests that the flow in them has a transverse as well as a streamwise component. Hence each lobe probably contains a pair of oppositely rotating corkscrew vortices (Fig. 12.11c).

What are the implications of the lobe-and-cleft struc-

ture for sedimentary processes at gravity-current heads? One is that, because of the greater velocity of the current, the rate and total amount of erosion beneath lobes should be greater than beneath clefts. Allen (1971) thus explained the occurrence of streamwise bands of flute marks (Sec. 11.2.4) beneath some turbidite sandstones. Fisher (1977) applied the same explanation to a pattern of erosional grooves radiating from a Hawaiian volcano erupting base surges. In the right circumstances, a turbidity current could plastically mould rather than erode a mud bed. The clefts ranging beneath the head should then correspond to ridges on the mud, formed where the fluid ascending at the margins of the lobes had dragged the mud up into the clefts. Dzulynski and Walton's (1965) longitudinal furrows and ridges probably arise in this way. The examples shown in Figure 12.15 were formed when a turbidity current made from a suspension of plaster of Paris drove over a soft mud bed in a laboratory tank. Note how closely the locally convergent marks resemble Simpson's (1972) pattern of moving clefts (Fig. 12.13).

## 12.8 Billows on the head

When the head of a laboratory gravity current is illuminated in a vertical streamwise sheet of light, a series of transverse vortices, or billows, is revealed along the top (Simpson 1969, 1972, Simpson & Britter 1979). Figure 12.16 shows examples from a laboratory experiment, and there is strong evidence for their occurrence

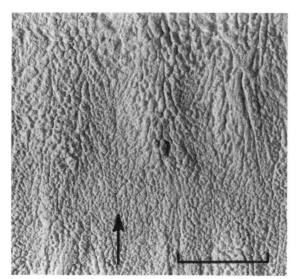

**Figure 12.15** Longitudinal ridges and furrows (lobe-and-cleft structure) preserved on underside of an experimental turbidity-current deposit made from plaster of Paris. Scale bar is 0.05 m.

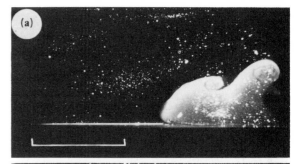

**Figure 12.16** Slit-lighting views (2 s apart) of a salty gravity current advancing from right to left into stagnant water. Scale bar is 0.05 m. From Simpson (1969) by permission of the Royal Meteorological Society. Photographs courtesy of Dr J. E. Simpson, University of Cambridge.

at field scale (Simpson & Britter 1980). These billows, growing in wavelength and amplitude to something like the thickness of the head as they are carried backwards (relative to the head), express the mixing of dense fluid out of the head of the current and into the ambient medium. Here is another example of Kelvin–Helmholtz instability (Sec. 11.2.2). Their presence on a gravity-current head shows that the head is being continually supplied with dense fluid, and that the average speed in the body of the current therefore exceeds that of the head (Britter & Simpson 1978, Simpson & Britter 1979).

Figure 12.17 shows a horizontal gravity current which appears to be steady because a velocity $-U_H$ equal and opposite to the velocity of the head has been superimposed on it. Neglecting the small amount of ambient medium driven under the head, three regions can be identified in the current. The head is of height $h_H$ and velocity $U_H$ relative to the ground. In the body behind the thickness is $h_B$ and the mean ground velocity $U_B$. The envelope of the billows leads downstream to a turbulent wake of thickness $h_W$ and mean velocity $U_W$ relative to the ground. How are the discharges in these regions related to the discharge represented by the billows? In the frame of reference in which the head is at rest, we have for each unit width of the flow

$$\text{discharge into head from body of flow} = h_B(U_B - U_H) \quad \text{m}^3\,\text{m}^{-1}\text{s}^{-1} \tag{12.13}$$

and

$$\text{discharge in wake} = h_W(U_H - U_W) \tag{12.14}$$

The discharge per unit width from the ambient medium through the billows to form the wake must therefore be the difference between Equations 12.14 and 12.13. Calling this quantity $q\,\text{m}^3\,\text{m}^{-1}\,\text{s}^{-1}$, we have

$$h_B(U_B - U_H) + q = h_W(U_H - U_W) \tag{12.15}$$

which, multiplying through by $1/h_B U_H$, becomes

$$\frac{U_B}{U_H} = \frac{h_W}{h_B}\left(1 - \frac{U_W}{U_H}\right) - \frac{q}{h_B U_H} \tag{12.16}$$

Equation 12.15 is a statement of continuity. Experimentally, Simpson and Britter (1979) find that $U_B/U_H$ is very nearly constant at approximately 1.16, the body of the current continually overtaking the head. The rate at which the dense fluid is being discharged into the medium is therefore $h_B(U_B - U_H)$. This may be non-dimensionalized by the discharge $h_B U_H$ to yield $(U_B - U_H)/U_H$, which is very nearly constant at approximately 0.16.

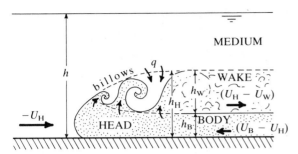

**Figure 12.17** Definition sketch for discharges into and out of an arrested gravity-current head.

Simpson and Britter (1979) established that the relative thickness of the wake is a function of the relative depth of the current. Experimentally, $h_W/h_B$ is about 0.6 for $h_B/h = 0.3$, where $h$ is the depth of the ambient medium (Fig. 12.17), rising to 3–4 in the 'deep-water' case. These experiments strongly suggest that it is the dynamics of the head, responding to the depth of the ambient medium, that determines the steady current following behind.

## 12.9 Gravity-current heads on slopes

What is the character of a gravity current made by suddenly releasing a dense fluid from a continuous and constant supply onto a slope? This is an important sedimentological question, because very few natural sediment-driven gravity currents occur on horizontal surfaces. An oceanic turbidity current could well find itself on a slope of 5–10° in the upper reaches of a submarine canyon, and would encounter slopes of less than 1° only on an abyssal plain, deep-sea fan or deep-sea cone. The surface beneath a *nuée ardente* or base surge could in places slope as steeply as 30–40°. The downslope component of the weight of the flow will then contribute significantly to the motion, in addition to the pressure forces.

Britter and Linden (1980) experimented on gravity currents initiated as above on slopes as steep as 90° from the horizontal. At slopes of 0.5° and larger, the head is of constant velocity, but gradually increases in size as it advances down the incline, due to (1) entrainment directly into the head, and (2) the addition of dense fluid from the steady current following behind in excess of the rate of mixing of fluid out of the head by billows. Figure 12.18 illustrates these enlarged heads. At moderate to large slopes the head advances at only about 0.6 of the mean velocity of the following current. In this case the downslope component of the weight

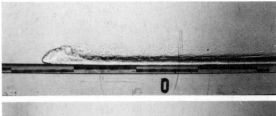
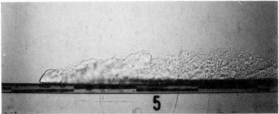
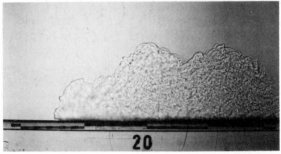

**Figure 12.18** Changes in shape of gravity-current head with bed slope (value in degrees given on each photograph). Current in each case 0.35–0.4 m long. After Britter and Linden (1980) *J. Fluid Mech.* **99**, 531-43, by permission of Cambridge University Press. Photographs courtesy of Dr R. E. Britter, University of Cambridge.

of the dense fluid is counteracted by increased mixing into the ambient medium. On slopes of less than about 0.5°, the head velocity gradually decreases with increasing distance of flow (Britter & Linden 1980). A decrease is also observed for currents on horizontal surfaces, once beyond the influence of initial conditions.

## 12.10 Dissipation of sediment-driven gravity currents

### 12.10.1 Causes of dissipation

The analyses and laboratory experiments so far described give us invaluable physical insights into the character of gravity currents in natural environments. Nonetheless, these studies all represent rather restrictive conditions in comparison with the real world. It has generally been assumed either that the motion is both uniform and steady (body flow alone), or that it is steady but non-uniform (head flow on horizontal or sloping boundary).

With these restrictions in mind, what factors will limit the life of a sediment-driven gravity current formed on a sloping boundary from a source of dense fluid which operated for a finite time?

One factor bound to curtail its life is the constant overtaking of the head by the body. Suppose that the current is of streamwise length $L$ when the last of the dense fluid is emitted by the source. Let the steady velocities of the head and body of the current be respectively $U_H$ and $U_B$. The rear of the body will have caught up with the head after the time $t = x/U_B = (x - L)/U_H$, where $x$ is the distance travelled by the rear of the current. It follows that $x = LU_B/U_B - U_H)$. As $U_B/U_B - U_H)$ is of order 5–10 for small slopes, a finite gravity current could become dissipated, simply through the feeding of the head, once it had travelled a distance an order of magnitude times its own starting length. The snake has (almost) eaten its tail!

Additional causes of dissipation emerge on inspecting Equation 12.4 for body flow and Equations 12.8 and 12.11 for motion of the head. Evidently the loss of either excess density or thickness or both will curtail the life of the flow. Excess density can be reduced through (1) mixing in the lobes and clefts, (2) mixing at the interface between the body of the flow and the wake above, and (3) loss of sediment through deposition. The first two rates of loss are not particularly important, but the third may be significant for currents laden with coarse particles, such as turbidity currents and base surges. Such currents may also experience significant rates of loss of height, as the 'ceilings' appropriate to the various grades of sediment present sink progressively nearer the bed.

### 12.10.2  Dissipation through sediment loss

To decide whether deposition will actually occur, with consequent losses of excess density and possibly thickness, we must compare the sediment load actually present in the gravity current with the load that can theoretically be supported by the forces due to the motion of the current. The idea behind this enquiry is that sediment can be deposited only if it is present in excess of the theoretically transportable load. Hence no sediment-related dissipation is expected if the actual load is at all times equal to or less than the theoretical value.

Let us therefore represent the potential for sediment-related dissipation by the loading factor $M = m_a/m_t$, where $m_a$ is the actual and $m_t$ the theoretical sediment load, each measured as dry mass per unit area. We shall

simplify the analysis by treating the current as a uniform slab of instantaneous thickness $h$ and velocity $U$ which contains sediment particles of a single size and solids density $\sigma$ at a uniform fractional volumetric grain concentration $C$, leading to a uniform constant current density of $\varrho_2$ (Fig. 12.19). It follows that

$$m_a = \sigma Ch \quad \mathrm{kg\,m}^{-2} \qquad (12.17)$$

It is more difficult to calculate $m_t$ because we must first separate the bedload and suspended-load components in the total sediment transport rate, and then divide each by the appropriate transport velocity to obtain the load (Sec. 4.4.1). From formulae corresponding to Equation 4.11

$$m_t = m_b + m_s \quad \mathrm{kg\,m}^{-2} \qquad (12.18)$$

$$= \frac{J_b}{U_b} + \frac{J_s}{U_s} \qquad (12.19)$$

where $m_b$ and $m_s$ are respectively the theoretical dry-mass bedload and suspended load, $J_b$ and $J_s$ the dry-mass transport rates respectively as bedload and suspended load, and $U_b$ and $U_s$ the respective transport velocities. For the total dry-mass sediment transport rate we shall use Bagnold's (1966) semi-empirical formulae, amounting to

$$J = \frac{\sigma}{(\sigma - \varrho_1)g}\left(0.148 + 0.01\frac{U}{8}\right)\frac{f_0}{V}\varrho U^3 \quad \mathrm{kg\,m}^{-1}\mathrm{s}^{-1}$$
$$(12.20)$$

and

$$\frac{\text{suspended transport rate}}{\text{bedload transport rate}} = 0.068\frac{U}{V} \qquad (12.21)$$

where $J$ is the total-load transport rate, $\varrho_1$ the density of the plain fluid, $U$ the current velocity, $V$ the terminal fall velocity of the sediment particles in the current, and $f_0$ the Darcy–Weisbach friction coefficient at the bed.

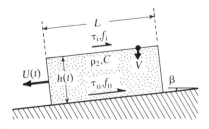

**Figure 12.19**  Slab model of a turbidity current.

Equation 12.4 will serve to give us $U$. The first term in the second bracket of Equation 12.20 relates to the efficiency of bedload transport, and the second term to the efficiency of transport in suspension (Sec. 4.4.4). We may take it that $U_b \simeq 0.2U$ and $U_s \simeq U$, at least for the finer grades of quartz sand in water.

Too many variables influence the loading factor for every one to be explored. As an example, however, consider a turbidity current 50 m thick containing sediment particles of solids density 2650 kg m$^{-3}$, terminal fall velocity 0.01 m s$^{-1}$ in the current, and fractional volume concentration 0.1, on a uniform slope of 0.5° in sea water, and let the Darcy–Weisbach friction coefficients be 0.02 and 0.01 respectively for the bed and interface. The proposed fall velocity corresponds to very fine sand at the low temperatures prevailing in ocean deeps. From these data Equation 12.17 yields $m_a = 1.325 \times 10^4$ kg m$^{-2}$ and Equation 12.4 gives $U = 12.49$ m s$^{-1}$. The density of the current turns out to be $\varrho_2 = 1188$ kg m$^{-3}$. Substituting into Equation 12.20, and then using Equation 12.21, we find that $m_b = 5.685 \times 10^1$ kg m$^{-2}$ and $m_s = 9.620 \times 10^2$ kg m$^{-2}$, making $M = 13.01$. Hence the current is grossly overloaded with sediment, the amount that available forces can support being an order of magnitude less than the actual load. Such a current should experience dissipation through loss of sediment alone.

### 12.10.3  Travel time and distance

In the light of this result, which stresses the irrelevance of the theoretically transportable load, we can regard our turbidity current as dissipating simply through loss of height at the rate $V$, the terminal fall velocity of the particles present from the top or 'ceiling' of the slab downwards. For how long and over what distance will this dissipating current travel?

To answer this question we must set up an equation of motion for the slab-like current. Let the current be of length $L$ when the last of the dense fluid emerges from whatever was the source. As the current is losing thickness, it must be decelerating, whence for each unit width of the slab Newton's second law requires us to write

downslope component
of immersed weight
of slab

| | | | | |
|---|---|---|---|---|
| − | total frictional resistance of slab | − | mass of slab times acceleration | $= 0$  N m$^{-1}$ |

$$(12.22)$$

Referring to Figure 12.19 the first term of this equation is

$$(\varrho_2 - \varrho_1)ghL \sin \beta \quad \text{N m}^{-1} \qquad (12.23)$$

the second is

$$\tau L = \frac{(f_0 + f_i)}{8} \varrho_2 L U^2 \qquad (12.24)$$

where $\tau$ is the sum of the shear stresses at the bed and the fluid interface, and the third is

$$\varrho_2 hL \frac{dU}{dt} \qquad (12.25)$$

where $dU/dt$ is the acceleration of the current (negative in the present case) and $t$ is time.

Equation 12.22 can be solved numerically with some difficulty, but if the third term is small compared to the other two, we can obtain an approximate solution by treating the slowly decelerating slab as though its motion at all times was steady. Consider the decay of the turbidity current whose loading factor we calculated. In this case the roughly equal first and second terms of Equation 12.22 are an order of magnitude larger than the third, and we may therefore go for the approximate solution by treating the slowly decelerating slab as though its motion at all times was steady. Consider the decay of the turbidity current whose loading factor we calculated. In this case the roughly equal first and second terms of Equation 12.22 are an order of magnitude larger than the third, and we may therefore go for the approximate solution. Now the thickness $h(t)$ of the slab declines as

$$h(t) = h_0 - Vt \quad \text{m} \qquad (12.26)$$

where $h_0$ is the thickness at the start of dissipation and $V$ the terminal fall velocity of the sediment in the current, equal to the rate at which the top of the slab approaches the bed. Substituting this expression into Equation 12.4 for the uniform steady flow of gravity current on a slope, the current velocity $U(t)$ becomes

$$U(t) = \left(\frac{8 \sin \beta}{f_0 + f_i}\right)^{1/2} \left(\frac{\Delta\varrho}{\varrho_2} g\right)^{1/2} (h_0 - Vt)^{1/2} \quad \text{m s}^{-1} \qquad (12.27)$$

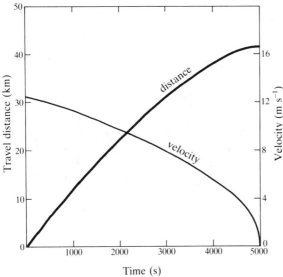

**Figure 12.20** Travel distance and velocity of a turbidity current treated using slab model (conditions stated in text).

As $U = dx/dt$, where $x$ is the distance travelled by, say, the front of the current, the time–distance function for the current is the integral of Equation 12.27

$$x = \left(\frac{8\sin\beta}{f_0 + f_i}\right)^{1/2} \left(\frac{\Delta\varrho}{\varrho_2} g\right)^{1/2} \int_0^{t_{\max}} (h_0 \big| - Vt)^{1/2} \, dt \quad \text{m}$$

$$(12.28)$$

where $t_{\max}$ is the total duration of the current, that is, the time when $Vt = h_0$. Equation 12.28 has the solution

$$x = \frac{2}{3V}\left(\frac{8\sin\beta}{f_0 + f_i}\right)^{1/2} \left(\frac{\Delta\varrho}{\varrho_2} g\right)^{1/2} [h_0^{3/2} - (h_0 - Vt)^{3/2}]$$

$$(12.29)$$

By setting $Vt = h_0$ in this equation we obtain the maximum distance of travel. This distance is inversely proportional to $V$ and increases as the power $3/2$ of the initial thickness. Increases in the excess density and bed slope also raise the travel distance, but their influence is less strong.

Figure 12.20 is a plot of Equations 12.27 and 12.29 for the turbidity current whose loading factor was calculated above and with $L = 5000$ m. The current flows for roughly 1.39 h and covers, according to the approximate solution, a distance of about 42 km. From it accumulates a sand bed with an average thickness of

1 m, assuming for the deposit a fractional volumetric grain concentration of 0.6. As the current path is 8.4 times longer than the current itself, this corresponds to an average instantaneous deposition rate of approximately $1.7 \times 10^{-3}$ m s$^{-1}$ (roughly 6 m h$^{-1}$).

It is readily admitted that the slab model just described is oversimplified. It nonetheless yields some important insights into the probable behaviour of naturally occurring turbidity currents, *nuées ardentes* and base surges. The model points to sediment overloading as an important if not overwhelming cause of dissipation, to the high speeds and long travel distances to be expected of such currents and, finally, to the high sediment deposition rates likely to prevail. Our conclusions need slight modification only in the case of *nuées ardentes*. These are formed of incandescent particles from many of which gases are exsolved during flow, thus counteracting the weight of the debris by partly fluidizing it.

## 12.11 Sloshing gravity currents

Thus far we have tacitly assumed that gravity currents are topographically unconstrained, with always sufficient space in which to advance without altering course. What happens to a gravity current that appears on the flanks of a depression, either a small ocean-floor basin or a valley on the side of a volcano, much narrower than the potential travel distance of the flow?

The general principle of conservation of energy (Sec. 1.5.3) offers an insight. Imagine that the topographic basin shown in Figure 12.21 is much smaller in width than the potential travel distance of the turbidity current which has arisen in its right-hand margin. The energy principle states that the current, on reaching the opposite flank, will flow up the slope until its kinetic energy is all exchanged for potential energy and frictional heat (energy loss). But the potential energy gained will then be re-exchanged for kinetic energy, and the current will flow back across the basin, the exchange being continued on alternate flanks until dissipation is completed. Such a current can be described as reflected or sloshing.

Assuming the fluids to be of uniform and constant density, we can write for a current ascending either flank in Figure 12.21

$$\varrho_2 g(\Delta\varrho/\varrho_2)\,\Delta y + E_{\text{loss}} = \tfrac{1}{2}\varrho_2 U_B^2 \quad \text{J m}^{-3} \quad (12.30a)$$

and for a descending current

$$\tfrac{1}{2}\varrho_2 U_B^2 + E_{\text{loss}} = \varrho_2 g(\Delta\varrho/\varrho_2)\,\Delta y \quad (12.30b)$$

where $\varrho_2$ is the current density, $U_B$ the mean body velocity, $\Delta\varrho$ the density difference between current and medium, and $\Delta y$ the vertical distance by which the centre of gravity of the current alters in elevation during the energy exchange. The three terms of Equation 12.30 represent energy per unit volume. The ratio $(\Delta\varrho/\varrho_2)$ appears in the terms for the potential energy in order to adjust $g$ on account of the flow in water rather than air (you are more 'weightless' when swimming than in air, although less frictional in air). The heat dissipated represents energy loss and is represented by $E_{loss}$. This loss is difficult to predict, but it might amount to as much as one-third of the driving energy. Thus the current whose loading factor and travel distance were calculated above would be expected from Equation 12.30a to rise at least 50 m up the side of a basin 12 km wide if it has been released on the opposite flank. A powerful current might therefore make several passes across a small basin before being completely dissipated (Fig. 12.21). Depending on basin shape, and the obliquity of the starting current, the flow would either repeatedly reverse over much the same track, or zig-zag along the basin.

Just to convince you about this energy exchange, Figure 12.22 shows a gravity current which had flowed from the left against the end of a short tank. The mass of fluid subsequently collapsed, to flow from the right and towards its starting point. Further details of sloshing currents can be found in Van Andel and Komar (1969) and Woodcock (1976).

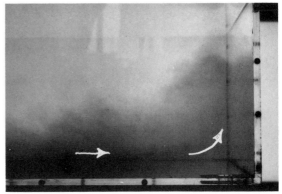

**Figure 12.22** Salty gravity current flowing from left and rising up vertical end wall of laboratory tank.

## 12.12 Turbidity-current deposits

Of the three kinds of sediment-driven gravity current discussed above – turbidity currents, *nuées ardentes* and base surges – our understanding of the relationship of the deposits to the physics of the flow is most complete in the case of turbidity currents. That understanding has come largely from studies of turbidity-current deposits preserved in the rock record.

Rock formations that accumulated through turbidity-current action are monotonous vertical repetitions of mainly decimetre-scale sandstone and mudstone beds with a lateral persistence measured in kilometres or tens of kilometres at the least (Fig. 12.23). A major advance in the understanding of this kind of deposit was made

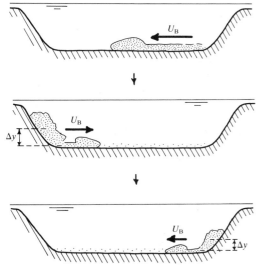

**Figure 12.21** Turbidity current sloshing in a restricted basin.

**Figure 12.23** Laterally uniform and extensive turbidite sandstones, Aberystwyth Grits (Silurian), Wales.

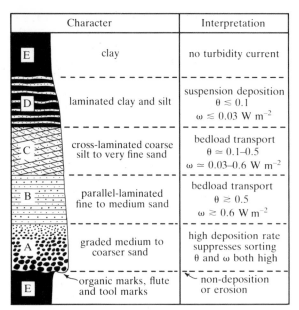

**Figure 12.24** Bouma sequence of grain sizes and sedimentary structures in a turbidite, and its hydraulic interpretation in terms of non-dimensional boundary shear stress ($\theta$) and stream power ($\omega$).

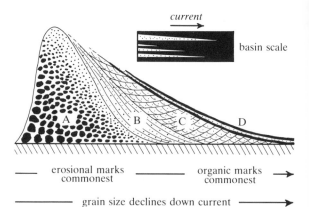

**Figure 12.25** Hypothetical variation of a turbidite sandstone along current path.

by Bouma (1962), who undertook a comparative study of numerous formations, finding that the sandstone beds were similar vertically in texture and sedimentary structures regardless of age and setting. The resulting major generalization, known as the Bouma sequence, appears in the left-hand column of Figure 12.24. Each sandstone turbidite has a sharp base overlying mudstone but a gradational top. The base in many instances preserves organic markings that existed on or within the mud prior to the appearance of the current, but in most cases bears proofs of erosion in the form of either tool or flute marks (Secs. 8.6 & 11.2.4). The sandstone above is divisible vertically into a number of zones using textural and structural features. Division A – the graded division – lacks internal lamination, and is either normally graded (coarse up to fine) or vertically uniform in texture (there may even be a slightly finer deposit at the base). Normally, the division consists of medium sand and coarser debris. Division B is typified by parallel laminations with parting lineations, consisting generally of medium grading up to fine sand. Division C is distinguished by current-ripple cross-lamination, in many beds involved in a later soft-sediment deformation. Very fine sand and clean silt typify this level. Division D is distinguished by the interlamination of clean silt and mud, but in certain beds the silt layers prove on close examination to contribute to the tops of cross-

laminated units, and so belong to division C. Division E consists only partly of mud introduced by the current. Overall, then, each sandstone bed is graded from coarse up to fine, and grading should not be thought of as restricted to division A.

Very few turbidity current sandstones in the rock record can be traced individually over large distances because of limitations of exposure, but general studies of turbidite formations point to the downcurrent changes in a single bed shown in Figure 12.25. Each bed is thickest close to its origin, thinning gradually down current. The Bouma divisions progressively disappear from the lowest upwards as the bed is traced down current. Division D seems to be restricted to the more distal reaches of the deposit. There is a tendency on the formation scale for the ratio of sand to mud to decrease down current, as the thinner and less extensive sandstone beds progressively drop out (Lovell 1970).

What would the bed formed by a sloshing current look like? It should consist of a sequence of closely superimposed parts, with abrupt current reversals and changes of grain size and thickness between the parts, as if the strip shown in Figure 12.25 had been folded back on itself, depending on how many passes the current made over the basin floor. Imagine a turbidity current which sloshed five times across a small basin, to give a deposit which varied in average grain diameter $D$ and thickness $H$ with total distance travelled as in Figure 12.26a. Consider a site A close to the flank on which the current appeared. The bed accumulated here will consist of five parts, the second and third of which are similarly thick, but much thinner than the lowest part, and the fourth and fifth of which are similarly thick, but thinner than the previous pair (Fig. 12.26b). The parts will differ texturally as in Figure 12.26c, where $\Delta D$ is the difference between the beds of the $n$th and $(n + 1)$th

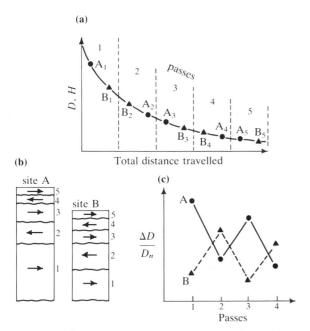

**(a)**

**(b)**

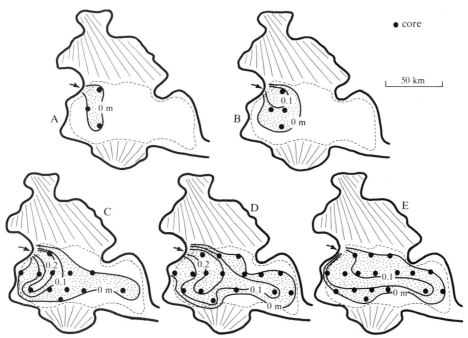

**Figure 12.26** Hypothetical character of deposit formed by a turbidity current sloshing in a restricted basin, at a station A near the margin on which the current was released and B near the opposite margin.

passes and $D_n$ the grain size at the $n$th pass. The patterns are different for the bed formed at a locality B, near the opposite flank of the basin (Figs. 12.26b & c). Mapped over a turbidite formation deposited in a small basin, these patterns would help to define basin size and shape.

These generalizations are remarkably consistent with our understanding of laboratory gravity currents and with our knowledge of turbidity-current deposits formed in the oceans within the last 10 000–20 000 years.

The erosional and/or sharp base of a turbidite bed is generally thought to represent the passage of the head of the current, and the overlying sandstone the deposit formed from the body of the flow. This interpretation is supported by Middleton's (1967) laboratory studies of gravity currents made with plastic beads. He observed that deposition began from the rear of the head. The deposits formed in these experiments were graded both vertically and in the downcurrent direction. The style of grading in Middleton's experiments varied with the concentration of the beads in the current.

The general vertical and lateral sequences of sedimentary structures inferred from ancient turbidites (Figs 12.24 & 12.25) are well supported by detailed studies of individual turbidites which have been traced widely over

**Figure 12.27** Thicknesses and distributions of Bouma divisions in a single turbidite correlated across abyssal plain in the Hispaniola–Caicos Basin, west-central Atlantic Ocean. After Bennetts and Pilkey (1976).

large areas in the modern oceans (Bornhold & Pilkey 1971, Bennetts & Pilkey 1976, Elmore *et al.* 1979, Van Tassell 1981). Figure 12.27 shows the areal distribution and thickness of the Bouma divisions in a turbidite that extends over the abyssal plain in the Hispaniola–Caicos Basin. This plain is comparatively small as abyssal plains go, but Bouma's division A is not noticeably more conspicuous in the huge Late Pleistocene turbidite described by Van Tassel (1981) from the Hatteras Abyssal Plain to the north. This turbidite, said to involve about 100 km³ of mud and sand, has an inter-rupted length of at least 500 km and a width of 100–140 km. Van Andel and Komar (1969) describe a turbidite formed by a sloshing current.

The right-hand column of Figure 12.24 suggests how the Bouma sequence preserved in these and older turbidity-current deposits can be interpreted physically. The interpretation rests heavily on our knowledge of the bedforms and internal sedimentary structures discussed in Chapters 4 and 6. This interpretation is amplified in Figure 12.28, which tentatively suggests in a represen-tative case: (1) the history of the leading edge of the depositional zone in the current (path connecting sta-tions 1, 2, 3 and 4), and (2) the evolution of the current itself as it passes each of the four fixed stations. Bouma's divisions B, C and D are readily understood in terms of bedforms in one-way flows, but the interpre-tation of division A remains difficult. There is no particular problem about the grading in this part of the turbidite, and indeed in the bed overall, in the light of our slab model when extended to a current carrying a range of particle sizes. What cannot be so readily ex-plained is the lack of lamination. As lamination reflects the capacity of a current to bring like particles together, we must suppose when division A was deposited that either turbulence-related sorting was suppressed by a high grain concentration near the bed, or the sediment

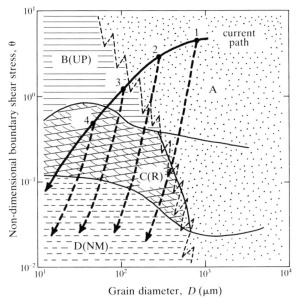

**Figure 12.28** Hypothetical evolution of a turbidity current at a sequence of four stations in its path, in terms of the grade and structure of its deposit. Path connecting the stations (continuous curve) illustrates the changing character of leading edge of depositional zone in current. Each broken line depicts evolution of the current as it passed a station. A, graded division (deposition rate too large for sediment sorting to occur); B(UP), division of parallel laminations (upper-stage plane beds); C(R), division of cross lamination (current ripples); D(NM), division of mud–silt laminae (no bedload movement).

deposition rate was so great that grain re-entrainment and consequent sorting were negligible. Reference again to the slab model suggests that the latter explanation is preferable. It remains to be discovered what is the critical rate for the suppression of lamination.

# Readings

Allen, J. R. L. 1971. Mixing at turbidity current heads, and its geological implications. *J. Sed. Petrol.* **41**, 97–113.

Bagnold, R. A. 1966. *An approach to the sediment transport problem from general physics.* Prof. Pap. US Geol. Surv., no. 422-I.

Benjamin, T. B. 1968. Gravity currents and related phenomena. *J. Fluid Mech.* **31**, 209–48.

Bennetts, K. R. W. and O. H. Pilkey 1976. Characteristics of three turbidites, Hispaniola–Caicos Basin. *Bull. Geol. Soc. Am.* **87**, 1291–300.

Bornhold, B. D. and O. H. Pilkey 1971. Bioclastic turbidite sedimentation in the Columbus Basin, Bahamas. *Bull. Geol. Soc. Am.* **82**, 1341–54.

Bouma, A. H. 1962. *Sedimentology of some flysch deposits.* Amsterdam: Elsevier.

Britter, R. E. and P. F. Linden 1980. The motion of the front of a gravity current travelling down an incline. *J. Fluid Mech.* **99**, 531–43.

Britter, R. E. and J. E. Simpson 1978. Experiments on the dynamics of a gravity current head. *J. Fluid Mech.* **88**, 223–40.

Britter, R. E. and J. E. Simpson 1981. A note on the structure of the head of an intrusive gravity current. *J. Fluid Mech.* **112**, 459–66.

Dzulynski, S. and E. K. Walton 1965. *Sedimentary features of flysch and graywackes.* Amsterdam: Elsevier.

Elmore, R. D., O. H. Pilkey, W. J. Cleary and H. A. Curran, 1979. Black Shell Turbidite, Hatteras Abyssal Plain, western Atlantic Ocean. *Bull. Geol. Soc. Am.* **90**, 1165–76.

Fisher, R. V. 1977. Erosion by volcanic base-surge density currents: U-shaped channels. *Bull. Geol. Soc. Am.* **88**, 1287–97.

Garratt, J. R. 1984. Cold fronts and dust storms during the Australian summer 1982–83. *Weather* **39**, 98–103.

Garvine, R. W. 1977. Observations of the motion field of the Connecticut River plume. *J. Geophys. Res.* **82**, 441–54.

Heezen, B. C., D. B. Ericson and M. Ewing, 1954. Further evidence for a turbidity current following the 1929 Grand Banks earthquake. *Deep-Sea Res.* **1**, 193–202.

Holyer, J. Y. and H. E. Huppert 1980. Gravity currents entering a two-layer fluid. *J. Fluid Mech.* **100**, 739–67.

Idso, S. B. 1976. Chubasio. *Weather* **31**, 224–6.

Irvine, T. N. 1980. Magmatic density currents and cumulus processes. *Am. J. Sci.* **280 A**, 1–58.

Komar, P. D. 1969. The channelized flow of turbidity currents with application to Monterey deep-sea fan channel. *J. Geophys. Res.* **74**, 4544–58.

Krause, D. C., W. C. White, D. J. W. Piper and B. C. Heezen. 1970. Turbidity currents and cable breaks in the western New Britain Trench. *Bull. Geol. Soc. Am.* **81**, 2153–60.

Lawson, T. J. 1971. Haboob structure at Khartoum. *Weather* **26**, 105–12.

Lovell, J. P. B. 1970. The palaeogeographical significance of lateral variations in the ratio of sandstone to shale and other features of the Aberystwyth Grits. *Geol. Mag.* **107**, 147–58.

Mellor, M. 1978. Dynamics of snow avalanches. In *Dynamics of snow avalanches*, B. Voight (ed.), 753–92. Amsterdam: Elsevier.

Middleton, G. V. 1966a. Experiments on density and turbidity currents. I. Motion of the head. *Can. J. Earth Sci.* **3**, 523–46.

Middleton, G. V. 1966b. Experiments on density and turbidity currents. II. Uniform flow of density currents. *Can. J. Earth Sci.* **3**, 627–37.

Middleton, G. V. 1967. Experiments on density and turbidity currents. III. Deposition of sediment. *Can. J. Earth Sic.* **4**, 475–504.

Moore, J. G. 1967. Base surges in recent volcanic eruptions. *Bull. Volc.* (2) **30**, 337–63.

Simpson, J. E. 1967. Aerial and radar observations of some sea-breeze fronts. *Weather* **22**, 306–27.

Simpson, J. E. 1969. A comparison between laboratory and atmospheric density currents. *Q. J. R. Meteorol. Soc.* **95**, 758–65.

Simpson, J. E. 1972. Effects of the lower boundary on the head of a gravity current. *J. Fluid Mech.* **53**, 759–68.

Simpson, J. E. 1982. Gravity currents in the laboratory atmosphere and ocean. *Annu. Rev. Fluid Mech.* **14**, 213–34.

Simpson, J. E. and R. E. Britter 1979. The dynamics of the head of a gravity current advancing over a horizontal surface. *J. Fluid Mech.* **94**, 477–95.

Simpson, J. E. and R. E. Britter, 1980. A laboratory model of an atmospheric mesofront. *Q. J. R. Meteorol. Soc.* **106**, 485–500.

Stith, J. L., P. V. Hobbs, and L. F. Radke 1977. Observations of a *nuée ardente* from the St Augustine Volcano. *Geophys. Res. Lett.* **4**, 259–62.

Van Andel, T. H. and P. D. Komar 1969. Ponded sediments of the Mid-Atlantic Ridge between 22° and 23° North latitude. *Bull. Geol. Soc. Am.* **80**, 1163–90.

Van Tassell, J. 1981. Silver Abyssal Plain carbonate turbidite: flow characteristics. *J. Geol.* **89**, 317–33.

Woodcock, N. H. 1976. Ludlow Series slumps and turbidites and the form of the Montgomery Trough, Powys, Wales. *Proc. Geol. Assoc.* **87**, 169–82.

# 13 To and fro

Wave motions − origin of wind waves − origin of tides − waves and wave currents in shallow and deep water − wave energy − mass transport currents − sediment transport due to waves and tides − rolling grain and vortex ripples − plane beds − sand waves − longshore bars and troughs − role of waves during storms − storm deposits.

## 13.1 Some introductory experiments

The preceding chapters may have conveyed the impression that we need refer only to unidirectional currents in order to give a physical account of sedimentary phenomena. We must emphasize that there is another important class of naturally occurring flows, characterized by rhythmical changes of speed combined with reversals of flow direction. These are the oscillatory currents associated with tides and with wind-generated surface waves. It is not immediately obvious that tides and wind waves belong in the same general class, but it turns out, as we shall see, that the tide can be analysed exactly as a type of wave. In addition to requiring some understanding of the origins of wind waves and tides, we shall in particular want to know what the currents due to them are like, what factors control these currents and the sediment transport they promote, and what bed-forms are adjusted to tidal and wave regimes. Some insight into these questions can be gained from simple experiments, from which it will become apparent that oscillatory flows can be much more complicated than unidirectional ones. But perhaps that is just part of their fascination.

The help of a companion and a domestic bath are necessary for these experiments (Fig. 13.1). A wave maker is also needed, so shape a piece of plywood or hardboard about 0.3 m wide so that when upright it fits fairly snugly across the deep end of the bath. Slit a piece of rubber tubing lengthways and slip this over the edge of the wood fitting the bottom of the bath. Now fill the bath with cold water to a depth of about 0.2 m and let the turbulence dampen out. Get your helper to intro-

duce the wave maker across the deep end of the bath, with the rubber-covered edge pressed firmly against the bottom. Your helper should then steadily flap the wave maker from side to side, beginning at a frequency of 1−2 flaps per second at an amplitude of a centimetre or less.

You will see that the to-and-fro motion of the wave maker creates undulations on the water surface that are (1) regularly spaced in a train and (2) moving away (progressive) at a steady speed, until they reach the opposite end of the bath where they break and are partly reflected. The spacing of the undulations is called the wavelength $L$. Their speed is described as the celerity or phase velocity $c$. By getting your companion to vary the rate but not the amplitude of flapping, you will produce waves of a range of periods $T$ and will confirm that $T = L/c$. Notice that by increasing the amplitude of flapping, the waves can be increased in height $H$, that

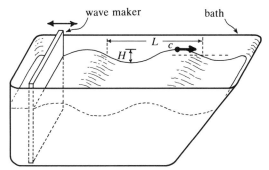

**Figure 13.1** Apparatus for making waves, and their properties.

is, the vertical distance between trough and crest. The wavelength and phase velocity, however, are little affected.

Do these travelling undulations generate currents? You will now need (1) some dry coloured paper torn into small shreds, and (2) a small quantity of well boiled wet tea leaves. The paper shreds sprinkled onto the waves float on the surface. As the waves pass by the pieces execute in the horizontal plane a short to-and-fro motion perpendicular to the wave crests. But as the fragments simultaneously move up and down, in response to the passage alternately of wave crests and troughs, the combined motion, and hence that of surface water particles, must be a virtually closed orbit in a vertical plane. Notice that the orbit is small in comparison with the spacing of the waves and also that the speed of the shreds is much less than the phase velocity. Notice also that the markers change only slowly in position. On increasing the wave height, you will find that the orbit of the markers increases in size. The scale of the orbit is in fact comparable with the wave height.

Does this orbiting current penetrate below the surface? Introduce some of the tea leaves, which will sink to the bottom. On starting up the wave maker, both the paper markers and the tea leaves will be found to exhibit a to-and-fro motion on the same period. But look for two differences between these motions. The tea leaves at the bottom will traverse a rectilinear path, rather than a curved orbit, and this path may be shorter than that of the paper on the surface. Presumably the orbits of water particles at an intermediate depth take an intermediate scale and form. Hence the waves can act at a distance by setting up currents within the water body over which they travel.

The possibilities of the experiments do not end here. Scatter some crystals of potassium permanganate over the floor of the bath and set the wave maker gently in motion. The flecks of colour shed from each crystal denote the same oscillatory bottom current as was indicated by the tea leaves. But you will also see that the whole patch of coloured water is at the same time slowly drifting away from the site of the crystal. The bottom drift may be towards either the shallow or the deep end of the bath, depending on the position of the crystal. Hence wave-generated currents comprise at least two components: (1) an unsteady periodic part, and (2) a much slower but steady one-way streaming called a mass-transport current. The latter could be of critical importance in determining the rate and direction of net sediment transport in wave and tidal systems.

Our experimental progressive waves were produced rather artificially. They nonetheless afford a useful insight into the character of the waves generated by the wind and by the gravitational effects of heavenly bodies.

## 13.2 Making wind waves

Figure 13.2 shows waves at sea. Their period is typically between 1 and 10 s. The causal relationship between the wind and such waves will have been noticed by the most casual observer. Waves on a pond advance with the wind and have crests perpendicular to the wind. Visits to the coast under a range of weather conditions will soon convince one that the tallest if not always the longest waves tend to occur on the windiest days.

How does the wind generate waves? Evidently mechanisms of energy transfer are involved, but exactly how they work is not wholly clear (Phillips 1977). There is an interplay between the turbulent wind blowing over the irregular moving surface of the water and the waves themselves, which modify the wind and, because they are moving at various rates, also each other. Figure 13.2 makes it clear that naturally generated waves comprise a spectrum of heights and wavelengths.

Energy reaches the waves through the action of the forces associated with the wind. As the air is viscous, the wind exerts a time-averaged shear stress on the water surface. This stress must vary with position on the waves due to the wave shape. Because the wind flow is compressed above the wave crests, but expanded above the troughs, Bernoulli's equation (Eqn. 1.12) tells us that a pressure difference must exist between trough and crest. The greater pressure in the trough as compared to the crest will tend to augment the wave height. But the

**Figure 13.2** Fairly regular waves at sea, period about 3 s. At least two distinct sets of smaller waves are superimposed on them.

wind is turbulent. From our discussion in Chapter 6, it can be thought of as composed of turbulence elements of a range of sizes. These parcels of air have a certain life and independence of behaviour, and are convected with the wind at a range of speeds somewhat less than the mean wind speed. They can be described in terms of a spectrum of sizes and a spectrum of frequencies. As these parcels pass over the water surface, they exert a corresponding fluctuating pressure upon it, now pulling the surface upwards and now pressing it down. Consequently, a spectrum of water surface undulations are formed which travel with the wind, and which may be further amplified as further turbulence elements of an appropriate size and persistence act on them. But as the surface undulations present a range of sizes, they must as we found experimentally be travelling at different rates, and so may interfere with each other. In particular, we might expect some of the lengthier waves to modify the shorter ones and extract energy from them.

Some limit to the wave growth is to be expected, if only because gravity is tending all the time to flatten the water surface. A loss of energy is represented by any waves that become sufficiently steep as to break. Finally, amongst undulations fashioned by the wind, there must be some whose phase and frequency are so related that in time they cancel out.

Observations made at sea during times of fairly constant wind show that wave period and length are augmented by increasing distance (fetch) and time of travel (Barnett & Wilkerson 1967). Wave height also tends to increase. This particular property of waves is of considerable interest to mariners, marine engineers and sedimentologists, because the energy of waves is a steeply increasing function of their height. Consequently, several empirical schemes have been proposed for predicting wave characteristics from wind speed, duration and fetch (Darbyshire & Draper 1963, Carter 1982). Table 13.1 shows Frost's (1966) simple scheme for equilibrium wave heights in open waters. Slightly larger heights are likely to apply nearer the coast, because of the effects of reduced water depth.

## 13.3 Making the tide

If you frequent sea coasts at all, you will be familiar not only with waves but also with tides (Doodson & Warburg 1941). The tide − what has been called the heartbeat of the ocean − is the rhythmical rise and fall of the surface of the sea caused by the changing distance of the Earth from the Moon and Sun. These vertical

**Table 13.1** Wave height related to wind conditions (adapted from Frost 1966).

| Wind speed at 10 m height (m s$^{-1}$) | Beaufort wind force | Description | Expected wave height in open sea (m) |
|---|---|---|---|
| 0 | 0 | calm | 0 |
| 1.5 | 1 | light air | 0.1 |
| 3.3 | 2 | light breeze | 0.2 |
| 5.3 | 3 | gentle breeze | 0.6 |
| 7.5 | 4 | moderate breeze | 1.0 |
| 9.6 | 5 | fresh breeze | 2.0 |
| 11.9 | 6 | strong breeze | 3.0 |
| 14.3 | 7 | near gale | 4.0 |
| 16.7 | 8 | gale | 5.5 |
| 19.1 | 9 | strong gale | 7.0 |
| 21.7 | 10 | storm | 9.0 |
| 24.1 | 11 | violent storm | 11.5 |
| 26.8 | 12 | hurricane | 14.0 |

movements create horizontal water currents, the tidal streams.

The vital thing to remember about tides and tidal currents is that they embody unsteadiness on several timescales. Consider the predicted tide in March 1984 for Immingham on the east coast of England (Fig. 13.3). The most obvious feature of the graph is that there are two high waters and two low waters every 25 h approximately (24.84 h exactly). At times, however, successive high waters and successive low waters are markedly unequal, a feature called the diurnal inequality. The next thing to notice is that the high-water and

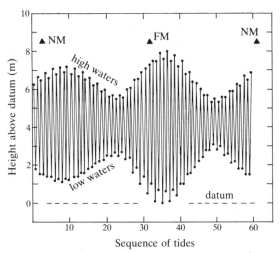

**Figure 13.3** Predicted tidal heights at Immingham, east coast of England, throughout March 1984. NM, new moon; FM, full moon.

low-water levels are gradually changing, as also is the tidal range, that is, the maximum difference between a high water and the following low water. The pattern, called the spring–neap cycle, recurs every 14.8 days. The spring tides are those marked by a large rise and tidal range. Neap tides achieve only a shallow range and rise. Looking at the whole of 1984, we find that the spring–neap cycle itself displays repeating changes (Fig. 13.4). The greatest spring tide reaches a maximum height at the equinoxes, while also displaying a monthly inequality (period 29.5 days). This inequality shows an interesting difference between the two parts of the year. The higher of the greatest spring tides in the month accompanies the full moon at about the time of the vernal equinox, but the new moon six months later. The inequality scarcely appears at the solstices. Neap high waters display the opposite pattern, with respect to both the monthly inequality and the longer-term changes.

Why should the sea rise and fall in this complicated rhythmical manner? The forces involved must relate to the gravitational attractions exerted by the Moon and Sun on the ocean, since the tides are observed to vary in strict accord with the Moon's phases and the movement of the Earth.

In Newton's gravitational theory, the attraction between particles of masses $m$ and $M$ with centres $R$ apart varies as $mM/R^2$, that is, as the product of the masses and inversely as the square of the separation. Suppose that $M$ is the Moon's mass and $R$ its distance from the centre of the Earth, whose radius $r$ is negligibly small compared to $R$. The average force on water particles of mass $m$ on the Earth would therefore be $F = mMG/R^2$ N, where $G$ (N m² kg$^{-2}$) is the universal gravitational constant. Now should all particles experience this force, no differential movement and therefore no tide could arise. Hence the tide-raising force is the difference between attractions as experienced at respectively the centre and the surface of the Earth, namely

$$\text{tide-raising force} = -r\frac{\mathrm{d}F}{\mathrm{d}R} = \frac{2rmMg}{R^3} \quad \text{N} \qquad (13.1)$$

The Earth–Sun distance varies on an annual cycle, being smallest at the winter solstice (perihelion) and greatest at the summer solstice (aphelion). The variation is monthly in the case of the Moon, from least at perigee to largest at apogee. Putting the appropriate distances and masses into Equation 13.1, and comparing each answer to the Moon at perigee, we obtain the relative forces shown in Table 13.2. The Moon clearly influences tides more than the Sun, the relative proximity of the Moon more than offsetting its small mass.

It is sufficiently close to the truth for several aspects of the tide to be explained on the assumption that the orbits of the Earth and the Moon lie in the same plane (Fig. 13.5a). When the Moon is aligned with the Earth and Sun, at full Moon (opposition) and new Moon (conjunction), the resultant tide-raising force is the sum of the Moon's and Sun's contributions. It is therefore a maximum, spring tides resulting (Fig. 13.3). The resultant tide-raising force is the difference between the two

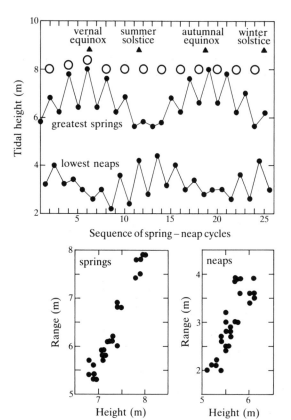

**Figure 13.4** Predicted greatest and least high waters (relative to tidal datum) in a spring–neap cycle at Immingham throughout 1984, and the associated tidal ranges. Full moon represented by open circles.

**Table 13.2** Relative magnitude of tidal forces (Moon at perigee = 1).

| Force due to Moon | |
|---|---|
| perigee | 1.00 |
| apogee | 0.67 |
| | |
| Force due to Sun | |
| perihelion | 0.385 |
| aphelion | 0.381 |

contributions during the quarters when the Moon lies perpendicular to the axis of the Earth and Sun (quadratures). The force is then a minimum, neap tides being seen. The Moon orbits the Earth every 27.4 days, but its phases reappear every 29.5 days, since the Earth is continuing its journey meanwhile. Spring and neap tides therefore recur twice every lunar month. The monthly inequalities, and the cycles of greatest springs and least neaps (Fig. 13.4), depend on the way in which the long axis of the Moon's elliptical orbit changes relative to the axis of the Sun and Earth (Fig. 13.5a). At the equinoxes, full moon and new moon occur in either perigee or apogee. The difference between successive spring tides is then at its maximum, in contrast to the equality of the neaps. At the solstices, however, when the Moon is in apogee and perigee at the quarters, it is the neaps which differ most.

The reason why broadly two high waters and two low waters occur daily (Fig. 13.3) is because the Earth is spinning on its axis, at a rate relative to the Moon of once every 24.84 h. As the resultant tide-raising force causes the ocean to bulge on both sides of the Earth (Fig. 13.5b), an observer on the Earth notices the passage of two bulges and two lows in each period of this duration.

The final puzzle is the diurnal inequality (Fig. 13.3).

We assumed above that the orbits of the Moon and Earth lay in the same plane. In fact, the Moon's orbit is set at about 5° to that of the Earth. Moreover, the Earth's axis of rotation is set at 23.5° to the plane of its orbit, giving the Moon's movement in latitude, or declination, a possible range of 57°. A suitably placed observer on the Earth would therefore experience alternately high and low high-water levels, as the Earth's rotation carried him from near the pole to near the equator of the tidal bulge (Fig. 13.5c).

Finally, there are numerous places on the Earth where the tidal regime is quite unlike that at Immingham, which reveals a typical semi-diurnal tide (Figs 13.3 & 13.4). The greatest differences are shown by the diurnal tides, marked by only one high water and one low water every 24.84 h, and by spring tides at the Moon's quarters. Defant (1958) further discusses these tides and also the well represented regimes intermediate between the semi-diurnal and the diurnal types.

## 13.4 Waves in shallow water

To an observer on the rotating Earth, the tidal bulge appears as a progressive wave with a period of 12.42 h and wavelengths measuring up to thousands of

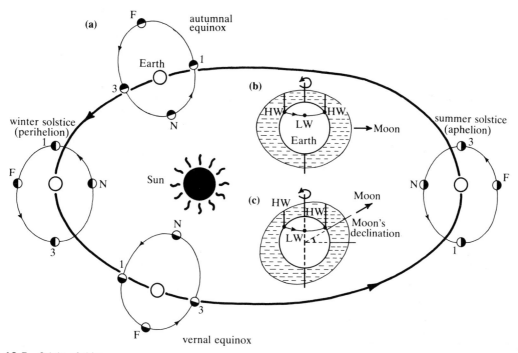

**Figure 13.5** Origin of tides.

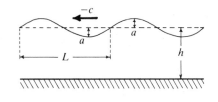

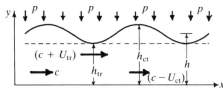

**Figure 13.6** Definition sketch for waves in shallow water.

kilometres. Yet the average depth of the oceans is only about 4 km. The wavelength of the tidal wave is consequently very large compared to the depth of water below, and the tide is therefore a shallow-water wave (long wave). What is its speed and what currents accompany it?

Consider uniform long waves of wavelength $L$ and amplitude $a = H/2$ travelling from right to left at phase velocity $-c$ over water of density $\varrho$ and uniform undisturbed depth $h$ (Fig. 13.6). To answer the questions posed, we must arrest the waves and make the motion steady. This we accomplish by means of our earlier device of superimposing on them a current of mean velocity $c$ from the opposite direction. The depth $(h + a)$ at a wave crest exceeds the mean depth. But the current velocity there must be less than the mean, say $(c - U_{ct})$, where $U_{ct}$ is the deficit. The discharge per unit width of flow is therefore $(h + a)(c - U_{ct})$ m$^3$s$^{-1}$m$^{-1}$. The opposite prevails at a trough, the discharge being $(h - a)(c + U_{tr})$, where $U_{tr}$ is the excess velocity. Continuity demands equal discharges beneath trough and crest, whence assuming $U_{ct}$ and $U_{tr}$ to be of equal magnitude $U$,

$$(h + a)(c - U) = (h - a)(c + U) \qquad (13.2)$$

which on multiplying out and rearranging becomes

$$\frac{c}{U} = \frac{h}{a} \qquad (13.3)$$

As the motion is steady, Bernoulli's equation (Eqn. 1.12) applies to the crest and the trough, both lying on the same surface streamline, subject to a uniform atmospheric pressure $p$. The constant being the same in each statement of the equation, and with the bottom as datum,

$$p + \tfrac{1}{2}\varrho(c - U)^2 + \varrho(h + a)g =$$
$$p + \tfrac{1}{2}\varrho(c + U)^2 + \varrho(h - a)g \quad \mathrm{N\,m^{-2}} \qquad (13.4)$$

where $g$ is the acceleration due to gravity, the left-hand terms represent the crest, and the right-hand ones the trough. Multiplying out, and as $p$ is also a constant,

$$cU = ga \quad \mathrm{m^2\,s^{-2}} \qquad (13.5)$$

Substituting for $a$ from Equation 13.3

$$c = (gh)^{1/2} \quad \mathrm{m\,s^{-1}} \qquad (13.6)$$

or alternatively for $c$

$$U = a(g/h)^{1/2} \qquad (13.7)$$

Equation 13.6 states that the phase velocity of a progressive wave in shallow water depends only on the depth, increasing as the square-root of the depth. The maximum horizontal component of the associated current, given by Equation 13.7, varies directly as the amplitude and inversely as the square root of the depth. On removing the imposed velocity, to regain a stationary frame of reference, we see that this component is directed with the waves at a crest and opposite in a trough.

The above analysis was made primarily to derive an expression for the phase velocity of a wave in shallow water (long wave), but in so doing we also derived Equation 13.7 for the mean velocity of subsurface water particles beneath the crest and trough. Instead of trough and crest, however, we could have chosen any two arbitrary corresponding points on the wave profile. We should then have discovered − try the analysis for yourself − that the mean water velocity beneath the wave varied as the local depth. Hence if we could specify the wave profile, we should also know how this velocity changed in space and time. A simple harmonic profile is plausible, in which case the velocity varies sinusoidally in both space and time, for example $U(t) = U_{ct} \sin(2\pi t/T)$, where $U$ is the value at time $t$ and $U_{ct}$ as before is the velocity at the crest (i.e. the amplitude of the sinusoidal velocity profile).

Equations 13.6 and 13.7 are only approximations to the truth, obtained because we assumed that (1) all points on the wave have the same phase velocity, and (2) the particle motion is symmetrical. A better approximation results from the supposition that each point on the wave has a phase velocity proportional to the depth below that point. This wave distorts as it travels, the crest advancing at a slightly greater phase velocity than the trough. Distortion in shallow water has important implications for the shape of the tidal curve and pattern of tidal velocities, and similar implications for shoaling wind waves. Finally, what is the quantitative meaning of Equation 13.7, which should describe tidal streams? Values representative of the North Sea are $a = 2.0$ m and $h = 30$ m, giving peak tidal currents of approximately $1.1$ m s$^{-1}$ (about 2 knots). Gravel, as well as sand, could be moved by such currents.

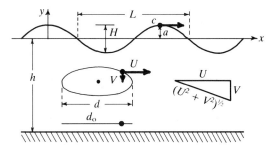

**Figure 13.7** Properties of waves and the associated fluid motion.

## 13.5  Waves in deep water

What is the phase velocity of a progressive wave on water so deep that the wavelength is small compared to the depth (a deep-water wave or short wave)? Using the previous device and assumptions (Fig. 13.6), but taking the level of the wave trough as datum, Bernoulli's equation for crest and trough affords

$$p + \tfrac{1}{2}\varrho(c - U)^2 + 2\varrho ag = p + \tfrac{1}{2}\varrho(c + U)^2 \quad \text{N m}^{-2}$$

(13.8)

As $a = H/2$, we obtain

$$H = \frac{2cU}{g} \quad \text{m}$$

(13.9)

One assumption was that the motion of a surface water particle is symmetrical, the horizontal velocity being a maximum of $U$ at a crest and $-U$ at a trough. Hence we may imagine the particle as travelling at a fixed speed $U$ round a vertical circular orbit of diameter $H$. As one whole orbit of length $2\pi(H/2)$ is completed per wave period $T$

$$U = \frac{\pi H}{T} \quad \text{m s}^{-1}$$

(13.10)

Now $T = L/c$ by definition. Substituting from this and from Equation 13.10 into Equation 13.9, we find

$$c = \left(\frac{gL}{2\pi}\right)^{1/2} \quad \text{m s}^{-1}$$

(13.11)

The phase velocity of a deep-water wave is therefore independent of depth and proportional only to the wavelength. Evaluating constants, a useful rule of thumb is that the phase velocity equals roughly 1.3 times the square root of the wavelength.

## 13.6  Wave equations

### 13.6.1  General equations

The above formulae for long and short waves are about as far as one can go with the mathematical description of waves without using advanced methods. By applying such techniques (Wiegel 1964, Kinsman 1965), however, a much more detailed understanding of waves becomes possible.

In the simplest of these advanced models (the Airy model), the wave form is described by the simple harmonic function

$$y = \frac{H}{2}\sin\left(\frac{2\pi x}{L} - \frac{2\pi t}{T}\right) \quad \text{m}$$

(13.12)

where the symbols appear in Figure 13.7. Notice particularly that $y$ is measured positive upwards from the undisturbed water surface, so depths below the surface are negative. Holding the distance $x$ in Equation 13.12 constant, we see how the water surface elevation at a fixed station changes with time $t$. The wave profile results from holding $t$ constant.

The phase velocity of waves on water of any depth $h$ is given by the wave dispersion equation

$$c = \frac{gT}{2}\tanh\left(\frac{2\pi h}{L}\right) \quad \text{m s}^{-1}$$

(13.13)

where $\tanh(2\pi h/L)$ is one of the hyperbolic functions (Fig. 13.8), as extensively tabulated by Wiegel (1964). For deep water $\tanh(2\pi h/L)$ tends to 1 and Equation 13.13 simplifies to

$$c = \left(\frac{gL}{2\pi}\right)^{1/2}$$

(13.14)

derived earlier as Equation 13.11. In shallow water

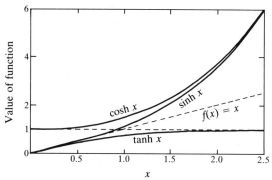

**Figure 13.8** Properties of hyperbolic functions relevant to wave theory.

$\tanh(2\pi h/L)$ tends to $2\pi h/L$, whence Equation 13.13 reduces to

$$c = (gh)^{1/2} \qquad (13.15)$$

making use of the definition $T = L/c$. This relationship we derived as Equation 13.6.

The wavelength of waves on water of any depth is given by the expression

$$L = \frac{gT^2}{2\pi} \tanh\left(\frac{2\pi h}{L}\right) \quad \text{m} \qquad (13.16)$$

which for deep-water conditions simplifies to

$$L = \frac{gT^2}{2\pi} \qquad (13.17)$$

and in shallow water reduces to

$$L = (ghT^2)^{1/2} \qquad (13.18)$$

arguing as for the celerity. It is useful to state here a little more exactly what we mean by 'deep water' and 'shallow water' with respect to the waves. The former condition is reached for $h/L > 0.5$ and the latter when $h/L < 0.04$. Between these limits there is nothing for it but to use the full formulae, Equations 13.13 and 13.16.

The equations representing the motion of water particles beneath waves are more complex and introduce the other chief hyperbolic functions. Referring to Figure 13.7, the horizontal component of the particle velocity in water of any depth is given by

$$U = \frac{\pi H}{T} \frac{\cosh\left[2\pi(y+h)/L\right]}{\sinh(2\pi h/L)} \cos\left[2\pi\left(\frac{x}{L} - \frac{t}{T}\right)\right] \quad \text{m s}^{-1} \qquad (13.19)$$

and the vertical component by

$$V = \frac{\pi H}{T} \frac{\cosh\left[2\pi(y+h)/L\right]}{\sinh(2\pi h/L)} \sin\left[2\pi\left(\frac{x}{L} - \frac{t}{T}\right)\right] \quad \text{m s}^{-1} \qquad (13.20)$$

The velocity of a surface particle varies directly as the wave height but inversely as the period. Reference to Figure 13.8 will show that, speaking generally, the particle velocity declines with increasing distance below the surface. The sine and cosine terms tell us that a particular velocity is repeated once every wavelength and period. Under deep-water conditions, Equations 13.19 and 13.20 define a circular orbit of diameter $d$ decreasing exponentially downwards as

$$d = H \exp\left[-2\pi(y+h)/L\right] \quad \text{m} \qquad (13.21)$$

where 'exp' stands for the base of natural logarithms. At $y = -L/2$, for example, $d$ is only 4% of $H$, and this is commonly taken as the practical limit of wave effectiveness (wave base).

Sedimentologically, the most useful forms of Equations 13.19 and 13.20 are those for the bottom. Here $V = 0$ but the greatest horizontal component of the particle velocity $U_{max}$ is

$$U_{max} = \frac{\pi H}{T \sinh(2\pi h/L)} \quad \text{m s}^{-1} \qquad (13.22)$$

in water of any depth, and

$$U_{max} = \frac{1}{2} H(g/h)^{1/2} \qquad (13.23)$$

in shallow water. The last formula we derived earlier as Equation 13.7. The corresponding equations for the orbital diameter $d_0$ at the bottom are

$$d_0 = \frac{H}{\sinh(2\pi h/L)} \quad \text{m} \qquad (13.24)$$

in water of any depth, and

$$d_0 = \frac{H}{2\pi h/L} \qquad (13.25)$$

for shallow-water conditions. These simplifications follow from the character of the hyperbolic functions (Fig. 13.8).

Figure 13.9 summarizes the general implications of these equations. Notice that, as implied by the formulae, water particle orbits are closed. The predicted

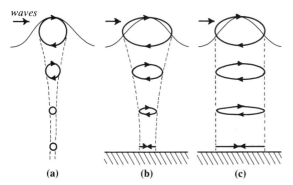

**Figure 13.9** Water particle orbits associated with (a) deepwater, (b) intermediate, and (c) shallow-water waves.

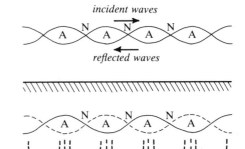

**Figure 13.10** Standing waves and the internal motion due to them. A, antinode; N, node.

orbital velocities and shapes agree well with laboratory waves (Wallet & Ruellan 1950, Morison & Crooke 1953) and moderately well with observed ocean waves (Thornton & Krapohl 1974). Referring to Table 13.1, Equation 13.22 implies that storm waves could stir sand as deep as the outer continental shelf.

### 13.6.2 Limiting steepness of progressive waves
Waves break if they grow too steep, the practical limit being defined by

$$\frac{H}{L} = 0.142 \tanh\left(\frac{2\pi h}{L}\right) \qquad (13.26)$$

where the usual deep-water and shallow-water simplifications can be made.

### 13.6.3 Standing waves
Progressive waves encountering an obstacle are partly or wholly reflected back on themselves, to form a system of standing waves. The tide forms a standing wave in restricted seas and estuaries of an appropriate size and shape. Partial reflection of progressive waves occurs not just at steep cliffs but also from beaches and submerged shoals, and has important sedimentological consequences.

The equation for a standing wave is formed by summing the equations for the constituent progressive waves. Consider the standing wave made by two equal progressive waves travelling in opposite directions. Those advancing in the positive x-direction comply with

$$y = \frac{H}{2} \cos\left(\frac{2\pi x}{L} - \frac{2\pi t}{T}\right) \quad \text{m} \qquad (13.27)$$

while those in the negative x-direction obey

$$y = \frac{H}{2} \cos\left(\frac{2\pi x}{L} + \frac{2\pi t}{T}\right) \qquad (13.28)$$

Using the trigonometric rule for the expansion of $\cos(A \pm B)$, we find that

$$y = H \cos(2\pi x/L) \cos(2\pi t/T) \qquad (13.29)$$

A standing wave therefore is stationary and has twice the amplitude of its constituent progressive waves. The wave appears repeatedly to turn itself inside out, the surface oscillating within the envelope shown in Figure 13.10. Antinodes (A) occur at wave crests and troughs and nodes (N) lie between. The direction of motion of a water particle reverses once each wave period. At antinodes the horizontal velocity component is zero but the vertical component is a maximum (Fig. 13.10). The maximum horizontal current is observed at nodes. Notice how the current is reversed every $L/2$ along the bottom.

### 13.6.4 Wave energy and power
How much energy do progressive waves possess? Evidently they have both kinetic and potential kinds. The latter arises from the fact that water, so to speak, has been emptied from the troughs and piled under the crests. The kinetic energy depends on the motion of the subsurface water particles. Hence the total energy must be the sum of the potential and kinetic energies, as evaluated over the length of the wave and the full water depth. It is conveniently expressed per unit length of wave crest.

Take the kinetic energy. Referring to Figure 13.7, the speed of an orbiting water particle equals $(U^2 + V^2)^{1/2}$, by Pythagoras' theorem. Putting $\varrho$ as the water density,

251

the kinetic energy in a unit volume is therefore $\frac{1}{2}\varrho(U^2 + V^2)$ J m$^{-3}$, whence the energy per unit length of wave crest is the total of this quantity contained in a unit slice bounded by the surface and bottom across a wave, i.e.

$$\frac{1}{2}\varrho \int_0^L \int_0^{-h} (U^2 + V^2)\, \mathrm{d}x\, \mathrm{d}y \qquad (13.30)$$

which, introducing Equations 13.19 and 13.20, turns out after some mathematics to be

$$\frac{1}{4}\varrho ga^2 L \quad \text{J m}^{-1} \qquad (13.31)$$

where $a = H/2$ is the wave amplitude. The energy is therefore proportional to the wave height squared.

To derive the potential energy, consider in Figure 13.11 a vertical fluid column of unit basal area extending upwards from the undisturbed water level. Its weight $\varrho gy_s$, where $y_s$ is the elevation of the surface, can be thought of as having been displaced upwards by its own height $y_s$ from the corresponding position in the wave trough. Hence the fluid column has gained potential energy equal to $\varrho gy_s^2$ J m$^{-2}$. The total potential energy of the wave is this quantity integrated over a unit slice through the wave crest, where $y_s$ is some function of $x$. For a sinusoidal wave

$$y_s = a \sin(2\pi x/L) \quad \text{m} \qquad (13.32)$$

whence the potential energy is

$$\varrho g \int_0^{L/2} [a \sin(2\pi x/L)]^2\, \mathrm{d}x \qquad (13.33)$$

which when evaluated becomes

$$\frac{1}{4}\varrho ga^2 L \quad \text{J m}^{-1} \qquad (13.34)$$

the same as for the kinetic energy (13.31). The grand total energy per unit length of crest is therefore $\frac{1}{2}\varrho ga^2 L$ J m$^{-1}$.

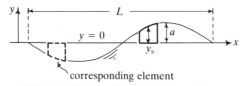

**Figure 13.11** Definition sketch for potential energy of a wave.

You can now see why sedimentologists and engineers are so keen on relating wave height to weather conditions, for wave energy is available for sediment erosion and transport. If we divide the grand total energy per unit length of crest by the wavelength, we obtain the areal density of the energy $\frac{1}{2}\varrho ga^2$ J m$^{-2}$. Multiplying this by the velocity at which the energy is being fed forwards with the waves, we further obtain the wave power per unit width of crest, corresponding to the power of a river. The appropriate velocity is the group velocity ranging from $c/2$ in deep water to $c$ for shallow conditions (Wiegel 1964).

## 13.7 Mass transport in progressive and standing waves

What caused the steady drift or mass-transport current noticed in the bath experiment? To answer this we must face the fact that the Airy wave-model gets round certain difficulties in the physical problem by means of simplifying assumptions. These assumptions are (1) the wave amplitude is negligibly small compared to other lengths, and (2) the fluid is inviscid. They are too severe for every aspect of wave behaviour to be revealed.

On relaxing the first assumption, we find that water particle orbits beneath progressive waves of finite height are very slightly open. The consequence, in an infinitely extensive and deep ocean, is at all depths a steady drift of water in the direction of the waves but at a rate small compared to local orbital velocities (Fig. 13.12a).

The introduction of viscosity means that boundary layers appear at the bottom and free surface. The bottom boundary layer is particularly important in sediment-transport problems. Its thickness can be shown to be $\delta = (\eta T/\pi\varrho)^{1/2}$, where $\eta$ is the fluid viscosity, and $\delta$ is just a few millimetres in value for large wind waves. The boundary-layer flow is oscillatory and

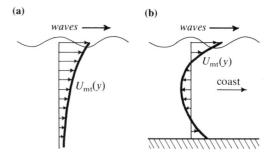

**Figure 13.12** Mass-transport current associated with waves in (a) an infinite water body, and (b) a water body restricted by a shoreline.

becomes turbulent at a sufficiently large Reynolds number. A vital consequence of the oscillatory motion, irrespective of the presence of turbulence, is that the instantaneous velocity has components both normal and parallel to the bed. Reynolds stresses (Sec. 6.2) are therefore created, and those associated with the normal fluctuations drive within the boundary layer a slow but steady drift in the direction of the waves (Longuet-Higgins 1958).

Hence the total mass-transport current is formed from both finite-amplitude and viscous effects. At the outer edge of the bottom boundary layer, the mass-transport velocity $U_{mt}$ is

$$U_{mt} = \frac{5}{16} \frac{(2\pi H)^2}{LT \sinh^2(2\pi h/L)} \quad \text{m s}^{-1} \quad (13.35)$$

for progressive waves, and

$$U_{mt} = -\frac{3}{16} \frac{(2\pi H)^2 \sin(4\pi x/L)}{LT \sinh^2(2\pi h/L)} \quad (13.36)$$

for standing waves. Like wave energy, the mass transport increases steeply with wave height. Figure 13.12b shows the sort of mass-transport velocity profile associated with waves approaching a coast. Note the onshore near-bed movement.

The sine term in Equation 13.36 tells us that the drift beneath standing waves reverses every $L/4$ along the bottom. The expected flow pattern – it could appear off a reflective beach – comprises two sets of stationary cells in which the fluid recirculates (Fig. 13.13). In the lower cells, involving the bottom boundary layer, the flow at the bed is towards the positions of nodes. The bottom current in the upper cells flows towards the positions of antinodes.

Laboratory work supports these predictions (Collins 1963), which are easily tested for oneself in the standing-wave case (Fig. 13.13). Fill with water to about half its

**Figure 13.14** Glass tank mounted on rollers in which to produce standing waves and wave ripples.

depth a rectangular glass tank measuring roughly $0.25 \times 0.25 \times 0.30$ m and place it lengthwise across two suitable pieces of round dowelling (Fig. 13.14). Grasping one of these rollers in each hand, very gently rock the tank from side to side until you find the resonant frequency of a standing wave with central node (Fig. 13.15a). Let the water become still and then sprinkle into the tank some very small potassium permanganate crystals (the bottom boundary layer is only about half a millimetre thick). On gently rocking the tank again at the resonant frequency, patches of colour will spread over the bottom towards the node (Fig. 13.15b). Here is proof of the lower cells. Now introduce the largest crystals available, so that they pierce the boundary

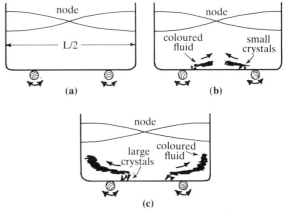

**Figure 13.15** Experimental demonstration of mass-transport currents due to standing waves.

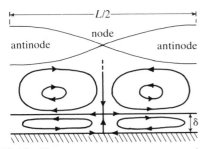

**Figure 13.13** Mass-transport currents associated with standing waves.

layer. On further rocking the tank at the resonant frequency, an upward-moving coloured stream will be seen at each antinode (Fig. 13.15c), thus marking the outer cells.

But we have not yet exhausted viscous effects. Paint a band of wood glue a centimetre or so wide round one of the wooden rollers used above and sprinkle the painted area with potassium permanganate crystals (Fig. 13.16a). Immerse the dried cylinder vertically in a large tank of stagnant water. Rapidly move the cylinder to and fro parallel to a fixed diameter, keeping the amplitude of the motion comparatively small (Fig. 13.16b). An outward-directed jet-like motion occurs at each end of the diameter (Fig. 13.17). The mass-transport currents to which these jets contribute (Fig. 13.16c) are driven by Reynolds stresses, and correspond to those created by standing waves (Fig. 13.13). The difference, however, is that the scale and spacing of the circulatory drifts are now controlled by the shape of the solid cylinder rather than the free surface. To make the water oscillate instead of the cylinder, all we need do is change the reference frame and 'unwrap' the cylinder (Fig. 13.16d). The near-bed motion is from troughs to crests, with the outer jets rising from the troughs. A similar streaming could therefore be provoked by irregularities on a wave-affected sea bed.

To summarize, the currents induced by real waves may comprise (1) an oscillatory component, (2) a weak

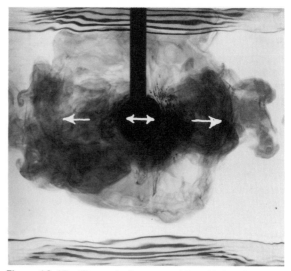

**Figure 13.17** Diametrically opposed jets (darker fluid only) formed by oscillating a cylinder along a diameter in stagnant water.

mass-transport current, due to finite-amplitude and viscous effects, that is either unidirectional (progressive waves) or recirculatory on a large scale (standing waves), and (3) a weak recirculatory component due to bed irregularities. Note that the first and second components invariably coexist.

## 13.8 Sediment transport due to wind waves and tides

### 13.8.1 Setting particles in motion

It is much more difficult to predict theoretically the entrainment of sediment particles by wave currents than by uniform steady flows (Sec. 4.2). This is because particles beneath waves experience various time-dependent forces. These forces are significant for wind waves, with their comparatively small periods, but as a first approximation may be neglected for the longer-period tide, to which Figures 4.3 and 4.4 may be applied. A different set of thresholds has been developed for sediment entrainment by wind waves, mainly on the basis of laboratory experiments (Rance & Warren 1969, Komar & Miller, 1973, 1975, Dingler 1979). There are two reasons for this. One is the significant role of the time-dependent forces mentioned, and the other the peculiar character of the bottom boundary layer. This layer as we saw is extremely thin, which means that it will sustain quite large wave-imposed velocities before becoming turbulent. It is a matter of observation that

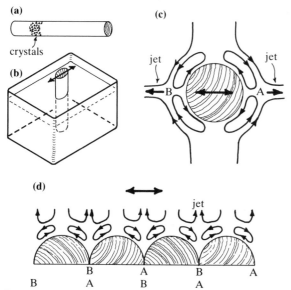

**Figure 13.16** Mass-transport currents due to the oscillation of a circular cylinder along a diameter in stagnant water, and its transformation into the pattern of currents associated with wave ripples.

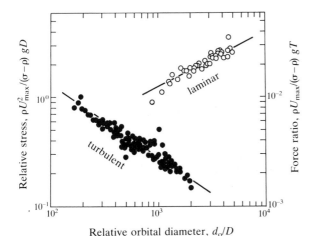

**Figure 13.18** Experimental threshold curves for the entrainment by waves of quartz-density solids in water. Data sources listed in Allen (1982a).

sediment entrainment and transport occur in both the laminar and turbulent forms of the wave-generated bottom boundary layer.

Empirically (Fig. 13.18), the flat-bed laminar entrainment threshold for mineral-density sands can be stated as

$$\frac{\varrho U_{max}^2}{(\sigma - \varrho)gD} = 0.405 \left(\frac{d_0}{D}\right)^{1/2} \qquad (13.37)$$

where $U_{max}$ is the maximum wave-induced orbital velocity just outside the boundary layer, $\sigma$ and $\varrho$ are respectively the sediment and fluid densities, $g$ the acceleration due to gravity, $D$ the sediment particle diameter, and $d_0$ the diameter of a water-particle orbit just outside the boundary layer. Recalling the quadratic stress law (Eqn 1.27), the left-hand term will be seen to express non-dimensionally the force acting on one grain (see also Eqn 4.7). The right-hand term is the relative orbital diameter. In the turbulent case (Fig. 13.18), the corresponding empirical threshold is

$$\frac{\varrho U_{max}}{(\sigma - \varrho)gT} = 280 \left(\frac{d_0}{D}\right)^{-2/3} \qquad (13.38)$$

where $T$ is the wave period. The left-hand side now is the ratio of fluid acceleration to gravity forces in the boundary layer. Quartz sands become entrained under turbulent conditions roughly when $D \gtrsim 500\,\mu m$.

### 13.8.2 Sediment transport

How good are wind-wave and tidal currents at trans-

porting sediment? As these currents are time periodic, the answer depends on (1) the kind of sediment, and (2) the questioner's timescale. As regards the sediment, the following discussion relates largely to bed material, that is, sand and gravel.

According to the Airy model, wave currents are simple harmonic and at a fixed station can be represented by a sine curve. Suppose for the moment that real wind waves and tides afford precisely such currents. Let us further suppose that the instantaneous bed-material transport rate due to such currents increases as the cube of the difference between the applied current and the entrainment velocity, that is, $\Delta U^3 = (U - U_{cr})^3$, where $U$ is the mean current velocity and $U_{cr}$ the threshold (Sec. 4.4.5). Figure 13.19a shows how the velocity varies over one wave period at a fixed station. This sinusoidal curve is divided by the abscissa into two exactly equal but displaced halves. Similarly, the curve for $\Delta U^3$ also falls into two displaced but exactly equal halves. Thus sediment transport is confined to those intervals when $U(t) > U_{cr}$, reaching its greatest instantaneous value once in each half-cycle, when $U(t) = U_{max}$. At certain instants, then, the sediment transport rate is substantial. But as the transport direction in one half-cycle is the

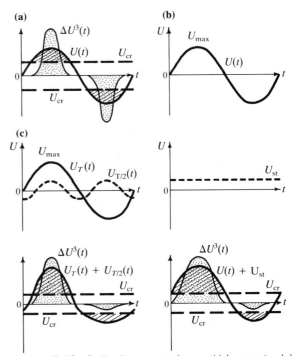

**Figure 13.19** Bedload transport due to tidal currents. (a) Symmetrical oscillatory current. (b) Symmetrical oscillatory current combined with a small steady current. (c) Distorted tidal wave in shallow water.

opposite of that in the other, and as the two halves are precisely equal, the net total transport and the net transport rate over the whole wave period must be zero. By the net total transport (kg m$^{-1}$ in dry-mass terms) we of course mean the integral of the instantaneous transport rate over one wave period. The net transport rate (kg m$^{-1}$ s$^{-1}$) is the net total transport divided by the wave period.

Hence on a timescale large compared to the wave period, a sinusoidal wave-related current affords no net sediment transport, although the same current at an instant may induce intense bed movement. Hence to obtain a non-zero transport in the long term, we must find some way of unbalancing the velocity distribution between the two parts of the wave cycle.

One way is to combine the sinusoidal motion with a small steady undirectional current (Fig. 13.19b), something that must commonly happen naturally (e.g. waves on a tide, tidal currents plus a wind-drift current). The upper graph shows the sinusoidal current $U(t)$ and the middle one the steady stream $U_{st}$. In the lower graph we see the currents added together. The two parts of the cycle are now unequal and the area under the curve for $\Delta U^3$ is significantly larger in one part than the other, the difference being the net total transport over one wave period. The net transport rate is therefore also non-zero. Evidently the net total transport and net transport rate will each grow as $U_{st}$ increases relative to $U_{max}$, the amplitude of the sinuosoidal current. You can see that if $U_{st}$ is large enough, the transport becomes effectively unidirectional.

The second way of unbalancing the velocity cycle is through wave distortion. We saw above that the crest of a shallow-water wave advances slightly faster than the trough, so that the profile steepens in the travel direction. This effect, noticeable from both shoaling wind waves and the tides, can be modelled so far as the induced currents are concerned by adding two sinusoidal velocity curves, one with one-half the wave period. In the upper graph of Figure 13.19c the curve $U_T(t)$ represents the basic wave and the curve $U_{T/2}(t)$ the distortion-related velocity component. In the lower graph we see the combined curves and the resulting sediment transport. The two parts of the wave cycle are now unequal in a different way than with a steady current (Fig. 13.19b). Wave distortion lengthens one part of the cycle, but produces the larger peak velocity in the shorter portion. There is no net discharge of water, as may be found by integrating an expression for the combined velocities over one wave period. On the other hand, the net total sediment transport and the net transport rate are clearly non-zero, and directed in the sense of the larger peak velocity. This effect results from (1) the non-linear dependence of the transport rate on velocity, and (2) the operation of a non-zero threshold condition. As with the mass-transport current, the additional current due to wave distortion increases as the square of the (basic) wave height.

We have explored these two sources of imbalance separately, although in natural environments they are probably combined. The net sediment transport rate, however, is likely to be generally small in comparison with the peak instantaneous rate. Wave and tidal environments should therefore be thought of as places of considerable sediment reworking for little resultant transport.

### 13.8.3 Some effects peculiar to the tide

As well as transporting sand and gravel, many tidal waters are turbid with suspended mud. What sequences of deposits might we then expect? Consider combined mud and sand transport under the pattern of tidal currents sketched in Figure 13.20, where $U_{crs}$ is the threshold for sand transport and $U_{crm}$ that for mud deposition. Assuming that some mud remains uneroded, a sequence of alternating sand and mud layers should result. The increments of sand will be unequal, however, because of the imbalance in the velocity distribution through time. If the imbalance was sufficiently severe, a sequence consisting of two mud layers

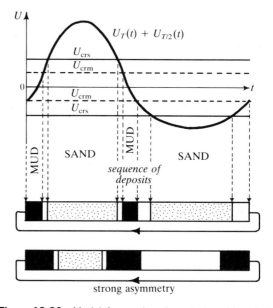

**Figure 13.20** Model for mud and sand deposition during one cycle of an asymmetrical tide.

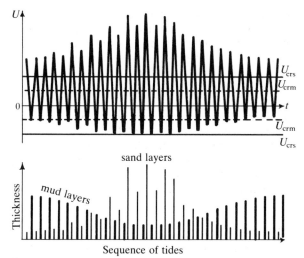

**Figure 13.21** Sequence of sand and mud deposits (thickness of mud layers exaggerated) due to a hypothetical spring–neap cycle of strongly asymmetrical semi-diurnal tides.

followed by one sand layer could result, where in practice the mud layers may not be separable. Given the right circumstances for preservation, each tide might therefore yield between two and four recognizably distinct sediment units.

It is interesting to speculate for a whole spring–neap cycle. Choosing a semi-diurnal tide of Immingham type (Fig. 13.3), and a strongly distorted velocity cycle, the hypothetical sequence of layers shown in Figure 13.21 might result. Here the sand layer thickness grows with the maximum of $\Delta U^3$ whereas the thickness of mud layers increases with the length of the interval of mud deposition. The layers of each kind therefore show a pattern of thickness in time, but with a phase difference of one-half the spring–neap period. At neaps, for example, relatively thick mud layers accompany comparatively thin sand deposits. Notice the diurnal inequality evident in the sand thicknesses.

These sequences of sand and mud layers were predicted on the basis of given tidal currents. Fossil tidal deposits show similar thickness patterns from which, using appropriate sediment-transport formulae, past tidal regimes can be reconstructed quantitatively (Nio *et al.* 1983).

## 13.9 Wave ripples and plane beds

### 13.9.1 *External and internal features of ripples*
What happens once waves have entrained coarse

sediment? Can wind-waves generate a power-related sequence of bedforms corresponding to the sequence of unidirectional currents (Sec. 4.8)? Further experiments with the wave tank provide a partial answer.

Half fill the tank (Fig. 13.14) with cold water and carefully spread clean fine-grained quartz sand over the bottom in an even layer a few centimetres thick. Drop in well boiled wet tea leaves to mark the bottom flow. Rock the tank from side to side on the rollers, at first very gently but afterwards with increasing vigour. The motion of the tea leaves will give some idea of the orbital diameters and velocities. As expected (Sec. 13.6.3), these quantities are largest beneath the node half-way along the tank. The sand will be seen not to move until the wave amplitude reaches a critical value. Try to measure the wave period and nodal orbital diameter for this condition. Assuming simple harmonic motion, use these data to calculate $U_{max}$. Does your estimate plot on one of the threshold curves in Figure 13.18? Slightly increase the amplitude and watch the moving sand. The grains become grouped into low symmetrical ridges perpendicular to the bottom current, which rock to and fro. These structures Bagnold (1946) called rolling-grain ripples (Fig. 13.22); their wavelength is substantially less than the diameter of the water particle orbits. If the grain motion is allowed to continue, or if the wave amplitude is increased, you may be lucky enough to see the occurrence of a sudden change in the ripples brought on by an increase to a critical steepness. A pebble or shell dropped into the tank has the same effect. The new

**Figure 13.22** Wave (vortex) ripples in roughly diamond-shaped patches initiated around shells and stones, Norfolk coast, England. Flatter rolling-grain ripples occur on surface between. Trowel is 0.28 m long.

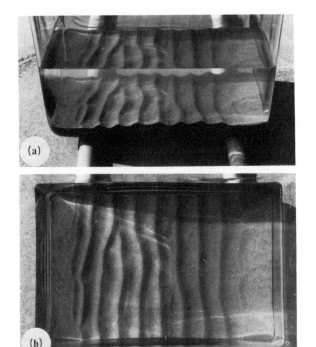

**Figure 13.23** Experimental wave ripples (bed 0.32 m long) in (a) profile and (b) plan, produced by rocking tank.

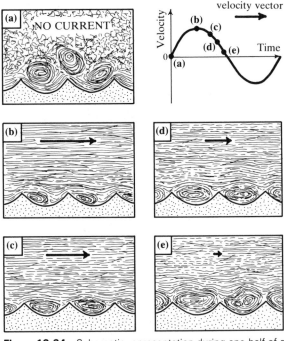

**Figure 13.24** Schematic representation during one-half of a cycle of flow patterns associated with vortex ripples. Largely after Bagnold (1946).

structures, tending to spread in diamond-shaped patches, have a wider spacing and are called vortex ripples (Bagnold 1946) or orbital ripples. As you can see from the tank (Fig. 13.23), as well as from the field (Fig. 13.22), vortex ripples are symmetrical and trochoidal in form.

You can judge for yourself why the terms 'orbital' and 'vortex' are applied to these structures. Their wavelengths can be seen from the tank to be comparable with the water particle orbital diameter. The stream affecting vortex ripples is heavily charged with sand, and the grains, with each reversal of bottom current, become involved in a separated flow alternately to one side of the crest and then the other. As each newly formed separation vortex can grow only by displacing the mature vortex representing the previous current stroke, sand-charged water is continually being 'pumped' upwards from the bed (Fig. 13.24). Bagnold (1946) and Longuet-Higgins (1981) give further details of this important process for dispersing bed grains upwards into the water column.

Vortex ripples abound in shelf, coastal and lacustrine environments and are even recorded from river backwaters and floodplains. Natural examples (Fig. 13.25) closely resemble the experimental structures (Fig. 13.23)

and possess an almost perfect symmetry. In further sharp contrast to current ripples (Figs 4.9 & 4.10), they are regular and little curved in plan, even when comparatively short-crested. Vortex ripples range in wavelength from about 0.04 m in very fine sand to 1–2 m in very coarse sand and fine gravel. Ripples towards the lower end of this size range are known from stormy continental shelves in depths up to 204 m. Large-wavelength forms in biogenic and other coarse debris occur as deep as 140 m (Conolly 1969, Newton *et al.* 1973, Yorath *et al.* 1979, Hamilton *et al.* 1980).

As wind waves shoal, generating an increasingly asymmetrical pattern of currents, the ripples they produce also grow more asymmetrical but without losing regularity (Fig. 13.26). Thus they remain recognizably wave-related forms (Reineck & Wunderlich 1968). Notice also that strongly asymmetrical wave ripples can occur in much coarser sediment (Fig. 13.26f) than current ripples (Fig. 4.26).

The shape of vortex ripples might lead one to expect a symmetrical internal lamination. The falseness of this inference is shown by field studies (Newton 1968), which reveal laminae inclined only in the direction of wave propagation (Fig. 13.27), that is, in the direction of the expected net sediment transport. Cross-lamination

**Figure 13.25** Wave (vortex) ripples in (a) very fine-grained quartz sand (lens cap 0.045 m across), (b) fine sand (trowel 0.28 m long), (c) medium sand (geological hammer for scale), and (d) very coarse-grained pebbly shelly sand (bag 0.3 m square). Various localities on British coast.

sets generated by wave ripples range between complex erosional forms (Allen 1981a) and steeply climbing varieties (McKee 1938).

### 13.9.2 Instability of a plane granular bed

What in our tank experiments made the plane bed change into a rippled one? We can readily perceive the oscillatory component of the current, but we also know that, on its own, the component gives rise to no net sediment transport. Considering the ripples, we see that they are stationary on a timescale comparable with the wave period, but can only have arisen if grains had in one place been dug from below the level of the undisturbed bed and in another gradually heaped up above it. This can only have happened if on the bed there was a stationary spatial series of reversing net sediment transport paths. An additional component of flow similar to that in Figure 13.16d could have promoted the required net transports.

Do such drifts occur above rippled beds? Make some large wave ripples in the tank (Fig. 13.14) and let the water become stagnant. Take a few large crystals of potassium permanganate and quickly insert them using sharp-pointed forceps one-half to two-thirds of the way up the flanks of one of the ripples, so that the tops of the crystals are roughly level with the surrounding sand. On gently rocking the tank at the resonant frequency you will see oscillating patches of colour spread slowly towards the ripple crest. Here is proof of the lower mass-transport cell (Fig. 13.16d). Now sprinkle crystals

259

**Figure 13.26** Asymmetrical wave-related ripples in (a) very fine-grained quartz sand (scale 0.15 m long), (b & c) fine-grained sand (pencil 0.18 m long), (d) fine sand (trowel 0.28 m long), (e) coarse to very coarse sand (hammer 0.33 m long), and (f) very coarse pebbly and shelly sand (bag 0.3 m square). Waves propagated from top in (a), upper right in (b) and (c), bottom in (d), left in (e) and right in (f). Various localities on British coast.

Figure 13.27 Internal structure of active wave ripples from the inshore zone of the Baltic Sea. Waves propagated shorewards (to right). Each sample is about 0.15 m wide. Downturning of laminae at edges occurred during sampling of the loose sand. From Newton (1968) with permission of Elsevier Science Publishers. Photographs courtesy of Dr E. S. Newton, D'Appolonia, Houston.

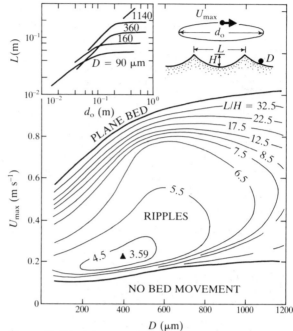

Figure 13.29 Existence and properties of field and laboratory wave ripples in terms of grain size and maximum near-bed water particle orbital velocity, based on 648 observations (main plot). The numbered curves record the outward limit of the occurrence of ripples of that $L/H$ value. Adapted from Allen (1979) and Miller and Komar (1980).

over the bed large enough to penetrate the bottom boundary layer. As proof of the outer cells, an expanding tree-like mass of colour will rise from each trough (Fig. 13.28). It thus appears that a granular bed affected by waves is inherently unstable, the irregularities on it being sufficient to set up a series of recirculatory drifts, which then interact with the grain transport to cause the

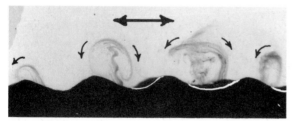

Figure 13.28 Tree-like jets and return currents formed experimentally above a bed of vortex ripples.

emergence of a preferred ripple wavelength. This demonstration was first proffered by Darwin (1884) and more recently by Kaneko and Honji (1979).

### 13.9.3 Bedform sequence

Field and laboratory work show that wind waves create a bedform sequence in which ripples at relatively low fluid stresses are succeeded by an upper-stage plane bed at a stress comparable to that for the bedform in unidirectional currents (Inman 1957, Kaneko 1980). Figure 13.29 shows these bedforms in terms of the maximum near-bed water particle orbital velocity and grain size (Allen 1979). On the basis of 648 laboratory and field observations, the relative ripple wavelength (wavelength divided by height) at each grain size is a minimum at a comparatively low orbital velocity, and is less in medium sands than in either finer or coarser grades. These steepest ripples are sharp-crested vortex ripples representing variation within a closely defined range of relative water particle orbital diameters. When $d_0/D$ is large (boundary layer nominally laminar)

$$L/d_0 \simeq 1.0 \qquad (13.39\text{a})$$

where $L$ is the ripple wavelength. At small values of $d_0/D$ (boundary layer turbulent)

$$L/d_0 \simeq 0.65 \qquad (13.39\text{b})$$

These relationships, combined with the entrainment thresholds and wave equations, may be used to predict ancient wave conditions from field measurements of sharp-crested vortex ripples (Komar 1974, Allen 1981b, Moore 1982, Sundquist 1982). Wave ripples become flatter as the orbital velocity is increased (Fig. 13.29) and also more rounded. Eventually they merge into a plane or very gently undulating bed associated with intense transport (Fig. 13.29).

Figure 13.29 also shows that ripple wavelength increases with the diameter of near-bed water particle orbits up to a limit set by grain size, above which the spacing is constant or declines (Inman 1957, Miller & Komar 1980).

## 13.10 Sand waves in tidal currents

Tidal waves change slowly enough to fashion some of the bedforms characteristic of one-way currents, especially current ripples and small dunes (Sec. 4.6). Nevertheless, one kind of transverse sandy bedform does seem to be peculiar to the tide. These large structures, called sand waves, are widely distributed subtidally in restricted seas (McCave 1971, Field et al. 1981), in estuaries and inlets (D'Anglejan 1971, Langhorne 1973, Bouma et al. 1980), in straits (Boggs 1974, Keller & Richards 1967), on shelf shoals (Mann et al. 1981), and on isolated shallow platforms (Jordan 1962, Hine 1977). Sand waves appear to be to the tide what the ripples just described are to wind waves. In shape sand

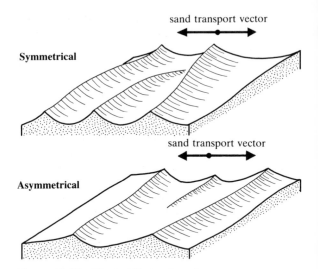

**Figure 13.30** Principal kinds of sand wave.

waves are regular long-crested ridges of sand varying in cross section from symmetrical, with sharp crests and broad rounded troughs, to strongly asymmetrical, with one face much shorter and steeper than the other (Figs 13.30 & 13.31). Typically, they occur in depths of several tens of metres or less, but can in places be found near the shelf edge. Their wavelength is ordinarily 50–300 m, with occasional sand waves reaching a spacing of about 1000 m. The height generally lies between 1 and 8 m, exceptionally reaching 20–25 m. On a global basis, sand-wave size increases with water depth, but there are local exceptions, suggesting that such factors as grain size and current strength may be influential. Sand waves become temporarily flattened at the crest during storms (Langhorne 1982).

Sand waves are on the whole much flatter structures than sub-aqueous dunes. The sides of the waves rarely dip more steeply than 10° overall and can be inclined as

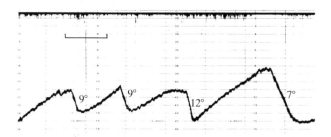

**Figure 13.31** Part of an echo-sounder trace across a field of asymmetrical sand waves with superimposed dunes in Start Bay, English Channel. Water surface is uneven black line at the top of photograph. Horizontal scale bar is 100 m long. Vertical scale appearing on chart is in metres. Overall slopes of leeward sides of sand waves recorded in degrees (ship's track perpendicular to sand-wave crests). Photograph courtesy of Mr D. N. Langhorne, Institute of Oceanographic Sciences.

little as 1°. These mild slopes preclude any general avalanching, except locally near the crests, but not the superposition of small bedforms. Side-scan sonar records and echo-sounder traces (Fig. 13.31) leave no doubt that most sand waves are decorated with dunes having wavelengths of the order of 5–15 m. These superimposed forms characteristic of unidirectional currents tend to be aligned parallel to the crest of the sand wave. Their sense of facing appears to depend mainly on the direction of the tidal stream when the record was made, but is also influenced by sand-wave shape and symmetry.

What is the meaning of sand-wave asymmetry? Evidence supporting the long-held opinion that the steeper flank faces in the direction of net sediment transport is given by Langhorne (1982) from one closely studied sand wave and by Ludwick (1972) from others. Pingree and Griffiths (1979) find mathematically that the observed facing of sand waves in British seas closely matches the likely pattern of net sediment transport due to tidal-wave distortion. Hence the shapes of sand waves and of wave ripples appear to carry the same implication regarding sediment transport.

The origin of sand waves is uncertain. One school treats them as simply very large dunes, implying the irrelevance of the periodic nature of tidal streams. It is just possible, however, that sand waves correspond dynamically as well as morphologically to wind-wave ripples (Allen 1982a). Consider the ratio of the thickness of the wave boundary layer to the bedform spacing in the two cases. For wave ripples it is of the general order $10^{-2}$ to $10^{-3}$, and for sand waves $10^{-3}$ to $10^{-4}$, each in terms of the true fluid viscosity. However, using a 'turbulent' viscosity (Sec. 1.9), the tidal boundary layer is of order 1 m thick, bringing the two ratios into agreement. In this important respect, then, the two bedforms seem to correspond, and so may depend on similar dynamics.

Sand waves are so large and difficult to study directly that only the bedding in the outermost layers of the modern forms is known by direct sampling. Consequently, various speculative models for sand-wave internal structure are current (Fig. 13.32). Relatively symmetrical forms may, as Reineck (1963) suggests, be characterized by opposed cross-bedding sets representing the reversal of dunes over the waves, in response to comparably effective ebb and flood streams. Strongly asymmetrical sand waves, shaped by markedly unequal tidal streams, may include long avalanche layers inclined in the direction of net sediment transport (McCave 1971, Allen 1982a). Amongst the laminae there could be convex-up erosion surfaces due to erosive

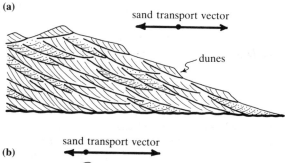

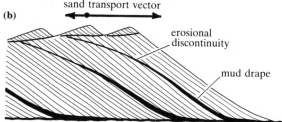

**Figure 13.32** Speculative models for the internal structure of (a) nearly symmetrical sand waves, and (b) strongly asymmetrical sand waves.

rounding of the wave crest during the subordinate flow, together with slack-water mud drapes. Sub-fossil tidal sands in The Netherlands contain cross-bedding sets and drapes (Fig. 13.33) showing a clear and simple pattern of sediment thicknesses and types explicable in terms of the spring–neap cycle (Fig. 13.21) (Visser 1980, Terwindt 1981). Similar cross-bedding patterns mark older tidal sands (Allen 1982b, Allen & Homewood 1984). However, to judge from Langhorne's (1982) study, many asymmetrical sand waves could be much more complex internally than these examples.

## 13.11 Longshore bars and troughs

Lying off many a sandy shore, not far below mean water level, is a belt of parallel shore-aligned bars and troughs which attain a total width of as much as 1500 m (Greenwood & Davidson-Arnott 1979). The depth at the outer edge of a belt of these longshore bars rarely exceeds 5–10 m and the bars themselves are no more than 1–2 m tall. Their spacing tends to increase from 10–50 m for the innermost structures outwards to 75–300 m at the deep-water edge. The bars in plan vary between straight and sinuous to cusped. Storms may cause the innermost bars in some belts to change from straight to cusped or sinuous. As a general rule, the widest-spaced bars occur off the coasts with the severest wave climate.

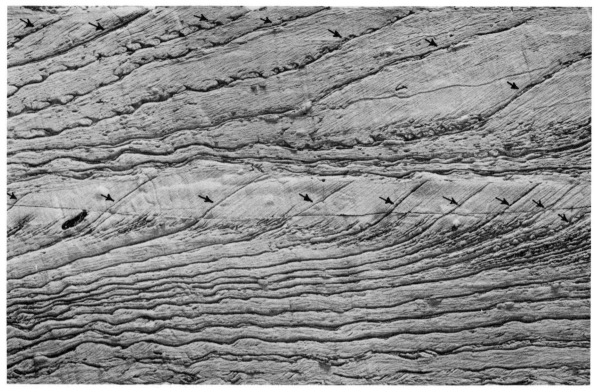

**Figure 13.33** Two cross-bedded units (0.75 × 1.6 m) of tidal origin observed in vertical section from the Oosterschelde, The Netherlands. Dominant transport from right to left. Mud drapes arrowed. Sequence in lower set records change from neaps to springs, and in upper set from springs to neaps. From Terwindt (1981) by permission of the International Association of Sedimentologists. Photograph courtesy of Dr J. H. J. Terwindt, University of Utrecht.

What causes these belts? Dependence in some way on wind-waves seems clear, because some of the best-developed bars occur in tideless waters. Of the available explanations, perhaps the most appealing is the idea that the bars represent the response of the sandy bottom to mass-transport currents created by wave reflection at the beach (Lau & Travis 1973), as described by Equation 13.36 and illustrated in Figures 13.13 and 13.15. Bars should arise beneath the nodes formed by the incident and reflected waves, that is, at a spacing of one-half the wavelength of the surface waves. Bar spacing will increase with the scale of the waves and, on account of shoaling effects, from shallow to deep water. Storm waves of 5–10 s period, for example, should yield bars spaced a few tens of metres apart, in agreement with experience.

## 13.12 Waves and storm surges – back to the beginning

We earlier derived Equation 1.56 for the wind-induced current in a water body. What is the interplay between the wind current and waves during a storm, when each is likely to be most effective in moving sediment? This question cannot yet be fully answered, but it is clearly an important one for understanding processes in lakes and shallow seas.

Consider a shelf sea bounded by a sandy coast in an area acted on by an onshore-moving storm. The main effects of the storm waves will be (1) to disperse upwards and set in to-and-fro motion large quantities of sediment, particularly at the shore where the waves break, and (2) to induce a small net onshore bottom sediment transport. Very little is known of the time-averaged wave-induced sediment load $m_w$, but plausibly

$$m_w \propto (\varrho U_{max}^2)^n \quad \text{kg m}^{-2} \qquad (13.40)$$

where $U_{max}$ is the maximum orbital velocity of a near-bed water particle, and $n \geqslant 1$ an exponent. As $U_{max}$ for constant waves declines with increasing water depth (Eqn 13.22), we should expect $m_w$ also to decline with increasing depth and distance from shore. But in Section 1.11 we found that the wind-induced bottom current flows offshore. As it considerably exceeds in magnitude the shoreward mass transport, an offshore sediment transport should result, at the rate

$$J \propto (\varrho U_{max}^2)^n U_{sm} \quad \mathrm{kg\,m^{-1}s^{-1}} \qquad (13.41)$$

where $U_{sm}$ is a characteristic wind-induced bottom current. But as $U_{sm}$ also declines with increasing depth and distance from shore, the sediment dispersed into the water at the shore will become deposited (Eqn 4.19) as it is fed outwards into greater depths.

Hence we may imagine that a turbidite-like bed of mainly shore-derived sand gradually spreads outwards over the shelf (Fig. 13.34). Its base is likely to be erosional, and the lower part may include a lag of shell and other debris scoured up from nearby on the shelf, since the storm waves will affect the deeper bottom before the sand can spread to reach it from the shore. According to Figure 13.29, parallel lamination will

characterize the bed in the shallows where $U_{max}$ is large, and wave-ripple lamination will typify the deposit in the deeper waters beyond. A rippled top to the bed, followed by bioturbation, should develop everywhere as the storm declines and fair-weather mud deposition is resumed. The reality may be somewhat more complex, however, as we have ignored effects due to the Earth's rotation, and assumed that the wave and wind currents act along the same line.

Turbidite-like sand beds closely resembling the predicted sequence and interpreted as storm layers are being increasingly described from the rock record (Brenchley *et al.* 1979, Wright & Walker 1981, Buller & Johnson 1982, Dott & Bourgeois 1982). A few modern counterparts have been studied (Aigner & Reineck 1982). The mechanics of storm sedimentation are little known, however, and represent one of the many challenges of physical sedimentology.

## Readings

Aigner, T. and H.-E. Reineck, 1982. Proximality trends in modern storm sands from the Helgoland Bight (North Sea) and their implications for basin analysis. *Senckenbergiana Maritima* **14**, 183–215.

Allen, J. R. L. 1979. A model for the interpretation of wave ripple-marks using their wavelength, textural composition, and shape. *J. Geol. Soc. Lond.* **136**, 673–82.

Allen, J. R. L. 1982a. *Sedimentary structures*, Vol. I. Amsterdam: Elsevier.

Allen, J. R. L. 1982b. Mud drapes in sand-wave deposits: a physical model with application to the Folkestone Beds (early Cretaceous, southeast England). *Phil. Trans. R. Soc. Lond.* A **306**, 291–345.

Allen, P. A. 1981a. Wave-generated structures in the Devonian lacustrine sediments of SE Shetland, and ancient wave conditions. *Sedimentology* **28**, 369–79.

Allen, P. A. 1981b. Some guidelines in reconstructing ancient sea conditions from wave ripple marks. *Marine Geol.* **43**, M59–67.

Allen, P. A. and P. Homewood, 1984. Evolution and mechanics of a Miocene tidal sand wave. *Sedimentology* **31**, 63–81.

Bagnold, R. A. 1946. Motion of waves in shallow water. Interaction between waves and sand bottoms. *Proc. R. Soc. Lond.* A **187**, 1–16.

Barnett, T. P. and J. C. Wilkerson, 1967. On the generation of ocean wind waves as inferred from airborne radar measurements of fetch-limited spectra. *J. Marine Res.* **25**, 292–321.

Boggs, S. 1974. Sand-wave fields in Taiwan Strait. *Geology* **2**, 251–3.

Bouma, A. H., M. L. Rappeport, R. C. Orlando, and M. A. Hampton, 1980. Identification of bedforms in Lower Cook

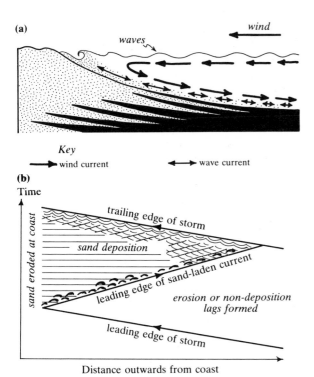

**Figure 13.34** Model for storm sedimentation. (a) General pattern of currents. (b) Events during a storm, and the possible sequence and thickness (proportional to time) of resulting storm bed. Symbols for sedimentary structures as in Figure 12.28 (vortex ripples shown as trochoids).

Inlet, Alaska. *Sed. Geol.* **26**, 157–77.

Brenchley, P. J., G. Newall, and I. G. Stanistreet, 1979. A storm surge origin for sandstone beds in an epicontinental platform sequence, Ordovician, Norway. *Sed. Geol.* **22**, 185–217.

Buller, A. T. and S. O. Johnson, 1982. Storm-influenced marine sandstones in the Ordovician Lower Havin Group, Nord-Trøndelag. *Norsk Geol. Tidsskr.* **62**, 211–17.

Carter, D. J. T. 1982. Prediction of wave height and period for a constant wind velocity using the JONSWAP results. *Ocean Engng* **9**, 17–33.

Collins, J. I. 1963. Inception of turbulence at the bed under periodic gravity waves. *J. Geophys. Res.* **68**, 6007–14.

Conolly, J. R. 1969. Western Tasman Sea floor. *N. Z. J. Geol. Geophys.* **12**, 310–42.

D'Anglejan, B. F. 1971. Submarine sand dunes in the St Lawrence Estuary. *Can. J. Earth Sci.* **8**, 1480–6.

Darbyshire, M. and L. Draper, 1963. Forecasting wind-generated sea waves. *Engineering* 5 April, 482–4.

Darwin, G. H. 1884. On the formation of ripple-marks. *Proc. R. Soc. Lond.* **36**, 18–43.

Defant, A. 1958. *Ebb and flow.* Ann Arbor: University of Michigan Press.

Dingler, J. R. 1979. The threshold of grain motion under oscillatory flow in a laboratory wave channel. *J. Sed. Petrol.* **49**, 287–94.

Doodson, A. T. and H. D. Warburg, 1941. *Admiralty manual of tides.* London: HMSO.

Dott, R. H. and J. Bourgeois, 1982. Hummocky stratification: significance of its variable bedding sequences. *Bull. Geol. Soc. Am.* **93**, 663–80.

Field, M. E., H. C. Nelson, D. A. Cacchione and D. E. Drake 1981. Sand waves on an epicontinental shelf: northern Bering Sea. *Marine Geol.* **42**, 233–58.

Frost, B. A. 1966. *The relation between Beaufort Force, wind speed and wave height.* Sci. Pap. Meteorol. Office, no. 25.

Greenwood, B. and R. G. D. Davidson-Arnott, 1979. Sedimentation and equilibrium in wave-formed bars: a review and case study. *Can. J. Earth Sci.* **16**, 312–32.

Hamilton, D., J. H. Sommerville and P. N. Stanford, 1980. Bottom currents and shelf sediments, southwest of Britain. *Sed. Geol.* **26**, 115–38.

Hine, A. C. 1977. Lily Bank, Bahamas: history of an active oolite sand shoal. *J. Sed. Petrol.* **47**, 1554–81.

Inman, D. L. 1957. *Wave generated ripples in nearshore sands.* Tech. Mem. US Beach Erosion Board, no. 100

Jordan, G. F. 1962. Large submarine sand waves. *Science* **136**, 839–48.

Kaneko, A. 1980. The wavelength of oscillation sand ripples. *Rep. Res. Inst. Appl. Mech. Kyushu Univ.* **28**, 57–71.

Kaneko, A. and H. Honji, 1979. Double structure of steady streaming in the oscillatory viscous flow over a wavy wall. *J. Fluid Mech.* **93**, 727–36.

Keller, G. H. and A. F. Richards, 1967. Sediments of the Malacca Strait, southeast Asia. *J. Sed. Petrol.* **37**, 102–27.

Kinsman, B. 1965. *Wind waves.* Englewood Cliffs: Prentice-Hall.

Komar, P. D. 1974. Oscillatory ripple marks and the evaluation of ancient wave conditions and environments. *J. Sed. Petrol.* **44**, 169–80.

Komar, P. D. and M. C. Miller, 1973. The threshold of sediment motion under oscillatory water waves. *J. Sed. Petrol.* **43**, 1101–10.

Komar, P. D. and M. C. Miller, 1975. On the comparison between the threshold of sediment motion under waves and unidirectional currents with a discussion of the practical evaluation of the threshold. *J. Sed. Petrol.* **45**, 362–7.

Langhorne, D. N. 1973. A sandwave field in the outer Thames Estuary. *Marine Geol.* **14**, 129–43.

Langhorne, D. N. 1982. A study of the dynamics of a marine sandwave. *Sedimentology* **29**, 571–94.

Lau, J. and B. Travis, 1973. Slowly varying Stokes waves and submarine longshore bars. *J. Geophys. Res.* **78**, 4489–97.

Longuet-Higgins, M. S. 1958. The mechanics of the boundary-layer near the bottom in a progressive wave, an appendix to the paper by Russell and Osorio, 1958. *Proc. 6th Conf. Coastal Engng* 184–93.

Longuet-Higgins, M. S. 1981. Oscillating flow over steep sand ripples. *J. Fluid Mech.* **107**, 1–35.

Ludwick, J. C. 1972. Migration of tidal sand waves in Chesapeake Bay entrance. In *Shelf sediment transport: processes and patterns.* D. J. P. Swift, D. B. Duane and H. O. Pilkey (eds), 377–410. Stroudsburg: Dowden, Hutchinson and Ross.

McCave, I. N. 1971. Sand waves in the North Sea off the coast of Holland. *Marine Geol.* **10**, 199–225.

McKee, E. D. 1938. Original structures in the Colorado River flood deposit of the Grand Canyon. *J. Sed. Petrol.* **8**, 77–83.

Mann, R. G., D. J. P. Swift and R. Perry, 1981. Size classes of flow-transverse bedforms in a subtidal environment, Nantucket Shoals, North American Atlantic shelf. *Geo-Marine Lett.* **1**, 39–43.

Miller, M. C. and P. D. Komar, 1980. Oscillation sand ripples generated by laboratory experiments. *J. Sed. Petrol.* **50**, 173–82.

Moore, P. S. 1982. Ripple-mark analysis of a fine-grained epeiric-sea deposit (Cambrian, South Australia). *J. Geol. Soc. Aust.* **27**, 71–81.

Morison, J. R. and R. C. Crooke, 1953. *The mechanics of deep water, shallow water, and breaking waves.* Tech. Mem. US Beach Erosion Board, no. 40.

Newton, R. S. 1968. Internal structures of wave-formed ripple marks in the nearshore zone. *Sedimentology* **11**, 275–92.

Newton, R. S., E. Seibold and F. Werner, 1973. Facies distribution patterns on the Spanish Saharan continental shelf mapped with side-scan sonar. *Ergebnisse 'Meteor' Forsch.* C **16**, 55–77.

Nio, S. D., C. Siegnethaler and C. S. Yang, 1983. Megaripple cross-bedding as a tool for the reconstruction of the palaeohydraulics in a Holocene subtidal environment, S.W. Netherlands. *Geol. Mijnb.* **62**, 499–510.

Phillips, O. M. 1977. *The dynamics of the upper ocean.* Cambridge: Cambridge University Press.

Pingree, R. D. and D. K. Griffiths, 1979. Sand transport paths around the British Isles resulting from $M_2$ and $M_4$ tidal interactions. *J. Marine Biol. Assoc. UK.* **59**, 497–513.

Rance, P. J. and N. F. Warren, 1969. The threshold of movement of coarse material in oscillatory flow. *Proc. 11th Conf. Coastal Engng.* Vol. 1. 487–91.

Reineck, H.-E. 1963. *Sedimentgefüge im Bereich der südlichen Nordsee.* Abh. Senck. Naturf. Ges., no. 505.

Reineck, H.-E. and F. Wunderlich, 1968. Zur Unterscheidung von asymmetrischen Oszillationsrippeln und Strömungsrippeln. *Senckenbergiana Lethaea* **49**, 321–45.

Sundquist, B. 1982. Palaeobathymetric interpretation of wave ripple-marks in a Ludlovian grainstone of Gotland. *Geol. För. Stockh. Förh.* **104**, 157–66.

Terwindt, J. H. J. 1981. *Origin and sequences of sedimentary structures in inshore mesotidal deposits of the North Sea.* Spec. Publn. Int. Assoc. Sed., no. 5, 4–26.

Thornton, E. B. and R. F. Krapohl, 1974. Water particle velocities measured under ocean waves. *J. Geophys. Res.* **79**, 847–52.

Visser, M. J. 1980. Neap–spring cycles reflected in Holocene subtidal large-scale bedform deposits: a preliminary note. *Geology* **8**, 543–6.

Wallet, A. and F. Ruellan, 1950. Trajectories of particles within a partial clapotis. *Houille Blanche* **5**, 483–9.

Wiegel, R. L. 1964. *Oceanographical engineering.* Englewood Cliffs, NJ: Prentice-Hall.

Wright, M. E. and R. G. Walker, 1981. Cardium formation (U. Cretaceous) at Seebe, Alberta – storm-transported sandstones and conglomerates in shallow marine depositional environments below fair-weather base. *Can. J. Earth Sci.* **18**, 795–809.

Yorath, C. J., B. D. Bornhold and R. E. Thomson, 1979. Oscillation ripples on the northeast Pacific continental shelf. *Marine Geol.* **31**, 45–58.

# Index